ICONS OF EVOLUTION

ICONS OF EVOLUTION

An Encyclopedia of
People, Evidence, and
Controversies

VOLUME 1

Edited by Brian Regal

Greenwood Icons

SETON HALL UNIVE
University Libraries So. Orange NJ 07

GREENWOOD PRESS
Westport, Connecticut · London

Libray of Congress Cataloging-in-Publication Data

Icons of evolution: an encyclopedia of people, evidence, and controversies / edited by Brian Regal.

p. cm.—(Greenwood icons)

Includes bibliographical references and index.

ISBN 978–0–313–33911–0 (set: alk. paper)—ISBN 978–0–313–33912–7 (v. 1: alk. paper)—ISBN 978–0–313–33913–4 (v. 2: alk. paper)

1. Evolution (Biology)—Encyclopedias. I. Regal, Brian.

QH360.2.I33 2008

576.803—dc22   2007033049

British Library Cataloguing in Publication Data is available.

Library of Congress Catalog Card Number: 2007033049
ISBN-13: 978-0-313-33911-0 (set)
        978-0-313-33912-7 (vol. 1)
        978-0-313-33913-4 (vol. 2)

First published in 2008

Greenwood Press, 88 Post Road West, Westport, CT 06881
An imprint of Greenwood Publishing Group, Inc.
www.greenwood.com

Printed in the United States of America

The paper used in this book complies with the Permanent Paper Standard issued by the National Information Standards Organization (Z39.48-1984).

10 9 8 7 6 5 4 3 2 1

Table of Contents

Preface and Acknowledgments

The purpose of this two-volume compilation is to give the reader a set of clear, concise, readable, understandable, and fascinating explanations for the evidence for evolution and why that evidence is so important for understanding how life on earth came to be and how it spread and diversified. It shows the reality behind some of the most famous symbols of the evolutionary process and evolutionary studies. The book is geared toward readers new to the subject and to those who have some knowledge of it but want to expand their understanding of the topic with some more technical explanations. Readers may be surprised that some of these icons are not quite what they thought they were. There are many misconceptions about these icons, claims made for and against them. For example, was Charles Darwin an atheist? Was *Archaeopteryx* the first bird? What *is* a missing link? What is the significance of all those horse fossils? What does "survival of the fittest" really mean? What ever happened to Peking Man? And others.

As this book is a compilation of topics, a sort of "best of," readers are encouraged to argue over which topics should have or should not have been included. This particular collection—by no means a definitive one—tends to lean heavily toward geology and anthropology with little discussion of the important role of plants in evolution and, besides peppered moths, little mention of insects or fish. The arrangement of the chapters will also probably not satisfy everyone. They were shuffled and reshuffled a number of times before reaching this line up. I tried to follow a roughly chronological order so that the individual chapters would have a certain logical progression with a balance between individuals, ideas, and artifacts. There is also overlap between different chapters. This is only natural as these icons do not exist in a vacuum but are part of a much wider continuum. The reader will also no doubt notice that not all scientists and historians always agree on things. This should not be seen as a problem but as evidence of the robust and dynamic nature of the scientific enterprise and historical analysis. As new discoveries are made, ideas change. This is how scholars operate. As with everything in science and history nothing is ever final; there are always new tomorrows which bring the possibility of new knowledge. To do otherwise would be to stagnate and never learn anything new. The book includes a chronology of events in evolution history, an extensive bibliography so that readers can go to original sources and see where the proofs are, and short biographies of the authors.

The topics covered in these volumes are not the only icons involved here. In their research into evolutionary mechanics and their writings on the history and social implications of evolution many of the authors in this collection are considered icons themselves, and others will be thought of that way in the future. When choosing authors I wanted a mix of well-known researchers, writers, and teachers and up-and-coming new scholars. They have written their chapters each from a different prospective and in different styles. They alternate between historical narratives, scientific dispositions, and philosophical discourses. While each chapter is designed to stand as a separate unit, they can be taken together as an entire story.

Volume 1 begins with Olof Ljungström's discussion of why evolution is the cornerstone of biology and acts as an overview of the history of evolutionary thought. This helps give some context for the rest of the chapters. This is followed by chapters on the early giants: Bowdoin Van Riper on Charles Darwin, Lisa Nocks on T.H. Huxley, and Dawn Digrius on Gregor Mendel. Peter Bowler then explains what "survival of the fittest" means and why it is so misunderstood. Then come some of the most famous icons of human evolution: Marianne Sommer on Neanderthal Man, Pat Shipman on Java Man, and Brian Regal on the disappearance of Peking Man. David Rudge and Christine Janis follow with discussions of evolution in action with the peppered moth and the horse series. Finally comes Anne Katrine Gjerløff on the Taung Child of Africa with the volume rounded out by a discussion of the history of monkey trials and the ongoing assault against the teaching of evolution from some quarters.

Volume 2 shows that evolution is not just rocks and fossils but involves interpretation and analysis of those fossils as well as a way of studying the descent of living organisms by techniques only dreamed of in the nineteenth century. After David Fastovsky takes a look at our old friends the dinosaurs and Bent Lindow gives a history of flight and the fossil *Archaeopteryx*, Olof Ljungström explains the fossil record and how it is organized and what it tells about the history of life. Bowdoin Van Riper explains another cornerstone of evolutionary studies, radiometric dating. He explains how we know how old the fossils are. The infamous hoax

Piltdown Man is discussed and a crack is taken at figuring out who the culprit is. Then Jeffrey Schwartz and Adam Wilkins explain how modern biologists figure out how to arrange living things into an orderly fashion and how the modern notion of evolution and genetics mix to form the current view of how evolution works. Rounding out the list of icons are the life of Louis Leakey by Anne Katrine Gjerløff and the most famous human fossil since the Neanderthal Man, Lucy, by Holly Dunsworth, and Christopher Stringer explains how humans got out of Africa. Finally, comes Intelligent Design and the newest assault on evolution, which, it turns out, is not so new after all.

I would like to thank Kevin Downing who first brought this project to me and allowed me to transform it from an enormous one-author job into an enormous multi-author job. That decision allowed me to tap some of the finest authors and thinkers—scientists, historians, and philosophers—on evolution in the world. It was a great honor and privilege to work with them. They did an outstanding job and put up with me constantly harassing them to get things done, despite the fact that they all had many other commitments and projects they were working on. Also thanks to everyone at Greenwood Publishing for all their hard work.

It should be known that there is another book called *Icons of Evolution*; however, it is a work geared specifically to refute evolution and to show that scientists are wrong in their conclusions. Many of the topics that book attacks are supported in this one with copious amounts of evidence. That leaves the reader with an interesting project on their hands: Compare the two and see where they stand.

—Brian Regal
New York, June 2007

Introduction

Where did life come from? How does it change? These are important questions. They are questions the science of evolution seeks to answer. The answers it provides satisfy some, angers and confuses others, and startles us all. This book presents some of the most famous and well-known answers to these questions: twenty-four icons of evolution in all. It is important for everyone, not just scientists, to understand the workings of evolution as it is a process so intimately linked to our lives and our very survival.

Evolution is the term used to label biological change on earth. It is the natural process which is responsible for the wide diversity of life. Evolution, or transmutation as it has also been called, was not invented or created or constructed any more than gravity was. It began with the appearance of the first living things and will continue for as long as life exists here. It is not one process but the result of a combination of several processes: change in the environment, genetic variation, and the adaptation, or lack of adaptation, living things make to their environments. As each generation is produced it contains certain differences—variations—from the genetic makeup of the previous generation. As these variations accumulate over time they sometimes allow the organism, and the population they are part of, a slightly better ability to adapt to and survive any changing environmental conditions than they previously had. As the variations continue to build up that group can reach a point where they are more different than similar to the original population they were part of. At that point they cease being the original population and become a new species.

While questions about the immutability of life go back to antiquity, the real discussion of the topic began in the seventeenth century as part of the Enlightenment. This attempt to work out rational answers to natural phenomena was prompted in part by recent discoveries in geology and biology as well as a growing fossil record. The discussion burst upon the Western consciousness in 1859 when Charles Darwin published *On the Origin of Species*. What Darwin did was to give a workable explanation for how transmutation operated, the core of which was the process he called natural selection.

The idea was controversial right from the start and Darwin and his work were both praised and condemned. While sides formed, they were not always easily categorized. The reaction was oddly mixed: some scientists rejected evolution while some theologians approved. Even those who did accept the basic premise of evolution were wary of the idea of natural selection. As a result of these genuine questions and doubts about the idea it did not become widely accepted at first: In fact Darwinism (not evolution itself but the concept of natural selection) went in and out of fashion for decades. It came into acceptance slowly as more and more evidence supporting it accumulated. Mountains of evidence, both figurative and literal, grew during the nineteenth century giving the concept an increasingly strong empirical base. The earth gave up evidence in the form of readable geologic formations which included a wide array of fossils—from plants to fish to mammals to the fantastic dinosaurs—which had very different living relatives and in some cases no living relatives at all and which showed the course of biological change over time.

As the twentieth century appeared a new experimentalist approach came over biology and paleontology. This along with the rediscovery of the pioneering work on heredity done by Gregor Mendel led to the science of genetics. This was of enormous consequence for evolutionary studies as now there was another way to explain how evolution worked to go with the fossil record. The nuclear age brought on knowledge of atomic structure and radioactive decay as a way to work out the age of the rocks and fossils in an increasingly sophisticated and reliable way. These discoveries showed the world was indeed an ancient place with an age more than enough to accommodate the time spans needed for evolution to operate. By the end of the twentieth century evolution had gone beyond an abstract theory and moved into the realm of an established fact accepted by biologists as the underlying mechanism of life on earth.

The course of evolution—the living things it produces, and the people who study it—form a story which is alternately fascinating, inspiring, intriguing, dramatic, tragic, perplexing, and exciting. It is a story billions of years in the making and one which none of us can afford not to understand.

Chronology

200s	B.C.E. Aristotle calls the idea of organisms striving to perfection "orthogenesis."
800s	B.C.E. Greeks think fossils are the relics of their ancestors, the heroes and Titans.
1500s	Legends of wild men and monstrous races abound in Europe.
1654	Archbishop Ussher calculates the day of creation to be 4004 B.C.E.
1667	Nicolaus Steno demonstrates that fossils are the remains of once-living creatures.
1677	Sir William Hale's *Primitive Origination of Mankind* is published.
1698	Edward Tyson creates the first description of a chimpanzee in London.
1709	Johann Jakob Scheuchzer describes the *Homo diluvinii testis*.
1735	Karl von Linné (Linnaeus) publishes the *Systema Naturae*.
1749	Count de Buffon publishes the *Histoire Naturelle*.
1750s	Johann Gottlob Lehmann and Giovanni Arduino work out the theory of the superposition of geologic strata.
1775	Johann Friedrich Blumenbach proposes the idea of five races of man.
1795	Erasmus Darwin publishes *Zoonomia*.
1800–1850	The hierarchical scale for geologic time is worked out, though not the actual dates for the various ages.
1802	William Paley publishes *Natural Theology*. Jean-Baptiste Lamarck proposes the law of acquired characters.
1807	The Montmarte fossil is discovered outside Paris.
1809	Charles Darwin is born.

1812	Georges Cuvier and Alexandre Brogniart work out the concept of extinction.
1820	Georges Cuvier proposes the ideas of catastrophism.
1822	Gideon Mantell finds the teeth of the *Iguanodan*. Gregor Mendel is born in Silesia (now the Czech Republic). The Reverend William Buckland visits Paviland cave and discovers the Red Lady.
1824	William Buckland makes the first scientific description of a dinosaur.
1830	Charles Lyell publishes *The Principles of Geology* and promotes the uniformitarian view of geology. The Engis skull is found in Belgium. The Devonian controversy rages.
1831–1836	Darwin's voyage on the *H.M.S. Beagle*.
1833	William Jardine publishes *Natural History of Monkeys*.
1838	The first recognized horse fossil (*Hyracotherium*) is discovered then described by Richard Owen.
1838	William Lawrence publishes *Lectures on Physiology, Zoology, and the Natural History of Man* based on his lectures.
1841	Richard Owen names large fossil reptiles "dinosaurs."
1844	Robert Chambers publishes *The Vestiges of the Natural History of Creation*.
1848	The Gibraltar skull is found.
1856	The Neanderthal Man is found in Germany.
1857	Hermann Schaaffhausen unveils Neanderthal Man at the meeting of the Natural History Society of Prussian Rhineland and Westphalia.
1858	Alfred Russel Wallace contacts Charles Darwin, asking him to comment on his new explanation of transmutation.
1859	Charles Darwin publishes *On the Origin of Species*.
1860	The Oxford University debate between T.H. Huxley and the Reverend Wilberforce.
1861	The Berlin specimen of *Archaeopteryx* is discovered in Solnhofen, Bavaria, Germany.
1862	Ernst Haeckel introduces the tree-of-life image and coins the term *phylogeny*.
1863	T.H. Huxley publishes *Man's Place in Nature*. Charles Lyell publishes *The Antiquity of Man*. James Hunt starts the Anthropological Society of London to promote the polygenetic approach to human diversity.
1864	Herbert Spencer coins the term "survival of the fittest."
1865	John Lubbock publishes *Prehistoric Times* and arranges the first modern scientific names for geologic periods.

1866	Gregor Mendel publishes his article on pea plant variation and heredity.
1866	The La Naulette skull is found in Belgium.
1868	The first Cro-Magnons are found at Les Eyzies, France.
1871	Charles Darwin publishes *The Descent of Man.*
1875	Francis Galton puts forward his idea of inheritance working through units he calls "stirps."
1876	T.H. Huxley visits O.C. Marsh at Yale University to view Marsh's horse fossil collection.
1876	S. George Jackson Mivart publishes *Man and Apes.*
1879	Ernst Haeckel publishes *The History of Creation.* The U.S. government establishes the Bureau of American Ethnology.
1880	August Wiesmann differentiates between somatic and reproductive cells. Gabriel de Mortillet modifies Lubbock's system, adding Acheulian, Aurignacian, Solutrean, and other ages to the geologic time scale.
1888	Galley Hill Man is found along the banks of the Thames River. Madame Blavatsky publishes *The Secret Doctrine.*
1890– 1895	Eugene Dubois finds Java Man.
1890s	William Bateson studies variation in heredity.
1890s	Western naturalists realize China is a trove of possible hominid fossils.
1900	Hugo de Vries and others "rediscover" the work of Gregor Mendel.
1903	Ernest Rutherford and Frederick Soddy discover that radioactive decay generates heat.
1903	Max Schlosser publishes *Fossil Mammals of China.*
1905	John William Strutt does the first experiments in radiometric dating.
1906	The first Nebraska Man is found outside Omaha.
1908	George McJunkin finds buffalo fossils outside Folsom, New Mexico. The La Chapelle-aux-Saints Neandertal skeleton is found. Marcellin Boule calls it "bestial."
1912	Piltdown Man is found at Sussex.
1921	Otto Zadansky finds a hominid tooth at Zhoukoudian, outside Peking.
1921– 1930	Roy Chapman Andrews, sponsored by Henry Fairfield Osborn and the American Museum of Natural History, leads the Central Asiatic Expedition to Mongolia in search of the cradle of man.
1922	The second Nebraska Man is found.

1925   T.H. Morgan begins his experiments on heredity with
        fruit flies at Columbia University.
1925   Henry Fairfield Osborn proposes the Dawn Man theory.
        The Scopes monkey trial is held in Dayton, Tennessee.
        Raymond Dart announces his discovery of Taung Child
        (*Australopithecus africanus*) in South Africa. The photo of
        François de Loys's ape is circulated.
1926   Otto Zadansky makes the Peking tooth public. A.W.
        Grabau dubs it Peking Man, and Davidson Black writes
        the descriptive paper. The Folsom points are verified to
        show that humans entered the Americas 10,000 years ago.
1931   The National Academy of Science proclaims the earth to
        be at least 1.6 billion years old, but no more than 3 billion
        years old. Pei Wenzhong finds the Peking Man skullcap.
1935   Jia Lanpo takes charge of the Zhoukoudianzhen site and
        tries to maintain it in the face of World War II.
1940s   Neo-Darwinism, or the modern synthesis, is developed.
1941   Franz Weidenreich attempts to ship Peking Man to the
        United States for safekeeping, but it is lost in transit.
1949   William Libby develops carbon-14 dating at the
        University of Chicago.
1950s   H.B.D. Kettlewell does his research into peppered moths.
        Radiometric dating begins.
1953   Piltdown Man is exposed as a fake. James Watson and
        Francis Crick work out the structure of DNA.
1964   Louis Leakey announces his discovery in Africa of *Homo
        habilis.*
1966   Willi Hennig develops cladistics as a way of organizing
        living things by their physical characteristics.
1967   Alan Wilson begins the genetic study of human evolution.
1974   "Lucy" is found in Ethiopia.
1980s   Christopher Stringer proposes the Out-of-Africa II
        population replacement theory. Milford Wolpoff proposes
        the Out-of-Africa I multiregional theory. *Homo ergaster* is
        described.
1987   The *Edwards v. Aguillard* case is decided by the U.S.
        Federal Supreme Court. The court finds any explanation
        for biological origins that is religiously based cannot be
        taught in the public school system as science.
1980s   The concept of intelligent design reappears.
and 1990s
1999   Kansas Board of Education removes references to
        evolution from state-approved textbooks. A series of

court cases go back and forth until evolution studies are
reintroduced back into the school curriculum in 2006.

2006   The *Kitzmiller v. Dover* case is won by the pro-evolution
          parents of Dover school district children. The judge
          declares Intelligent Design to be religious and not
          scientific.

Evolution as a Paradigm

Olof Ljungström

What is a basic definition of the theory of evolution these days? Well, one is "descent with modification." At the heart of it lies the assumed common ancestry of all life. The history of this descent has at times been described using the metaphorical figure of a branching "tree." Modern genetics provides us with understanding of the mechanisms of reproduction and inheritance. It is combined with the mechanism of "natural selection," briefly that the genes of an individual, which confers an adaptive, reproductive advantage, will be passed on to a greater extent than the traits of other individuals of the same species. Together they form what has been labeled "the modern synthesis" of the theory of biological evolution. Evolution is also seen to operate on two different levels; *macroevolution* being the process of descent from a common ancestor of different species reconstructed through the fossil record, and *microevolution* observable as changes in gene frequency in a population from one generation to the next.

The theory of evolution is the framework within which science currently operates, even as attempts are made to fault, fine tune, or expand upon it. Its opponents are obliged to come up with a better alternative if they want to supplant it. As a scientific theory it is provisional. Its saving grace is that it is not fixed for all eternity. The theory of evolution does not offer a cut-and-dried scenario of the origin of life and the exact history of life, that of man included. What it holds out is rather a powerful model for explaining organic change, and a framework which allows for the explanation of what's being observed with more consistency than any hitherto formulated alternative.

It can be asked whether evolution as a concept qualifies as a paradigm. And what is meant by one? "Paradigm" was introduced into the

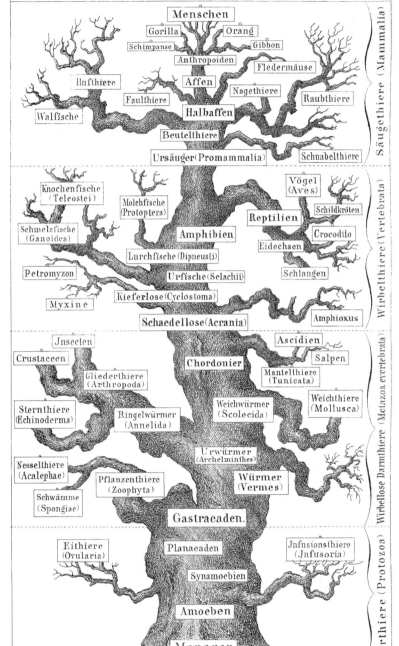

Menschen

Gorilla — Orang
Schimpanse — Gibbon
Anthropoiden
Fledermäuse
Ilnfthiere — Affen — Nagethiere
Faulthiere — Raubthiere
Walfische — Halbaffen
Beutelthiere — Schnabelthiere
Ursäuger(Promammalia)

Säugethiere (Mammalia)

Knochenfische (Teleostei)
Molchfische (Protoptera)
Vögel (Aves)
Schildkröten
Schmelzfische (Ganoides)
Reptilien
Crocodile
Amphibien
Eidechsen
Lurchfische (Dipneusti)
Petromyzon
Urfische (Selachii)
Schlangen
Kieferlose (Cyclostoma)
Myxine
Schaedellose(Acrania)
Amphioxus

Wirbelthiere (Vertebrata)

Jnsecten — Ascidien
Crustaceen — Salpen
Gliederthiere (Arthropoda) — Chordonier
Mantelthiere (Tunicata)
Weichthiere (Mollusca)
Sternthiere (Echinoderma)
Ringelwürmer (Annelida)
Weichwürmer (Scolecida)
Urwürmer (Archelminthes)
Nesselthiere (Acalephae)
Würmer (Vermes)
Pflanzenthiere (Zoophyta)
Schwämme (Spongiae)
Gastraeaden.

Wirbellose Darmthiere (Metazoa evertebrata)

Eithiere (Ovularia)
Planaeaden
Jnfusionsthiere (Jnfusoria)
Synamoebien
Amoeben
Moneren

Urthiere (Protozoa)

Haeckel del. Lith.Anst.v. J.G.Bach.Leipzig

The Tree of Life (facing page)

The choice of an icon to represent evolution in general is not a facile matter. It should be well known enough to measure up against the breadth and complexity of the phenomenon. On balance, the entire metaphorical figure of the branching tree of life was selected. There are older symbolic trees in human history; the old Norse of Scandinavia inhabited the Tree of the World, Trees of Life abound everywhere from the shamans of Siberia to the Jewish Kabala. The evolutionary Tree of Life is a metaphor representing the general shape of the relationship between different species over time, i.e., it is a metaphor for history of life. Like all metaphors it is problematic as it stands in the place of something that is not quite it. The tree of evolution is not evolution, it is not even the research being done in order to understand evolution and the history of life; it represents them. The tree here selected is one of the more famous nineteenth-century representations of branching evolution taken from a widely successful popular work by one of Charles Darwin's first converts, the German zoologist Ernst Haeckel's 1874 *Anthropogenic oder Entwicklungsgeschichte des Menschen* (*Anthropogenic or The Evolution of Man*, 1903; illustration from the copy in the author's private collection).

---

history of the sciences in 1962 by the philosopher of science Thomas Kuhn in his *The Structure of Scientific Revolutions*. It is used to describe a discontinuous model for how science works. The traditional view of the history of science was formulated already in the first half of the nineteenth century by the French positivist philosopher Auguste Comte, and the Cambridge mineralogist-philosopher William Whewell, who happened to coin the term "scientist" in the process. It stressed science as a continuous and cumulative process. How it supposedly worked was summed up in a quote from Newton, who had asserted that if he had been privileged to see further than other men, it was because he had been standing "on the shoulders of giants," the OTSOG hypothesis.

In opposition to this view, Kuhn analyzed the dramatic shift within sixteenth-century astronomy away from the model of a geocentric cosmos (earth as the center of the universe) to a heliocentric (sun-centered) one. In Kuhn's view this constituted a revolution by virtue of a shift of paradigm. Paradigms operate on a level above theories, where these interact and support each other. The shedding of one paradigm for another occurs when enough anomalies have cropped up within an existing paradigm, i.e., things inexplicable in accordance with present theory. At some point the hitherto dominant paradigm is discarded, and a new paradigm is formulated better able to provide explanations. The old and the new paradigms are incompatible. What is a fact in the former may not be one in the latter, what is a valid principle in one is invalidated in the other, and the set of options for theoretical explanations will be entirely different.

So, might evolution be called a scientific paradigm? As often, the answer can be yes and no. It occupies such a central position within a wide scientific field that it would seem to qualify as one. This is not least because of its seeming incompatibility with in particular religiously motivated views of the origin of the organic world. And there certainly can be said to exist an incompatible before and an after the advent of evolutionary thinking. On the other hand the formulation of evolution was such a drawn-out process, the lack of suddenness makes the label "revolution" seem less appropriate.

Postulating Darwinian evolution alone as the paradigm, and the true scientific revolution, would not seem to make matters clearer either. Historically the concept of evolution is a rather nebulous and unspecified set of divergent concepts and attitudes, of which Darwinian evolution forms a specific alternative. To give an example, historically the origin of life is part of this wider sphere of interests. It is not necessarily so in Darwin's original theory of evolution by natural selection, as it is mainly concerned with the mechanism behind adaptation and organic change, but not with the origin of life, beyond the fact that life on earth began at some point in time. The reason supporters of the theory of evolution are requested to provide a definite answer to the origin of life is due less to Darwin, than to that wider set of theories of evolution and the history of life stretching back to the eighteenth century. A review of this ensemble might show in what way evolutionary thinking represented an unreconcilable break with previous views. So this is where we now turn.

TELEOLOGY AND THE ARGUMENT OF DESIGN IN NATURE

> In His great goodness, God created the cat, that man could caress the tiger.
>
> – Alexandre Vialatte

In the mid-eighteenth century a pious French reverend, the Abbé Pluche, took a keen interest in the study of nature in order to argue, in a quite popular multivolume work, how it revealed the inimitable wisdom of God in the wonderful way in which everything was designed for a purpose, fit together, and displayed perfect balance. Of course, in his enthusiasm the pious French Abbé occasionally went overboard even in the eyes of his contemporaries by exhorting his readers to contemplate how wonderfully purposefully the tides allowed ships to enter and exit harbors.

Quaint anecdote aside, the Abbé Pluche's writings are a good example of eighteenth-century notions of how God was assumed to reveal Himself in nature. Intentional or not, today's intelligent design share at least a superficial resemblance to what for the eighteenth century has been termed the "argument of design." A major difference is of course the lack of two centuries of hindsight. Most typical of the eighteenth century was the insistence that everything was not only the product of divine "fiat", but had also been designed with a specific goal or purpose in mind. This view is commonly described as "teleological" (Greek *telos*, goal). It is like the quote above puts it: The purpose of the cat is to let man caress it as a proxy for the beautifully dangerous tiger.

There is a handy term for describing the pious eighteenth century mix of keen nature study, and reverence for the ingenious ways in which the Almighty had crafted everything. It was called *physico-theology*, after the 1713 popular treatise by William Derham. It was not an unsophisticated view of nature, or one lacking in shrewd observations, like the natural balance between how predators cull their prey no more harshly than to maintain their numbers (Bowler, 1984). However, while the default position within the Western world for most of its history has been that God had created all things, this did not necessarily imply that everything had always remained the same. In order to understand the shifts in our thinking about the natural world, it's necessary to briefly review the options for historical thinking available.

OF PROGRESS AND DEVELOPMENT

Intimately tied to the concept of evolution is the idea of progress. Darwinian evolution's lack of progressivism has always been one of its aspects hardest to sell. Even today progress is virtually taken for granted, something we less think about than use to "think with" when looking at the world. But progress was not originally part of the European historical imagination. Instead its legacy consisted of the histories of classical antiquity and the Bible, neither of which are in fact very progressive. Biblical history provided a macro-historical linear framework for the history of the world, human, organic and inorganic alike, with a clear beginning and end. Greek and Roman history provided micro-historical moral lessons and models of exemplary behavior with a focus on virtues and choices of action in war, diplomacy, and politics. They complemented each other nicely.

Historically, progress was for very long the less explored alternative in comparison with a view of mankind and nature as undergoing a process of degeneration. It was nurtured by the extraordinarily high regard in which Europeans held the works of classical antiquity. The self-imagery of the Renaissance was as the rebirth of a glorious past, but should not obscure to us that this recapture of lost greatness only gave the modern age a tenuous claim of, at best, equality. Renaissance and seventeenth-century Europe was not necessarily a happy place, rife with wars of religion, plague, and an economic stagnation. One of the more popular attitudes toward the present became the theory of "senectus mundus," of the aged world. The world and man in their present state were not simply fallen, as per Scripture, but had been skidding ever since. Compared to ancient man, living nearer to the time of the Garden of Eden and God's proximity, present-day humanity was stunted, dull, and full of vice. This was the intellectual view of the time. Thomas Hobbes in his *Leviathan* (1651) concluded that the only way to stop the natural state of anarchic war between everyman was through a social contract handing the ruler absolute power. Those serious about religion, especially among seventeenth-century Puritan Protestants, most decidedly felt they were living at the end of days. Considering the rotten state of the world, it would come none too soon in the opinion of many.

In comparison with this melancholy and pessimistic age, European spirits picked up remarkably in the eighteenth century, as this is where both progressivism and explicit theories of evolution were formulated. If one is to indicate a start for this shift, there was a curious literary and artistic debate in late seventeenth-century France. It was the first stirring of the today positively charged concept of being "modern," intimately tied to ideas of progress. It made its unassuming debut when the "Ancients" maintained that the perfection of the art and literature of the Greek and Romans was such that there could be no question about equaling, much less topping it. The "Moderns" of course held that present humanity had at least a shot at equality.

This newfound self-confidence was expressed by the philosophical movement of the Enlightenment in the eighteenth century. Enlightenment philosophy held out the promise of improvement and progress of human society by the application of reason to its problems. At the time this was a far more radical proposition in autocratic France than in the British Isles. Taking England as their social model, the French philosophers targeted royal absolutism, the organized religion, but above and beyond all the society of privilege (literally meaning "private law"), where noble birth was next to the only way for a man to get ahead in life, regardless of ability. In opposition they offered the principle of division of political power (Montesquieu), and the ideals of meritocracy and reason.

The extent of the optimism about the general direction of human history and society found its finest expression in the 1795 *Esquisse d'un tableau historique des progrès de l'ésprit* (sketch of a historical tableau of the progress of reason) by the mathematician Condorcet. This is not only due to its incredibly upbeat message, but also the circumstances under which it was written. Condorcet was telling his audience that human reason has a limitless ability for improvement, history is a record of its progress, and the future opens even brighter vistas, but he finished his work in the jails of the revolution, where he also died shortly after (murder or suicide). This general trend forms a backdrop to the formulation of specific theories the inorganic and organic worlds alike as the outcome of processes of development.

TOWARD A NATURAL HISTORY OF PROGRESS

The natural history in the eighteenth century had already shed the authority of classical antiquity. The likes of Pliny the Elder and Aristotle were up against the hitherto unsuspected vastness of the world and the multitude of beings inhabiting it. As European overseas exploration, trade, and conquest picked up, there was an increasing need to make sense of it all, to describe, collect, and bring back new plants and animals from all over the globe. The natural history of the Greeks and the Romans as well as the medieval scholastic philosophers had been cast in a hierarchical vision of the natural world commonly referred to as "The Great Chain of Being." Codified in the thirteenth century this symbolic representation of the world kept at least some of its currency all the way into the nineteenth century. It represented all of creation as links in a chain stretching hierarchically from unconscious base matter, up through vegetative nature, through the animal kingdom before reaching man. Humanity was positioned above the animals, but tied to the same world of earthly matter as they. Above man stretched further links in the chain in the form of beings of pure spirit, the angels, leading all the way up to God. These were however not part of the historical framework of fall and redemption like man, the protagonist of the great drama of creation. The aspect of the Great Chain of Being still with us, in a sense, is the interest in "missing links." The original formulation of the theory implied that a complete chain must be in existence. A seeming gap in the hierarchy meant an intermediary being must be presumed to exist somewhere. It is in the sense of presumed intermediary, transient forms between two kinds of organisms, that an aspect of the Great Chain can be said to remain with us.

Typical of the natural history of the eighteenth century was the tabular arrangements for the classification of all things. This grand stock-taking of creation was in itself an older idea, and a pious one. The plan of God for his creation was assumed to be discernible in the details. Upon obtaining the complete picture, this would be revealed. From the Renaissance onward cabinets of curiosities were assembled with a mind to allow a grand synthetic view of creation. These cabinets evolved into larger collections and eventually the various museums of natural history, ethnography, and so forth established from the late eighteenth century onward.

The culmination of this kind of tabular systematization was the binominal system of classification proposed by the Swedish botanist Carolus Linnaeus (Carl von Linné, 1707–1778) in his 1735 *Systema Naturae*. It was widely accepted and is still in use due to its convenience for naming new species, even if modern research has overridden many calls made by Linnaeus. His greatest forerunner was the English clergyman John Ray, who in the latter half of the seventeenth century tackled the basic problem of classification. The foundation for the work done by Ray, and later Linnaeus, lay in Aristotle's *History of Animals*, and the use of basic Aristotelian logic as a guide to classification. From Aristotle they carried over the nomenclature of "class," "genus," "species," and "genera," in descending order of generality. What they wanted was a classification of all beings according to the true composition of the natural world, the plan of God, a "natural system" (Young, 1992). The problem lay in working out the basic plan of creation that such a system could be based upon. It would otherwise simply be an "artificial system."

Other botanists besides Ray (the Swiss Bauhin, the Frenchman Tournefort, the German Rivinus, and others) proposed their own taxonomic solutions. Linnaeus in no way had to formulate the problems for which he suggested his solution. Apart from the voracity of his appetite for botanical examination, and, at least initially, a considerable measure of youthful overconfidence, Linnaeus's contribution was twofold: He proposed a system of botanical classification based on the composition and anatomy of the sexual parts of flowers (the stamen and the pistils), where earlier systems had used the shape of the leaves, or of the flowers themselves, and he proposed his binomial Latin nomenclature. The former could only be applied to plants of course, but it was the kingdom of plants which had caused the bulk of the problems of classification. (Animals were easy by comparison, and already Ray proposed a binomial system for their classification.) Linnaeus eventually used his naming principles for the classification of all living organisms. The ease of use of the nomenclature was a considerable selling point. To choose an example from the mammals (so named by Linnaeus from the Latin *mammae*, breast, for those who suckle their young): the "dog family," the *canidae*, is divided into the genus *canis*, dog, and the genus *vulpes*, fox. The *canis* subdivide into the species *familiaris* and *lupus*, by their common names dogs and wolves. For practical naming purposes the common dog is of the genus and species of *canis familiaris*. Man, of course, is of the genus of *hominidae*, species *homo*, and the genera *sapiens*, "the thinking man."

Including man into his system of classification was a bold step on Linnaeus's part. Man's connection to the natural world, while not in dispute, was at the time taking a far second to his position as a spiritual being. Linnaeus also made a quick and dirty further subdivision of man into four "varieties" on the level below that of species, one for each continent and with its own color; white, black, red, and yellow. It would seem almost inevitable that his fury for the classification of all life would include man as well, but it was perhaps not his most positive contribution, as it formed the basis for later more elaborate systems of racial classification.

Linnaeus's system seemingly brought the main concern of the eighteenth-century natural historians, that of classification, to a conclusion. He did so, at least in the case of plants, by focusing on their procreation in his "sexual system" of classification. Though he was attacked for it, he was in fact aware that the species he recognized were not abstractions of form, but represented collectives of individuals breeding together. In the application of the system this aspect was however left out. If the Linnean system of classification is a good example of the project of systematization in tabular form that so concerned eighteenth-century natural history, this is in part because it did not allow for organic change. The species were fixed and there to be discovered, described, and named.

Despite the optimism of men like Linnaeus, who from his home in Uppsala, Sweden, sent out dozens of disciples to virtually every corner of the globe (many of them never to return) to augment the university's collections, all was not well in the presumed static nature. Thorough as their observations were, and powerful as a model for explanation as physico-theology was, certain observations did not sit well. Nature provided wonderful examples of the utility of most things for man, but what on earth was one to make of parasites, intestinal worms for example? There was also the discovery of the dug-up remains and traces of long-dead mysterious creatures unlike those already known. Already John Ray had grappled with the issue. It was of course possible that they did represent extinct species, and biblical history had a ready explanation in the form of the deluge. Still, Ray balked at the possibility of extinction, and instead preferred to assume that these creatures still existed in some far-off corner of the globe. As the European scientific "global consciousness" (Pratt, 1992) expanded over the next century, the potential hiding grounds for these creatures shrank alarmingly. The idea that perhaps nature was not static and that new species arose under certain circumstances could not be laid to rest. If that was occurring, what was the process like?

A model of organic change was in fact proposed by Linnaeus's greatest contemporary rival the Count de Buffon, the director of the Jardin du Roi, the royal gardens in Paris, who in 1778 published *Les époques de la nature* (The epochs of nature) as an attempt to foot geology on a base of Newtonian physics. His theoretical stance in natural history did contain a definite consideration of organic change, and his geology was tied to it. Buffon proposed a theory of the history of the earth, which required a considerable amount of geological time. Supposedly the count conducted experiments at his country estate by letting white-hot iron cannonballs cool. By extrapolating the cooling time and the difference in mass between cannonball and planet, he arrived at an estimated age of our world of at least 75,000 years. That was the public figure, in private he conceded that it must be much higher.

From our present-day perspective of a world four billion years old, 75,000 may look unimpressive. However in comparison with the then-current framework of biblical chronology, allowing a mere 6,000 years of history, it was a lot. The most well-known biblical chronology was calculated by the Archbishop James Ussher (1581–1656) of Armagh, Ireland. According to him the first day of creation occurred on Sunday, 23 October 4004 B.C.E., Adam and Eve were driven from Paradise on Monday, 10 November 4004 B.C.E., and the Ark touched down on Mt. Ararat on 5 May 2348 B.C.E., which was a Wednesday. In comparison with that, Buffon's estimates opened up entirely new vistas of possibilities for thinking about geological and biological change.

As a natural historian Buffon established his intellectual and scientific credentials by taking on Linnean classification already in his first published work, his 1749 *Histoire Naturelle*. Buffon rejected what he considered an artificial system of classification based upon abstract relationships of form. For Buffon the relevant groupings in nature were populations maintaining themselves over time by breeding. From this followed his famous definition of species as a group of organisms capable of breeding.

Buffon was also a keen observer of the geographical distribution of organisms to the various continents and climatic zones. The problem of how plants and animals had managed to spread to their contemporary localities was not novel. Believers in Scripture dealt with it by simply assuming that they spread from the Ark on Mt. Ararat. Linnaeus had been slightly more ingenious, postulating that all species must at some point have inhabited a huge mountain on the Equator, as he had observed how climatic zones were mirrored in a change of climate depending on altitude. Buffon now made observations to the effect that most species are specific to certain regions and climates, and are in fact incapable of living under most other conditions, inaugurating the study of plant and animal geography, or biogeography. Regardless from where present-day animals had originally spread, it was inexplicable how they had managed to make their way to their habitats in their present shape. Something must have changed.

At this point Buffon might presumably have proposed an evolutionary solution. Instead he opted for the alternative of degeneration. Buffon assumed that all living species must be descended from an ancestral form defining the family on the superior level of classification, and these ancestral forms were still around. He knew of some 200 species of mammals and proposed 38 of these as ancestral. As they spread across the globe, they encountered different environments, and their bodies responded by degenerating. These changes were not assumed to be permanent, but reversible, should these degenerated new species be transported back to their original environment.

In this Buffon expressed a marked preference for Europe over the Americas, which he considered geologically younger and retaining more moisture. Over there plants, animals, and humans were all slightly soggy, smaller in stature, and weaker than the noble beasts of Europe. The European colonists he described as undergoing a process of stunting, a charge vigorously rejected by Americans like Benjamin Franklin and Thomas Jefferson present in the French capital from time to time in the last decades of the eighteenth century. Franklin even attempted to demonstrate the falsity of this theory at a dinner party. He asked the company present to stand up, and the half-dozen or so colonials present towered over the other guests.

THE IMPORTANCE OF GEOLOGY FOR EVOLUTION

As indicated by the example of Buffon, geology was of paramount importance for the development of any theory of organic change. Geology "sets the stage" for everything else. If the earth itself could be proven to have undergone change and development in a long or a short time span, this lent credibility to the concept of organic change. The great eighteenth-century authority on geology was the German Abraham Gottlob Werner (1749–1817), professor of geology at the Freiburg mining school. In 1786 Werner divided the earth's crust into primary, secondary, and tertiary strata. The topmost most modern deposits were termed "alluvium" but would later be relabeled as "quaternary." Of these the primary is formed by the bedrock of granite and gneiss, while the secondary consist of various strata of softer rock like limestone and sandstone, i.e. deposited rock, just like the tertiary strata consist of soft deposits like clay, sand, and chalk. Unlike the others, the primary strata contain no fossil. The secondary have an abundance of marine fossils, while the tertiary are where the fossil of various land-living animals can be found. Werner's great contribution was his thorough empirical study of the earth's crust, and the understanding of the history of geological formations it provided. Where he ran into opposition was over the mechanisms behind the observations.

Werner's geology hinged on erosion by water, washing matter into the oceans to be deposited on the seabed as sediment. It was a process that required very considerable amounts of geological time to function. Due to the insistence on the agency of water, Werner's and his followers' position became known as Neptunism after the Roman god of the sea. At the very end of the eighteenth century they were challenged by the school of Vulcanists, this time named after the god of fire and craftsmanship. The known active volcanoes are limited in number, but closer observations of these allowed geologists to identify extinct volcanoes, more plentiful than hitherto suspected. Suddenly volcanic activity was a contender as the main agent for the formation of the earth's crust. The most important geologist of this school was the Scotsman James Hooton, who found evidence to support the conclusion that the primary rocks were in fact a form of unerupted lava.

While the view of Hooton and the Vulcanists may have been closer to a modern understanding of the origin of the primary rocks, Wernerian geology provided most of a truly historic reconstruction of geological history. Hooton's steady-state view of the earth was of a machine operating in perfect cyclical balance. Between them both groups established a sense of geological time that had to be measured in millions of years or else the evidence made no sense. It also became clear that the history of the earth itself needed to be understood as a process. The inorganic world was by now established as having a long and eventful history possible to reconstruct from present-day observations. In light of this, the view of the organic world as fixed and unchanging also took a beating.

Here the question of evolution as a paradigm can again be raised. With regards to scriptural literalism, that would certainly be the case. However, it was not the exclusive religious viewpoint. The majority of the scholars proposing a long and complex history of change certainly considered themselves good Christians. What is more important to point out is the way in which observations and theories about the organic and inorganic world alike became intertwined and mutually supportive. Rejection of one meant the rejection of all, and in the face of mounting evidence and persuasive logic, nay-saying increasingly meant a personal sidelining from the debates of science. Science itself operates by consensus and once a sufficiently coherent system of explanation has been reached, research tends to shift to other sets of questions, regardless if a minority within the group is still challenging it, or if it is still challenged from without.

LAMARCK AND THE FIRST THEORY OF EVOLUTION

It fell to the self-styled philosopher-zoologist Jean Baptiste le Chevalier de la Marck (1744–1829), or Lamarck after the revolutionary fashion, to provide the first coherent theory of organic evolution. He struck upon it fairly late in his career. He had already made his mark within natural history by for instance establishing the distinction between vertebrate (having a spine) and invertebrate animals used to this day. In 1802 he also coined the term "biology." He can be regarded as part of a tradition of philosophical materialism interested in the problem of spontaneous creation of life, for want of a creator. This tradition included people like Julien de la Mettrie, who in 1748 published *L'homme machine* (Man the Machine), and the baron d'Holbach, author of *Système de la Nature* (1770) at the time hailed as the "Bible of atheism." Lamarck also took a keen interest in geology and chemistry, and what he took from them was the lesson that inorganic nature was anything but static. Considering that, what were the odds of organic nature being immutable?

Lamarck made some crucial assumptions about living organisms. First of all organisms were assumed to progress in development from simpler to more complex. Second they all progress upward in a predetermined linear fashion. This was evolution of a pre-programmed kind. Individual species were not progressing, but rather whole classes of creatures. What may feel counter-intuitive to modern readers is how Lamarck assumed that as a class of animals moved up the ladder of evolution, it left the rung below empty. Constantly pushing from below were new species of the lowest order, the worms, brought into existence through spontaneous creation. This was a fully historical theory of evolution as the present relationship between organisms was the assumed outcome of a real chain of events. Lamarck's theory left you with the conclusion that man himself was in fact the oldest species by virtue of being at the pinnacle of nature. Our ancestors started out as worms. The contemporary worms would one day become men. Destruction at the top was also continuous with successive generations of men extinguished and returned to the inorganic world from which life could go on generating spontaneously.

Lamarck went so far as to draw a visual representation of the hierarchy of evolution where the different positions represented both difference in level of organization and different histories. He did so in the first metaphorical tree of evolution. It represented the steps up the ladder which a class of animals were assumed to have traveled. It was not a tree of branching evolution assumed to represent the specific history of all life as a unique one-off event, but rather the general history of all life as a linear continuous process. Modern worms were not related to modern humans, and neither were any of the species in between, since life had not been created as a singular event, but as an ongoing process. All classes of animals traveled along individual trajectories, however, which mirrored each other.

In retrospect the part of Lamarck's theory of evolution that gathered most attention was his mechanism for adaptation, assumed to produce new species. According to Lamarck changes in environment changed an animal's "activities," "efforts," and "habits." Practice making perfect the animal acquired certain beneficial modifications. These were then passed on to its offspring. The next generation would be more proficient from birth than their parents, and push the new trend even further before passing it on in its turn. This has been called the "inheritance of acquired characteristics."

The prime example of how Lamarck thought it applied was the giraffe's neck. Originally short-necked, certain giraffes had, by stretching for the leaves on higher branches, slightly lengthened their necks; a very useful adaptation for the individual. This somewhat longer neck acquired through exertion was passed on to their young who in their turn kept stretching that neck even further, until the present-day long-necked giraffes were a reality.

The mechanism behind adaptation was not Lamarck's invention, bur rather a commonsensical view at the time. What was novel was the way it formed part of an explicit theory of evolution. It would in the nineteenth century become a central feature of theories of progressive evolution known as transformism within biology, as well as the preferred biological mechanism within Social Darwinism. Evolution through natural selection eventually provided a different answer to the problem of the giraffe's long neck, where the trait of a longer neck confers a reproductive advantage as some individuals can forage higher up in the trees, having more offspring sharing the beneficial trait. But this was the reason why later Social Darwinism would prefer Lamarckian adaptation. It was progressive and purposeful, where natural selection was open-ended and random. In a sense it put individuals in charge of their own evolution, in a way reconcilable with what was seen as the moral lesson of laissez-faire capitalism: Those animals that prospered where those who applied themselves. The lazy giraffe had remained short-necked and paid the price for it. Natural selection, on the other hand, made evolution random and beyond individual control as some individuals would get the longer necks and benefit purely by accident.

THE ADVENT OF PALEONTOLOGY

As is indicated by his changing name, Lamarck's career straddled the French revolution, Napoleon's empire, and the royal restoration. He started his career in Buffon's Jardin du Roi and ended it in the model institution of the revolution, the much-expanded Paris Museum for Natural History, established in 1793. This museum became the blueprint for such museums across Europe not least for the new lines of research initiated there. The Paris Museum for Natural History saw the inauguration of paleontology, the systematic study of fossils in order to reconstruct prehistoric life. This is treated in more detail in the chapter of this book dedicated to the fossil index. The key figure here was the comparative anatomist Georges Cuvier, generally identified as the founder of the science of paleontology, even if it was so named only much later by Cuvier's British supporter Richard Owen. As professor of comparative anatomy Cuvier benefited from effects of scale in the analysis of animal anatomy. The Paris museum was a combination of a number of collections and Cuvier could literally scan a larger material faster than any anatomist in prior history.

On the basis of this Cuvier undertook a complete revision of the classification of the animal kingdom. Linnaeus, and Ray before him, had used binomial systems, identifying class, genera, and species. Cuvier now realized that there were just a few basic structures to which all living organisms conform. Taking these as the basis of a system of classification he finally arrived at what was a natural system of classification. It still holds up rather well as his conclusions could eventually be reinterpreted in the light of Darwinian evolution, but reformulated as "phylum." Cuvier recognized four basic plans for all animals: the *vertebrata* (having an internal skeleton, and a spine, from fish to man), the *mollusca* (not possessing an internal skeleton), the *articulata* (arthropods, insects, shellfish, etc.), and the *radiata* (radially shaped starfish, sea-urchins, etc.). He referred to these as *embranchéments*, the major branches of life.

It was also Cuvier's contribution to have established what the historian of geology and paleontology Martin Rudwick has labeled "the reality of extinction" (1992). He established the prehistoric existence, and inferred extinction, of a large number of prehistoric mammals, many not unlike present species but in gigantic form, others even more curious and never seen in the flesh by man. His efforts were limited to the tertiary strata, but soon others using his methods began to delve into the older, secondary strata. This was where the today so-famous reptilian fossils named dinosaurs, "terrible lizards," so named in 1841 by Richard Owen, began turning up from the 1820s onward. At the same time botanists started working on the reconstruction of prehistoric plant life, using fossilized flora. What Cuvier had inaugurated was a general program for research geared toward finding, describing, and naming prehistoric life forms and their environment.

At this point, sometime around 1830, scientists had to their own satisfaction established that the earth was in fact extraordinarily old and that once upon a time now-extinct species of both plants and animals had lived on it. Geologists had a basic outline of geological time, and biologists and budding paleontologists knew that the flora and fauna of the various geological epochs were distinctly different. They also had the methods and concepts to begin the detailed reconstruction of what prehistoric life had looked like during the different geological ages. Even a logically consistent theory of evolution had been proposed by Lamarck. One might then expect a wide recognition of evolution.

Often the actual situation was much more complicated. The paleontologists were going about their task very much like natural historians, the animal- and plant-geographers of the present-day flora and fauna. And just like these, their research project demanded no loyalty to any specific theory about what was being observed. Paleontology was not by necessity Lamarckian at the time, which is portentous for how this scientific community eventually received Darwinian evolution. It was not necessary to advance any theory as to why and how in this cumulative phase of paleontology. For Cuvier, who had inaugurated this line of research, it was felt to be sufficient that earth had gone through a series of distinct geological ages, each populated with its peculiar fauna and flora. Each age ended with some form of major upheaval, a "cataclysm," after which the next set of beings took over the earth. It begged the question where these new species came from, but Cuvier felt it to be superfluous to speculate about this as opposed to describing them. The cyclical destruction could of course be seen as God's handiwork, sending disasters and then repopulating the earth. Cuvier might not have intended it, but his line of argument became known as "catastrophism," and regarded as a theory in support of the continued assumption of a benevolent Maker supervising natural development.

As for biological development, disregarding Cuvier's opposition, a minority of biologists did feel inclined to propose speculative scenarios in order to explain the new vistas of prehistoric life opening up. Lamarckism was picked up by Cuvier's and Lamarck's colleague the zoologist Etienne Geoffroy Saint-Hilaire, who used it as an evolutionary mechanism to propose that the embranchements of Cuvier actually developed out of one another. Cuvier and Geoffroy conducted a very loud and very public debate of the issues in the late 1830s. The outcome was inconclusive, but while Cuvier's non-speculative line was always stronger, enough French biologists in the decades to follow adopted a version of "transformist" evolutionism akin to Geoffroy's. There was thus both evolutionist biology in France before Darwin, and a simmering debate over the issue of species change before the publication of *Origin of Species*.

GERMAN ROMANTIC IDEALISM AND "NATURPHILOSOPHIE"

The Cuvier-Geoffroy debate was hailed by the old polymath Wolfgang von Goethe as the most momentous thing occurring in world politics at the time. Goethe might be best known as a poet, novelist, and playwright, but he was also the successful Prime Minister of the small principality of

Weimar and a philosopher and natural scientist of repute. Goethe strad-
dles the literary and philosophical currents of the neo-classicism of the
late eighteenth century and the highly idealist romanticism of the early
nineteenth century. In Germany the particular impact of romanticism on
natural history gave rise to idealist "Naturphilosophie" very accepting
of linear, teleological progressive development. Highly speculative, often
very obscure in its discourse it presented a view of the world where matter had soul and all things in nature were constantly striving toward
fulfilling their potential, to go from "potentiality" to "actuality." For the
philosopher Schelling nature was possessed of a soul striving to gain
consciousness, and man was the end result, having achieved it. The zoologist Lorenz Oken was a pantheist, considering nature divine in itself,
and proposing, among many things, a theory for the origin of life as having arisen in the shallow surf of the beaches of a primeval sea from an
"Ur-slime" of little globules of earth, light, and water.

It may all seem very outlandish and far removed from science. Still,
the other side of a man like Oken was as a practical organizer of the
first international scientific congress, a liberal political activist, and a
very fine empiricist observer of nature, zoology in particular. Somehow
the nitty-gritty of everyday scientific work coexisted with the lofty philosophical speculations. The weakness of the combination was in how little
the empiricism supported the speculation, and how little of the speculation actually helped explain data. Naturphilosophie also provoked a
pretty harsh reaction after a few decades. German scientists turned again
to the empiricism of "Baconian" induction, at times overcompensating
by trusting in "facts" alone, to the point of rejecting theory altogether,
and attempting to do science without offering interpretation.

What is interesting to point out here is that unlike in Britain and the
United States, traditional religious scruples were already sidelined in Germany and France. In both countries scientists as professional specialists
could engage in their inquiries without much sense of having to uphold
public propriety already in the first half of the nineteenth century. By comparison there was quite a conventional religious sensibility about British
"gentlemen of science." Specifically with reference to ideas about biological evolution French and German science had already grown accus-
tomed to more radical proposals than their British counterparts.

The realization that the shape of the history of life was in fact one of
branching originated with the German anatomist Carl Ernst von Baer.
One of the most powerful ideas supporting linear progressive evolution
was the assumption that the development of a vertebrate fetus recapitulated the entire development of the species. This could either be understood as a recapitulation of the actual process of development, or more
idealistically, as running through a "chain of being," i.e., before ending
up human, a fetus would run through the entire chain of hierarchic

development of vertebrates. The fetuses of other species would stop developing and hatch at their designated places in the hierarchy. The theory would even resurface into the late nineteenth century among those arguing for linear teleological evolution. Von Baer realized that embryological development, far from displaying a process of recapitulation, was instead a process of branching specialization. The implication was that species could hardly be ordered in a clear hierarchy like linear progressivism would have it (Bowler, 1984). This gradual realization of the branching history of life was even given form in a diagram published in 1858, just a year prior to Darwin's *Origin*, by Germany's leading paleontologist H.G. Bronn.

THE BRITISH PREHISTORY TO DARWINISM

Turning to Britain, as far as the geological and biological sciences were concerned roughly the same set of ideas held sway as on the continent. There was however one taboo remaining that the polite society of the typically British gentleman scientists would not even contemplate breaking; the special position of man as a spiritual being. The exception might have been Darwin, who already by the 1840s was well aware of the almighty stink his theory under development would cause upon publication. As it happened he could forecast it from the reaction to the anonymous publication of an ambitious work aimed at the educated public in general, Robert Chambers's *Vestiges of Natural Creation* (1844).

The book was not cutting-edge research, but synthesized the works of geologists, paleontologists, and zoologists over the previous decades. Their findings were presented as a grand panorama of prehistory. It also proposed that species evolved through transformation very much in the way Lamarck had seen it. Different species were not related but were going through a similar process of linear development. The account fell short of a satisfactory mechanism to explain why this was so. What really made scandal however, was the way in which the author refused to make halt at man. The author of *Vestiges* overstepped the invisible line which placed man outside the harsh realities of biology, by boldly stating that not only was man simply another animal, but his intelligence was a direct result of a process of animal evolution. The work was published anonymously, but we now know its author was the Scottish publicist and amateur scientist Chambers. At the time however, many of those with some inkling about the nature of the work Charles Darwin was engaged in suspected him instead. The scandal over *Vestiges*, a decade and a half before *Origin of Species*, is important for the history of Darwinism, as it convinced Darwin to postpone publication in order to work out his theory's full implications for the status of man. *Vestiges* had showed him an outline of the shape of the conflict it would cause.

That other great influence on Darwin was the "uniformitan" geology of Charles Lyell, who proposed it in an attempt to reform geology in the 1830s. The various schools of "diluvialism" (great floods as the main cause of geological change) and "catastrophism" (assuming unique cataclysmic events radically altered the crust and quickly formed new structures), were taken to task. On a solid base of evidence, like the formation of huge volcanoes on top of the known most recent strata of rocks, Lyell concluded that geological time was almost infinitely vast. This extended time frame allowed the conclusion that all observable structures had been formed by geological mechanisms still operating in the same way as ever, and with the same intensity. Lyell was actually mistaken in his postulation of a steady-state earth, where all mechanisms had always operated with the same intensity. The catastrophists assumed that prehistoric earth had been much warmer as it was cooling from an original molten state, allowing for more radical events involving more energy in prehistory, which was quite a valid theoretical proposition. This does not detract from the fact that Darwin derived a strong sense of how a mechanism for biological evolution should operate from Lyell's geology. At the heart of it should be a constant and unchanging mechanism still in operation.

DARWIN'S THEORY

Charles Darwin's career is the stuff of modern scientific mythology: his circumnavigation aboard the *H.M.S. Beagle* in the 1830s, the discovery of the Galápagos finches with their oddly shaped beaks, the 1859 publication of the *Origin of Species*, the subsequent scandal, and Darwin ending his days as a revered scientific household name. This poses peculiar problems when writing the history of his theory, as the amount of historical publications about it, both popular and professional, is staggering and has merited the label "the Darwin industry."

To briefly outline the history of its formulation, suffice it to say that there's a general agreement that Darwin had the basic mechanism of natural selection conceived at a fairly early date. The postponement of publication was made as to provide a firmer base for it. Darwin was after all very well read in both the philosophy and economic theory of his day. What Darwin ended up with was a theory of evolution that was wholly materialistic and open-ended. There was no remnant of the linear progressivism and teleology of the alternative theories of evolution. He broke with the idealistic notion of "type" and footed his theory on populations of actual individuals. There is no normative type which defines a species, with variations due to local ephemeral adaptation. The basic unit of analysis is populations of individuals breeding. The mechanism of the struggle for existence fueled the process of evolution, as Darwin proposed a model of nature as inherently competitive. From animal breeders he culled a considerable amount of useful knowledge about "artificial selection," how to breed for specific traits, but placed the impersonal mechanism of natural selection in the position of the breeder. Finally, from Lyell he brought the requirement to find a mechanism working uniformly, able to account for all observed phenomena as long as enough time was available. The theory made use of branching evolution, and assumed it to be basically directionless as all it entailed was a greater rate of survival and procreation for those individuals of a species better adapted to changed conditions.

In retrospect the area where his theory "failed," in the sense of Darwin's understandable inability to predict Mendelian genetics, was in its treatment of the problem of heredity. Darwin himself proposed that each parent contributed half the traits to their offspring, but that the individual contributions appeared blended together in it. This caused certain problems, most famously the criticism that if this was true even a beneficial individual adaptation within a population would quickly become blended into it as a drop of black in a bucket of white paint. In the history of science Darwin's "mistaken" theory of heredity was for a long time papered over as not worthy of real attention. Modern history of science has given it more of an airing for the simple reason that it is the aspect of Darwin's work which best reflects the extent to which his work was footed on basic problems of contemporary science.

The history of the actual publication of Darwin's theory of evolution is rather complex. Darwin had it basically worked out already around 1840, but was in no hurry to publish. He let trusted friends like Thomas Huxley and the botanists Hooker and Asa Gray in on it, slowly building a base of support. And then he was suddenly obliged to publish as to not lose priority. What happened was that in the summer of 1858 the young zoologist Alfred Russel Wallace sent Darwin an article for publication in some scientific journal. In it an astonished Darwin could read something virtually identical to his own theory of natural selection. With no knowledge of what Darwin had been fine-tuning all those years, Wallace had been working in New Guinea, and now felt sufficiently confident to publish his independently derived theory. At the time Darwin was halfway through a great manuscript titled "Natural Selection," but in no hurry to go to print. In the end it was Huxley who solomonically resolved the situation by having Darwin and Russel both publish an article on the subject (Bowler, 1984). Darwin then immediately set to work, at breakneck speed, on the one-volume book of his theory which in 1859 was published as *On the Origin of Species*.

THE RECEPTION OF DARWINISM

Contrary to what might be believed in retrospect, Darwin's theory of evolution by natural selection did not turn everyone into a Darwinist overnight. Best known, and hardly surprising, is the opposition against it from religious quarters in Britain and elsewhere. Less known are the limitations to its success within the fields of science, while its impact on the huge, heady, and hard-to-sum-up fields of philosophy, history writing, art, literature, and popular culture, might at times be underestimated. As for the sciences, they simply do not work in such a way that a theory will be instantaneously accepted for having the simple virtue of truth. A theory's correctness is rather the outcome of the thorough sifting of it by the collective of the scientific community. In the latter half of the 1860s Darwin's publication of his theory demanded their attention in order to make a ruling.

The reaction was mixed. In Darwin's native Britain and the United States it gained considerable acceptance within those sciences directly touched by it. The botanists may have cottoned on to it most of all, in particular Joseph Dalton Hooker in the United Kingdom and Asa Gray in the United States. Both had struggled with the mysteries of plant geography and found Darwinism helpful for explaining the distribution of species. Zoologists and comparative anatomists also stood to gain in their studies of animal "morphology" (the science of form). Darwinism's insistence on a common descent for the members of a species, over the older more abstract notion of type (blueprint, or the German "Bauplan"), coupled with natural selection, was very powerful in explaining the large amount of rudimentary organs and less than perfect anatomical homological adaptations. Sticking behind the idealistic types as basic forms for organic life was God, or so it was assumed, but some of the adaptations of these basic types were felt to be rather too impractical to be attributable to divine providence. If they were instead interpreted as the outcome of the process of non-directional materialistic natural selection, working on material at hand but unable to add to it, then the outcome made sense.

The group hardest to convince was the paleontologists, the students of the fossil record. One of the biggest hurdles for Darwinian natural selection lay in the fossil record itself. There were two basic interpretations: It was either complete enough to allow us, or too full of gaps to deny us, definite conclusions. At the time the fossil record seemed to indicate that radically different species had appeared and disappeared quite suddenly, with no transitory forms in between. If this was the actual shape of things in the past, it threw considerable doubt on Darwinian evolution, which assumed that slow and gradual change would lead to intermediary forms. Darwin consequently argued that this was simply an effect of our as-yet-imperfect knowledge of the fossil record. Intermediary forms would be forthcoming; as proved true, eventually.

The Darwinians, besides Darwin people like Russel and Huxley, "Darwin's bulldog," were very good at keeping the peace among themselves. This was necessary as they were in fact intellectually divided over the exact content of evolution. Even Darwin after formulating natural selection wavered between a conception of heredity as either hard or soft (Lamarckian acquired characteristics), while Wallace eventually argued that human evolution required divine intervention, and the staunch Huxley as a paleontologist never really found much use for the mechanism of natural selection.

Outside of the English-speaking world the reception was slower and sketchier. While *Origin* was definitely read, natural selection was not necessarily accepted. *Origin of Species* was translated into French in 1860, but it seems that translation did Darwin's argument a disservice by inserting it into a French debate over republicanism and religion. Later translations corrected this, but partly for patriotic reasons certain French biologists championed Lamarck as the true founder of the theory of evolution, relegating Darwin to a subsidiary role of having worked out some practical kinks.

The German situation was just as complex. Darwin was quickly seized upon by the group of German biologists, anatomists, and zoologists, politically mostly radical liberals, who in the 1850s set themselves up as the champions of scientific materialism in opposition to the remnants of the German tradition of idealistic Naturphilosophie and everything vaguely smelling of religious scruple in science. The group included Ludwig Büchner who wrote the program of the group, *Kraft und Stoff* (Force and Matter) in 1855, but also the anatomist Carl Vogt, a political refugee to Switzerland after the failed revolution of 1848, and the young Iena zoologist Ernst Haeckel, who would play the part of Darwin's German bulldog to Huxley's English.

These radical liberals and anti-religious materialists ran afoul of the rest of the German scientific community though. Their scholarly endeavors had most impact abroad, while their national impact was mostly with the German general public through popular accounts of evolution and natural history. The case of Haeckel is especially instructive. He was the disciple of the Berlin professor Rudolf Virchow, the inventor of the discipline of cellular pathology and discoverer of cellular partition summed up in the catch phrase *omne cellula, ex cellula* (every cell, from a cell). Virchow was an extraordinarily powerful scientific authority centrally placed in the Prussian, and from 1871 German, capital. He was also a confirmed liberal, to the point of becoming the leader of the largest liberal party of opposition to Bismarck's conservative government in the Reichstag.

When Haeckel clashed with Virchow over the status of Darwinism the political overtones were very strong. Virchow's charge against it was that it must lead to socialism, a claim vigorously denied by Haeckel. It was not the only problem Virchow and his supporters had with the theory. Another factor was the racist uses Haeckel and Vogt had put it to. This was no direct fault of Darwin's, who had in fact dismissed contemporary systems of racial classification as so much woolly thinking. In his native Britain evolution through natural selection was even hailed as an argument against racial determinism by the liberals of the British scientific community (Stocking, 1987). Upon publication Darwin's theory took on a life of its own and was put to uses he had often not envisioned.

Because that was the paradox of the situation; Darwinism was phenomenally successful with the German public, racist doctrines included, while professional scientists treated it with skepticism. As Virchow would have it, sticking to the tradition of Baconian inductive science (facts alone, no theories, and the conclusions should arrive by themselves), the Darwinian theory of natural selection would first have to be proven, before becoming a permissible premise for further research. This attitude was a known problem for Darwin, one that even Huxley made him the disservice of adopting: In order to consider the theory proven, artificial selection, i.e. breeding, must be shown to have produced a new species. It disregarded one of the major points of Darwin's conception of evolution as a gradual process requiring vast amounts of time and impossible to prove in such a manner.

The adoption of Darwinism was probably also slowed by the fact that most of what scientists did, the "practice" of science, made no demand on a clear stand on specific theories about evolution. This has been specifically demonstrated in the case of German science. Paleontologists, anatomists, and zoologists were preoccupied with the discovery, careful examination, and classification of all manner of organisms, living or extinct. Paleontologists were filling in blanks in the fossil record and reconstructing panoramas of prehistoric life. Some might adopt Darwinian evolution, others would not, but the work itself continued pretty much as it had been laid down by Cuvier and others already in the first half of the nineteenth century. Darwinian evolution was optional for this kind of work. The theory might be regarded as interesting, but as yet not proven, and active scientists had no need to either commit to or oppose it.

In view of this, it might then be asked: What was the significance of Darwin's theory at the time? Within the sciences the debates it sparked were important. Darwin was explicit in his materialism, in his shift toward population thinking and branching evolution through common descent, away from idealistic notions of design and archetypes. His stance on evolution as a random directionless process could hardly have hoped to shift the general preference for progressive evolution at the time. One could perhaps say that he blew the lid off a number of issues, in particular in British society, specifically the question of man's place in nature.

He also confirmed and presented a very well-substantiated and coherent argument for branching evolution. Even if many, or most, of his colleagues did not accept the specific mechanism of natural selection, it certainly confirmed the suspicion that branching evolution was the shape of the history of life, which had been growing for some time. The sturdy foundations laid by Darwin for his theory was also a boon. Opponents who dismissed it regularly found themselves unable to propose anything nearly as credible or comprehensive in its stead. Dismissing Darwinian evolution was not too hard, proposing a better alternative was.

What the debate over *Origin of Species* and the first Darwinists accomplished was to set the stage for the next generation of young scientists who would treat evolution less as a scientific problem to discuss, but as a point of departure for their own research. These "neo-Darwinists" came around in the 1880s and had a more dogmatic approach to the mechanisms of evolution. They were complemented and opposed by a dedicated group of self-professed neo-Lamarckians in the last decades of the nineteenth century.

Neo-Darwinism was launched by the German biologist August Weismann who, supported by the advances in cellular research made after Virchow, formulated a new theory of reproduction and "hard heredity". Hard heredity was crucial in his championing of Darwinian evolution through natural selection, as it completely ruled out all and every possibility of Lamarckian heredity of acquired characteristics. Weismann started out doing microscope studies of cell reactions, becoming convinced that all organisms transmit a kernel of immortal reproductive material to its offspring. He went on to deduce that these were located in the chromosomes (small structures easily stained, Greek *chroma*, color) within the kernel of the cell. This "germ-plasm" was impervious to outside influence and the parent simply contributed half each to their offspring without any possibility of modification of the germ-plasm itself. Thus having formulated a mechanism that would only work in conjunction with natural selection Weismann became the champion of a combative neo-Darwinism. The response to it was however not exactly a riotous triumph for natural selection. Weismann's hard stance rather forced those biologists working from Lamarckian assumptions to organize themselves into a competing neo-Lamarckian school which forcefully rejected Darwin and natural selection. In retrospect Weismann's insights into the inner structure of the cell was later incorporated within the emerging Mendelian genetics at the turn of the twentieth century.

In the meantime Lamarckists proposed a set of mechanisms to explain evolution of their own. What they had in common was the insistence on the process being non-random but in some way purposeful. One of the more popular alternatives has been labeled "orthogenesis," literally "evolution in a straight line." It was adopted not least among paleontologists in the United States in the decades prior to 1900. Orthogenesis considers evolution to be preprogrammed. The trajectory of the development of an individual species already lies inherent in it. It is a view which allows no room for the troublingly random and directionless mechanism of natural selection. As such it was one of the last refuges of idealistic teleology within the biological sciences.

SOCIAL DARWINISM AND THE POPULAR RECEPTION

As already stated, if the reception of Darwinian evolution was gradual within the sciences, it was more of an instant success within public culture. Though yet again, the actual mechanism of natural selection more or less shunted to the side in favor of other implications of evolution, and onto Darwin's theory was grafted a number of features not always reconcilable with it. The best-known example is how the philosopher and social theorist Herbert Spencer came up with the catch-phrase "survival of the fittest" to complement Darwin's "struggle for existence." Spencer might in retrospect appear to be the perfect exponent of the doctrine of Social Darwinism coming into fashion. Spencer's social theories of total individual freedom from government involvement and fierce competition between individuals was however assumed to operate within a Lamarckian framework. Thus Spencer provided a justification for the unregulated laissez-faire capitalism of Victorian society. Darwin's theory of evolution was grist to the mill.

The more important question that can been raised regarding Darwin and Victorian society is of course to what extent Darwin's work itself can be regarded as derivative of the society he was living in? This is one of these issues where historians' opinions have differed widely. Social historians, not least those inspired by Marxist theory, have not uncommonly stressed Darwinian evolution as simply the expression of and justification for the harsh capitalism of Victorian society. The influences they can cite are quite well known, as Darwin himself took a keen interest in the social and philosophical issues raised by his theory. He had a special debt to John Malthus's treatise of demographics, which helped him with thinking in terms of entire populations, and the Belgian Adolphe Quételet's attempt at a purely quantitative, statistical analysis of human society.

Traditional history of science for a very long time chose to deal only with the limited scientific aspects of Darwin's formulation of his theory of evolution to the point of next to denying Darwin as a member of his time and place. Both alternatives are of course radical simplifications, and the antithetic positions have been summed up as "externalism" as opposed to "internalism." In the end it is fair to say that taken to their extremes both positions are mistaken. There is no taking Darwinism out of its time and place, but neither can a theory with implications as complicated as Darwinian evolution, though powerful in its simplicity, be reduced as a mere reflection of collective, or class, interests.

Looking at the gamut of appropriations of Darwinian evolution for diverse needs and interests in the second half of the nineteenth century, one is struck by the tremendous impact of evolution in general, spanning a considerable range of social and political stances. This was part of the success of Darwinian evolution as a social and cultural phenomenon perhaps more than as a scientific theory. But since no watertight compartments existed between the popular reception and professional science, popular success would later be cashed in through the interest in evolution it sparked among later generations of young scientists.

Proportionally among the major political ideologies of the day, the greatest use of Darwinian evolution was of course made by liberalism. Darwinism was marshaled into justifying many things, from the necessity of unfettered industrialist capitalism, to the subjugation of non-European societies. It was all part of "the struggle for existence" and "the survival of the fittest." Kipling would wax eloquent about picking up "the white man's burden," in his great goodness allowing lesser endowed human races to piggyback on his presumed greatness.

Social Darwinism could also justify European great-power politics in the run-up to World War I. While liberalism in the first half of the nineteenth century had managed to stress peaceful national competition as a means toward the betterment of all, toward the end of the century the situation had become framed in much more aggressive terms, and social Darwinism was part of this shift. Where national competition previously was assumed to lift all, even if someone would pull ahead, the new situation was rather framed in social Darwinist terms of "conquer, or perish." The war of 1914 had been anticipated for a quarter of a century. It was assumed by many that a final showdown must inevitably occur to decide which of the major powers would go on to greatness, while others must diminish or fall. While some managed to work themselves up to an active longing for a war that would test their mettle and forge the character of a nation, even those who looked forward to it with trepidation had a hard time seeing a way to avoid it.

Turning away from international politics and back to the self-understanding of Western society, Social Darwinism was the basis for the first attempts to argue a conservative political agenda on the basis of science. It was the racist potential of Social Darwinism that attracted the Aryan supremacists like the French professor of law Gustave Vacher de Lapouge in the early 1880s, or the like-minded German publicist and anthropologist Otto Ammon in the 1890s. Hitherto the connection between science and liberalism had been exceptionally strong. It certainly was a contributing factor for the acceptance of social Darwinism among political liberals. From the late nineteenth century the catch phrases "survival of the fittest" and "the struggle for existence" would increasingly turn up in the arsenal of social thinkers intent on battling what they saw as the degenerative effects of modern society, political democracy, etc., in favor of a mythical European past of happy masters and servants, law and order, virtue instead of vice.

The main contender with liberalism among the political ideologies was of course the ascendant socialism, Marxist and otherwise. It is well known that socialism for the most part spurned Darwinian evolution. Where Darwin stressed competition, the anarchist theoretician Bakunin instead argued in favor of cooperation as the basic mechanism of human society. Hegelian idealist dialectics turned on its head had produced Marxist historical materialist dialectics and a progressive view of human history that was assumed to have the status of historical law, analogous to laws of nature. It had no real interest in things like natural selection. On the other hand, as Alfred Kelly (1981) has shown, among the German working class, the best organized and educated in the world in the late nineteenth century, judging from the workers' libraries own statistics you would be hard pressed to find a German worker who had actually read Marx and Engels, but they were absolutely lapping up Haeckel. They had no reason not to. Where the Nordic or Aryan supremacists worried over the competition from non-Europeans and the weak fertility of the upper and middle class as compared to the working masses (understood as not only socially but racially inferior), the working masses could lean back and confidently expect historical materialism as well as social Darwinism to guarantee them final victory. Socialists tended to pick up on Darwin's insistence on the way his theory operated on entire populations. They were this population, not the social elites they were confronting.

THE (RE-)DISCOVERY OF MENDELISM AND THE "MODERN SYNTHESIS"

The principles of heredity worked out by the Austrian abbot Gregor Mendel in the 1860s were so to speak rediscovered in 1900. Mendel's gardening experiments regarding how characteristics of various strains of garden peas are inherited is common currency these days, to the point where schoolchildren are sometimes allowed to repeat them. Mendel used the gardens of his monastery to conduct them and in 1865 had the results published in a local Austrian scientific journal where they were promptly forgotten. But unlike the popular misconception, he was not laboring entirely outside the fold of professional science. He was in contact with the prominent German botanist Karl von Nägeli, who asked him to instead try to work out other problems defying analysis. The situation was simply that the current state of science was unable to find a use for his results. Apparently the right conditions had come about in 1900. Most important, the view of heredity as a blending of the characteristics of the two parents was now considered deeply problematic. The germ-plasm theory allowed for a view of heredity as hard and biologists were becoming convinced that the hereditary characteristics lay preformed in it. At this point the biologists Carl Correns and Hugo de Vries realized that Mendel's work provided a useful model for discontinuous heredity with its dominant and recessive traits being passed on in unadulterated form through the generations.

Mendelism as it was put together after 1900 was in fact not part of the theory of evolution. Quite the opposite, it was viewed as an alternative capable of replacing it. Natural selection was regarded as an outdated nineteenth-century concept by this first generation of geneticists, who instead argued that discontinuous variation and the new concept of "mutation" were in themselves capable of explaining the history of biological variation (Bowler, 1984). Mendelism with its commitment to hard heredity was of course even more hostile toward Lamarckism. As far as Mendelism was concerned evolution was rather a non-adaptive process, where random mutations within the germ-plasm triggered change. Some Mendelists did go all the way to "orthogenesis," the concept of programmed evolution, assuming that there was a direction to the mutations occurring. Special attention was paid to the structure of the cell and it was established that the active material in heredity, i.e., the germ-plasm, was in fact located within the chromosomes of the cell and that at conception both parents contributed half each. The relevance of Mendel's laws of inheritance and the role of mutations in setting the range of variation of a species was confirmed experimentally using the *Drosophila melanogaster*, the banana-fly, which in the 1920s became one of the staple animals for conducting experiments on heredity.

In retrospect it might seem obvious how Darwinian natural selection and Mendelian laws of heredity would complement each other. The laws of hard heredity solved the old "paint bucket" problem of Darwin's. If a beneficial trait is introduced in a population obeying the Mendelian laws of heredity there is no risk of the trait becoming bred out of the population as it would with heredity of the blended kind. It took a while for this realization to sink in. Population genetics, as represented by scientific greats like T.H. Morgan, R.A. Fisher and J.B.S. Haldane, was conceived from 1900 onward as a highly mathematical enterprise, to the point of shutting out the field naturalists, preserving the original Darwinian emphasis on biogeography. Heredity became the research domain of laboratories studying mutations in pure strains of organisms, often downplaying adaptation and geographical factors facing the organisms in nature. Only when a proper meeting between Mendelian laboratory population genetics and Darwinian field science biogeography could be arranged would the modern view of evolution be worked out.

This occurred in the period roughly 1920–1940. It has been argued that a lot of the initiative for a synthesis came from the field naturalists like Ernst Mayr, one of these fieldworkers responsible for crafting the synthesis. It was to a great extent a matter of breaking the complex mathematics of population genetics down in such a fashion that the biogeographers could start to discern how these results could be applied to explain what they already knew from observation. A seminal work was Theodosius Dobzhansky's *Genetics and the Origin of Species* (1937). Dobzhansky brought with him a legacy of Russian genetics already with a strong element of Darwinian natural selection derived from Sergei Chetverikov early in the century. He was particularly well suited to mine the work of the geneticists for the kind of information useful to the biogeographers. His book was highly influential on Ernst Mayr's *Systematics and the Origin of Species* (1942), which appeared the same year as Julian Huxley's (T.H. Huxley's grandson) *Evolution: The Modern Synthesis*. The modern synthesis was finally transferred to paleontology by George Gaylord Simpson in *Tempo and Mode in Evolution* (1944). The new synthesis allowed for new collaborative efforts in the postwar era where mathematical formulas describing the mechanisms of heredity were tested against natural populations of organisms.

In some ways the process highlights the division within the theory of evolution between micro- and macro-evolution. The first can be studied in the laboratory, even studied in controlled experiments, while macro-evolution can only really be observed in retrospect. The modern synthesis has this double reliance on mechanisms of heredity and adaptive pressure according to the principle of natural selection. Its extension into paleontology was a matter of showing that it was a set of mechanisms capable of describing the already known state of affairs as well as offering more consistent models of explanation. But it remains a fact that macroevolution, as studied by paleontologists working with the fossil record, can never be experimentally either truly verified or falsified. It remains a matter of arguing for the most credible explanation.

On a final note, it can be argued that the modern synthesis does in fact represent a shift of paradigm in the sense of Kuhn, presented at the outset. In the postwar era it has taken the biological problem of heredity and evolution into a state of "normal science," as outlined in Kuhn's theory. Unlike the situation in the decades surrounding the turn of the twentieth century, where several theories of evolution and mechanisms of heredity were competing, scientists of today have a common conception of the basic principles of evolution and heredity. There is a shared terminology and a fundamental program of research stretching across all the biological disciplines, regardless of specialization. Ecologists and molecular biologists investigate different phenomena, in different ways, but share the conviction that regardless of what is being studied, it is the result of a biological evolution caused by spontaneous genetic change upon which the mechanism of natural selection has operated.

FURTHER READING

Bowler, Peter J. *Fossils and Progress. Paleontology and the Idea of Progressive Evolution in the 19th Century* (New York: Science History Publications, 1976).

Crook, Paul. *Darwinism, War and History* (Cambridge: Cambridge University Press, 1996).

Findlen, Paula. *Possessing Nature: Museums, Collecting, and Scientific Culture in Early Modern Italy* (Berkeley: University of California Press, 1994).

Kuhn, Thomas. *The Structure of Scientific Revolutions* (Chicago: University of Chicago Press, 1962).

Lovejoy, Owen. *The Great Chain of Being: A Study of the History of an Idea* (Cambridge, MA: Harvard University Press, 1936).

Oldroyd, David R. *Thinking About the Earth: A History of Ideas in Geology* (Cambridge, MA: Harvard University Press, 1996).

Rudwick, Martin J. S. *Scenes from Deep Time: Early Pictorial Representations of the Prehistoric World* (Chicago: Chicago University Press, 1992).

Stocking, George W. *Victorian Anthropology* (London: Free Press 1987).

Young, David. *The Discovery of Evolution* (London: Cambridge University Press, 1992).

Charles Darwin

Bowdoin Van Riper

Charles Darwin was neither the first to propose nor the first to popularize the idea that new species evolve, by natural processes, from existing species. Nor is he the sole architect of the modern theory of evolution. The idea that one species could "transmute" into another was raised by Darwin's grandfather, Erasmus Darwin, in his book *Zoonomia* (1795), by French naturalist Jean-Baptiste Lamarck in his *Zoological Philosophy* (1809), and by Etienne Geoffroy Saint-Hilaire in *Anatomical Philosophy* (1818–1822). An anonymous author, probably anatomist Robert Grant, argued for transmutation in the *Edinburgh New Philosophical Journal* (1826), and publisher Robert Chambers discussed it extensively in his anonymously published *Vestiges of the Natural History of Creation* (1844). The modern theory of evolution—though descended from Darwin's ideas rather than, say, Lamarck's—has also been shaped by developments in genetics and molecular biology that took place long after Darwin. The "modern synthesis" of genetics and evolutionary theory took place in the early 1930s: a century after the voyage of the *Beagle* and a half-century after Darwin's death. The evidence for evolution has also changed since Darwin's time. It now encompasses fields that did not exist in the mid-nineteenth century, such as molecular biology, and deeper levels of knowledge in fields that did, such as that provided by radiometric dating in geology.

No scientist, however, did more to define evolutionary theory than did Charles Darwin. He devised the conceptual core of the modern theory: the idea that species change over time, not because change is inherent in their nature but because the world they inhabit changes around them. He proposed the first plausible mechanisms for evolution: the idea that some inherited traits survive because they confer an advantage in the

Charles Robert Darwin in his old age. (Image © History of Science Collections, University of Oklahoma Libraries.)

struggle for survival (natural selection), and others because they confer an advantage in the competition to reproduce (sexual selection). He argued that the development of life on earth follows no clear path and has no predetermined goal, but is guided by directionless variations in heredity and climate. He rooted the idea of evolution firmly in the facts of natural history and presented a wide array of those facts in support of it. Finally, he established that his version of evolution was a powerful heuristic principle that ties together many distinct fields—that (as Ernst Mayr would later say) "nothing in biology makes sense except in the light of evolution."

Darwin's pivotal role in defining evolution has made him an icon in the most literal sense of the word: a scientist who has been transformed into a symbol. When we say "evolution" we are referring to Darwin's model of evolution rather than Lamarck's or anyone else's, and when we hear the word "evolution" it is *his* face—stern and graybearded like an Old Testament prophet's—that we see. Few other than biologists and historians of biology now remember Erasmus Darwin *or* the architects of the modern synthesis, and their names call forth no iconic images. The intimate ties between Darwin (the individual) and evolution (the idea) also run the other way. When Phillip Johnson writes of putting *Darwin on Trial* (1991), Michael Ruse of *Taking Darwin Seriously* (1998), or Kenneth Miller of *Finding Darwin's God* (2005), they are referring not to Darwin himself but modern evolutionary theory in general. Pro-evolution Web sites sell shirts and posters that portray Darwin as a rapper ("Darwin is My Homeboy"), a rock star ("Evolution Tour—Southern Hemisphere, 1835"), or a revolutionary (Darwin wearing the black beret of Fidel Castro's lieutenant, Ché Guevara). Bumper stickers designed to tweak conservative Christians proclaim "Darwin Loves You" and ask "What Would Darwin Do?" The organization Friends of Charles Darwin—originally formed to lobby for putting Darwin's picture on the British £10 note—even has an official song: "Charlie Is My Darwin," an adaptation of the Robert Burns lyric "Charlie Is My Darling," which paid tribute to Scotland's exiled would-be king Bonnie Prince Charlie.

The close association between Darwin and evolution reinforces, and is reinforced by, the popular image of scientists as lone geniuses whose theories appear in blinding flashes of insight. The "Darwin legend"—the sum of all the brief, once-over-lightly references to his life and career in textbooks, magazine articles, Web sites, religious tracts, films, and television programs—often portrays him in such terms. There, in rapid succession, he is "Darwin the lone explorer" in the Galápagos, "Darwin the lone scholar" agonizing over whether to publish his findings, and "Darwin the lone visionary" beset by critics and defended by his faithful bulldog, T.H. Huxley. The problem with this view is not that it is wholly wrong, but that it is partly right. Darwin *did* do significant things that nobody before him had done. Contrary to the legend, however, it is difficult to imagine a nineteenth-century scientist who was *more* connected with others, or one whose work was more a product of those connections. Darwin achieved what he did not by isolating himself from the world but by immersing himself in it. The ways he formed his ideas, the ways he developed them, and the ways he chose to present them were all shaped by that immersion. What follows is the story of that process.

THE MAKING OF A SCIENTIST, 1809–1831

Charles Robert Darwin was born on 12 February 1809, and came of age in the mid-1820s. He grew up, therefore, in a world where making a career out of science was rare and making a living at it was rarer still. The leading British universities—Oxford, Cambridge, and Edinburgh—had recently added a handful of chairs in various sciences to their traditional chairs in mathematics, but none had yet created an institutional base for science in the form of departments, laboratories, or degree programs. University medical schools treated medicine as a craft rather than a science, and focused on conveying practical knowledge to students rather than investigating or teaching the principles underlying it. Specialized schools of science and technology existed on the Continent, but not yet in Britain. The first government agencies charged with accumulating scientific knowledge—the Geological Survey, for example—would not be founded for decades. Medicine, law, the ministry, the army, the navy, and the foreign service were all considered respectable professions in the Britain of Darwin's youth. Scientific research itself was not yet recognized as a profession, however, and the few jobs that *did* require scientific knowledge—fossil collecting, surveying, mapmaking—were trades rather than professions and so beneath the dignity of a gentleman. Britons who made a career of science tended, therefore, to be either wealthy enough that they had no need to work or energetic enough that they could carry on two careers (a paid one and a scientific one) simultaneously. Darwin was both.

Both sides of Darwin's family were prosperous. His grandfather Erasmus and father, Robert, had both established successful medical practices. Robert had supplemented his with careful investments in land and in emerging industrial companies and by acting as a broker who brought together wealthy London investors and rising industrialists. The substantial wealth that Darwin's mother, Susannah, brought to the family was also a product of the Industrial Revolution. Her father, Josiah Wedgwood, was a third-generation potter who had made his fortune by applying the then-new techniques of mass production to the family's pottery works. The company's signature product was called "jasperware": a finely crafted style of pottery with a smooth, richly colored glaze and raised white decoration. Mass-produced and thus relatively inexpensive, it offered a combination of elegance and value that had never been possible before industrialization.

The Darwin-Wedgwood family's ties to the Industrial Revolution marked their wealth as "new money," and set them apart from "old money" families whose wealth came from land. Their political, religious, and intellectual affiliations also set them apart from Britain's traditional elite. Politically, they were Whigs rather than Tories, in favor of Parliament's independence from the king and relatively tolerant of calls for social and political reform. Religiously, they were Dissenters: Protestants (specifically Unitarians) but not members of the established Church of England. Both Erasmus Darwin and Josiah Wedgwood had been members of the Lunar Society of Birmingham, an informal scientific society that had included many of the sharpest minds of the late eighteenth century. Erasmus had written important treatises on science and medicine, along with poetry good enough to impress William Wordsworth and Samuel Taylor Coleridge. He sketched (but did not patent) a variety of inventions, and outlined a theory of evolution in his book *Zoonomia*. Josiah had spent his working life experimenting with new designs and glazes for his pottery, and won election to the Royal Society of London for his invention of the pyrometer, an instrument for measuring the temperature inside a kiln.

It would be stretching the point to say that young Charles Darwin grew up in a "scientific family," much less that his background predisposed him to an interest in evolution. What *is* clear is that he grew up in world where learning and innovation were valued, and traditional ways of thinking were accorded respect but not slavish devotion.

Sent to the University of Edinburgh at sixteen to study the "family business," Darwin soon made an alarming discovery: He hated medicine. The sight of blood, the suffering of patients, and the brutality of surgery performed without anesthetic appalled him, and the bitter infighting that divided the Edinburgh medical community depressed him. He went through the motions of keeping up with his studies, but preferred novels and natural history treatises to medical texts. Unable to tell his father that he could not face a career in medicine, he returned for a second year in the fall of 1826. The University of Edinburgh was more committed to science than most British universities, and Darwin took full advantage of its offerings. He took Thomas Hope's chemistry course in his first year, and Robert Jameson's natural history course in his second. Through the latter, he gained access to the superb natural history museum that Jameson had established. Over the course of many visits, he struck up a friendship with the curator, John McGillivray, and their conversations became a second, informal course in natural history. Darwin also joined the Plinian Society, an undergraduate natural history society, and through it met Robert Grant.

Grant, an Edinburgh graduate who taught in one of the private "anatomy schools" that offered additional training to medical students, was Britain's most gifted comparative anatomist. An expert on sponges and corals, he took on Darwin as a protégé and involved him in his ongoing research on *Flustra*, small marine animals that lived in colonies along the rocky coast of the North Sea. Darwin had been an enthusiastic natural history collector throughout his youth, but working with Grant was his first taste of real scientific research. He found the work thrilling, and his own small-but-significant discoveries about *Flustra* reproduction even more so. His delight turned sour, however, when Grant incorporated Darwin's discoveries into his own work and presented them in a paper with barely a mention of Darwin's name. The incident left Darwin feeling decidedly cool toward his one-time mentor. When Grant subsequently chose to share the theory of evolution he was developing, Darwin responded with stony indifference. He would later characterize the response as stunned silence and claim that the theory made no impression on him, but this seems unlikely. Over his two years in Edinburgh he had read widely and deeply on the "philosophy of zoology," including a careful study of his grandfather's *Zoonomia*. He was, if not yet comfortable with the idea of evolution, at least familiar with it.

The University of Cambridge, which Darwin entered in the fall of 1827 with the intention of taking holy orders and becoming an Anglican clergyman, was no hotbed of evolutionary ideas. Professor of geology Adam Sedgwick and professor of natural history John Stevens Henslow, who became Darwin's mentors, had no patience for it. They agreed with the evolutionists about what had to be explained—the history of life was long and complex, it showed a clear trend toward greater complexity, and individual species were well adapted to their environments—but not about how to explain it. Where Lamarck (for example) saw an "innate drive to complexity" that made each new generation of a species slightly more complex than the one before and eventually led to the formation of new species, Sedgwick and Henslow saw the sequential creation of ever-more-complex species whose form remain fixed from generation to generation. Where Lamarck saw adaptation in terms of organisms changing in response to their environment, Sedgwick and Henslow saw it as evidence of God's wisdom and benevolence. Sedgwick and Henlow agreed with the evolutionists on one critical issue. All saw the history of life as linear and progressive: a steady climb up a long ladder, of which humans occupied the top rung. Humans were, they all agreed, the natural and inevitable climax of the history of life.

Darwin, who became known as "the man who walks with Henslow," saw in the Cambridge botanist a mentor who (he believed) had Grant's breadth of knowledge but not his narrowness of spirit. Henslow broadened Darwin's knowledge of natural history generally and botany particularly, but also continued the practical education in how to *do* science that Grant had begun in Edinburgh. Darwin also continued his self-education in natural history while at Cambridge, reading widely and going on long collecting trips in the countryside with his cousin and friend William Darwin Fox. A middle-class fad for beetle collecting swept Britain in those years, and both pursued it with enough zeal to become competent amateur entomologists. Sedgwick became the third key player in Darwin's education. Darwin attended his geological lectures and, after his graduation in the spring of 1831, accompanied him on a multi-week geological reconnaissance of north Wales. The Welsh rocks, little studied before 1831, held the oldest fossils then known in Britain. Surveying them with Sedgwick was Darwin's first experience in *doing* geology as opposed to studying it, and like his work on *Flustra* he found it thrilling. Henslow may, in the summer of 1831, have represented the *man* that Darwin wanted to become—brilliant but kind, learned scientist and liberal clergyman—but Sedgwick had shown him the kind of research he wanted to do.

Later that summer, a letter from Henslow would unexpectedly give him his chance.

THE VOYAGE OF THE *BEAGLE*, 1831–1836

The letter from Captain Francis Beaufort, head of the Royal Navy Hydrographic Office, was slightly unusual but very clear: Lieutenant Robert FitzRoy, master and commander of His Majesty's Ship *Beagle*, needed a friend. It was not an idle request. FitzRoy was scheduled to leave on a two-year mission to survey the coasts of South America and give British mariners their first accurate charts of the area. The work would be difficult, exacting, and tedious, and FitzRoy literally feared for his sanity.

His family had a history of depression and suicide, and he had been given command of the *Beagle* when her previous captain, driven mad by stress and isolation on an earlier South American surveying mission, had retired to his cabin and put a bullet through his head. FitzRoy asked Beaufort if a "gentleman naturalist" could be added to the *Beagle* as a passenger, doubling the expedition's scientific staff and (more important) giving him someone to dine with and talk to who was outside the naval chain of command. Beaufort passed the request to a friend at Cambridge, who conferred with Sedgwick and Henslow. Recognizing the opportunity that the voyage represented, Henslow thought first of himself, then of a recent graduate named Leonard Jenyns, and then (since both he and Jenyns had commitments ashore) of Darwin.

The voyage of the *Beagle*, which Darwin joined after some timely lobbying of his father by Charles's uncle Josiah, was a naturalist's dream. A civilian passenger rather than a paid employee of the Royal Navy, he had no official role in the ship's primary mission. He was free to go ashore for days or even weeks at a time, as long as he was back aboard the ship when it was ready to leave for its next port of call. He enjoyed the social privileges of a naval officer—sleeping in a cabin, dining on privately purchased stocks of food and wine, being invited to social functions ashore—but had no official duties to distract him from his scientific work. The specimens he collected belonged to him, not the navy or the Crown, and he was free to dispose of them as he chose. The *Beagle* ultimately took him not only across the Atlantic and along the coasts of South America but across the Pacific to Tahiti, New Zealand, and Australia. He returned home in the fall of 1836 via the Indian Ocean, South Africa, and Ascension Island, having seen a greater diversity of environments in five years than most scientists of his era saw in a lifetime.

The ship's first stop, in the Cape Verde Islands off the coast of West Africa, gave him the opportunity to practice on his own the geological skills he had learned under Adam Sedgwick in Wales the summer before. He found that geology excited him as much as it had in Wales, and he continued making detailed geological observations for the remainder of the voyage: throughout South America, in the Galápagos Islands, and in Tasmania, Australia, and South Africa. Vertical motions of the earth's crust drew his particular attention, and his effort to understand them was the common thread that linked his observations of coral islands, volcanoes, earthquakes, and layers of fossil shells found high above sea level.

Darwin's view of geology was profoundly influenced by his reading of Charles Lyell's *Principles of Geology*, the first volume of which—published in 1830—he had received as a present from FitzRoy. Ten years older than Darwin, Lyell had already made his scientific reputation with his detailed work on the geology of southern France and Italy and his

unorthodox views about the history of the earth. Older scientists such as
Sedgwick and Henslow had deep respect for the former and deep suspicions about the latter. They rejected Lyell's claim that all geological phenomena, past and present, could be accounted for by geological forces
of the same type and the same (low) intensity as those in action today.
They insisted that deep gorges must have been carved, and high mountains raised up, by intense forces that had acted in the past but were
unknown in the present.

Sedgwick and Henslow were even more troubled by Lyell's idiosyncratic view of the history of life, which denied that it was characterized by
steady progress from simplicity to complexity. Belief in such progress was
one of the few points on which creationists like Sedgwick and Henslow
agreed with transmutationists like Lamarck and Grant. Indeed, for Lyell,
that was precisely the problem. A belief in organic progression could too
easily, in Lyell's view, become belief in transmutation. He preferred the
alternative, which might be called non-progressionism or steady-statism.
The essence of Lyell's view of the history of life was that there had been
simple organisms *and* complex ones on earth at every stage of its history.
Species had appeared and species had gone extinct, but the *overall* complexity of life on earth remained constant. The fact that the fossil record
seemed to show otherwise was, Lyell contended, a reflection of the
record's incompleteness. The fossils of simple organisms were common
in all parts of the geologic past, but fossils of simple organisms were
found only in the relatively recent present. This idea (for which he comprised a constituency of one) was novel at best and tortuous to the point
of willful self-deception at worst. When Henslow sent Darwin the second and third volumes of *Principles of Geology* but advised him not to
believe all of it, he probably had non-progressionism in mind,

Henslow's warning proved fruitless. Trained as a lawyer, Lyell was a
master at expressing complex ideas in clear and fluid prose. The first
volume of *Principles* was a sustained argument for Lyell's theory that
small geological forces acting over sufficiently long periods of time could
account for even the highest mountains or the deepest gorges. The second was a sustained argument for the reality of non-progressionism, the
incompleteness of the fossil record, and the limitations of Lamarck's
theory of transmutation. Both arguments were backed up by detailed
descriptions of supporting geological and paleontological evidence, and
the third volume consisted of a guided tour of earth history informed by
Lyell's point of view. Reading *Principles* made Darwin an enthusiastic
convert to Lyell's views, and gave him a theoretical framework for the
geological observations he made on the voyage. As early as the Cape
Verde Islands, even before he had finished Volume 1, Darwin had begun
to identify himself not just as a geologist but as a *Lyellian* geologist. He
would continue to do so throughout his life.

Geology was Darwin's principal scientific interest during his years on the *Beagle* expedition, but it was far from his only interest. His fascination with the living world and his enthusiasm for collecting specimens from it continued unabated, and the expedition gave him unmatched opportunities to pursue it. Over the course of five years he got not just a tour of the world, but a tour of places that were geographically isolated both from one another and from the world he knew: the Cape Verde Islands, South America, the Galápagos, Tahiti, New Zealand, Australia, the Cocos, Mauritius, South Africa, and Ascension Island. By the time he returned to home to England, he may have had a better sense of the exuberant diversity of life on earth than any scientist before him.

Darwin—trained by Grant, Sedgwick, and Henslow—saw it all with the eyes of a scientific observer rather than a gentleman-tourist. He sought out patterns, drew comparisons, and noted the ways in which different species of plants and animals related to their environments and to each other. His collections, particularly in places like South America and the Galápagos, were extensive and systematic. His observations, collected in notebooks which he carried with him everywhere, encompassed a wide range of scientific subjects: rocks, fossils, soil types, plants, animals, weather, climate, the appearance and customs of people, and so on. When port calls or encounters with homeward-bound ships made it possible, he sent specimens and reports (in the form of letters) back to Henslow, who shared them with fellow scientists in Cambridge and London.

The *Beagle* expedition was Darwin's graduate school. It turned him loose in unexplored scientific territory, and gave him his first experience with doing professional-grade scientific research on his own. It gave him his first published scientific communications (the letters to Henslow) and ensured his welcome into the scientific community when he returned. Most important, it gave him the beginnings of what today would be called a research program: a set of questions he wanted to investigate and a body of data with which to begin. Darwin boarded the *Beagle* in December 1831 as a young gentleman with a deep interest in nature. He left her in September 1836 as a scientist.

THE MAKING OF A CAREER: PUBLIC LIFE, 1836–1845

The decade after Darwin returned from the *Beagle* expedition was the most productive of his life. After a few weeks at home, he moved to Cambridge and then (in early March of 1837) to London, the center of the British scientific community. He quickly became a full-fledged member of that community and a regular participant in the meetings of the geological, zoological, and Linnean societies. He renewed old friendships in

Cambridge and established new ones in London. He met Lyell for the first time within weeks of his return, and Lyell introduced him in turn to with many of the leading scientists of the day. Darwin turned to his colleagues, old and new, for help in "working up" the mass of specimens and observations that he had brought home from the expedition. Experts in fields where he saw himself as an amateur, they undertook the technical work of identifying, classifying, and describing the specimens for eventual publication.

Richard Owen, a brilliant anatomist and paleontologist who Darwin had met through Lyell, analyzed the fossil mammals from South America. Eminent ornithologist John Gould took charge of the birds, Thomas Bell the reptiles, George Waterhouse the insects, and William Martin (keeper of the museum at the Zoological Society) the living mammals. Henslow agreed to handle the plants. The results of their work appeared, between February 1838 and October 1843, in a five-part report titled *The Zoology of the Voyage of the Beagle*. Partly financed by a £1,000 government grant, it eventually included nearly 600 pages of text and 166 black-and-white and color illustrations. Owen, Gould, and Bell wrote on their particular specialties, with Waterhouse covering the living mammals and Leonard Jenyns—who had turned down the opportunity to sail on the *Beagle*—the fishes. A planned volume on insects, which Darwin had intended to write himself, was dropped (along with a volume on plants, to have been written by Henslow) because of scheduling pressures. Even with those pressures, however, Darwin had contributed introductions to two volumes, supplied observations about habitat and geographic distribution throughout, and acted as general editor of the entire project.

The results were, frequently, surprising. Owen's analysis of the fossil mammals from South America showed that *Toxodon*, though sized and proportioned like a rhinoceros, was taxonomically closer to the capybara, a large (forty- to fifty-inch) living rodent native to the continent. *Mylodon* and *Megatherium* were extinct giant cousins of the modern sloth, and the giant *Glyptodon* was kin to the modern armadillo. The vertebrae and fragments of leg bone that Darwin had attributed to an elephant turned out to belong, according to Owen, to a giant species of llama. These revelations fundamentally changed the way Darwin thought about the South American fossils. They showed that the differences between South American and African mammals he had observed on the voyage were not new. The mammal populations of the two continents had been radically different for millions of years, and extant species of South American mammals were far more similar to extinct South American species than to extant African ones.

John Gould's analysis of the birds Darwin had collected brought similar revelations. Virtually all of the birds Darwin had brought home from the Galápagos were unique to the islands, and many were unique to specific islands. Darwin had suspected this of the Galápagos mockingbirds while he was there, but Gould surprised him by reporting that the three types he had collected were not three varieties of a single species but three distinct species. The island-to-island differences in the mocking-bird population were, in other words, even greater and (to an expert like Gould) more striking than Darwin had recognized. Gould also reported that fourteen types of small bird that Darwin had collected in the Galápagos and variously identified as finches, wrens, grosbeaks, and blackbirds were in fact fourteen *different* species of finches. Learning this raised the question of whether those species, too, were unique to particular islands, and forced Darwin to confront an unpleasant truth: He had not recorded *where* in the Galápagos they had been collected. Fortunately, both Captain FitzRoy and Darwin's own assistant, Syms Covington, had carefully labeled their own bird collections. Cross-checking theirs with his own, Darwin was able to assign each species to its island of origin. The results confirmed his suspicions: Most of the fourteen species of Galápagos finches, so different in appearance that he had assigned them to different families, were found only on specific islands.

The *Zoology* was a detailed, technical work aimed primarily at scientists. Darwin's own *Journal of Researches*, which he completed in 1837, was broader in scope and addressed to a broader audience. Originally released in 1839 as the third volume of the official report of the *Beagle* expeditions (edited by FitzRoy), it took on a life of its own when London publisher John Murray reissued it as a stand-alone title in 1845. Its blend of scientific observation and travel narrative lands drew a wide audience, and established Darwin to the public as a graceful and talented writer.

Darwin still saw himself as a geologist at this point in his career, and he was an extremely productive one. He was an active and respected member of the Geological Society of London, and served in the demanding, high-profile office of Secretary from 1838 to 1842. Well over half the thirty-seven papers and three of the four books he published in the decade after the *Beagle* expedition dealt with geology. His geological work built on the observations he had made on the expedition, bolstered by further research and correspondence and refined through conversations with Lyell and other colleagues. The books—*On the Structure and Distribution of Coral Reefs* (1842), *Geological Observations on Volcanic Islands* (1844), and *Geological Observations of South America* (1846)—reflected his intellectual kinship with Lyell and his interest in changes in the elevation of the land surface.

All of this—thirty-seven papers, four books written and another edited, dozens of scientific meetings, countless hours doing the work of the Geological Society—constituted Darwin's public scientific career in the decade after his return to England. He also, of course, had a private life. He met and married his cousin Emma Wedgwood in 1839, moved them to a county house in the village of Down (later Downe) in 1842, and presided over a growing household that included five children born in the first six years of their marriage and would eventually include ten. What neither his colleagues, his closest friends, nor his family realized was that Darwin was also carrying on a second scientific career. In the pages of his notebooks and the privacy of his own mind, he was working relentlessly to answer the questions that the *Beagle* expedition had raised in his mind about the origin of species.

THE BIRTH OF A THEORY: PRIVATE LIFE, 1836–1845

The questions had begun to form while Darwin was still on the *Beagle*, but the conclusions that Owen and Gould had drawn from the *Beagle* specimens had intensified them. Why do the living and fossil mammals of South American resemble one another so closely? Why are marsupials common in Australia but almost unknown in other parts of the world? Why are islands often populated by species found nowhere else in the world? Why are particular species of finch found on particular islands in the Galápagos? Why, if each species is created to be perfectly adapted to its environment, do species tend to resemble nearby species from different environments more than they do distant species from similar environments?

Darwin already had an idea about what the answer might be when he opened the first in a series of private notebooks on transmutation in the spring of 1837. If new species descended from existing species, rather than being created from scratch, the patterns of geographic distribution he had seen on the expedition would make sense. Living species would resemble fossil species from the same area because the fossil species were almost certainly their ancestors. A species divided into two isolated populations—on island and mainland, or two different islands—might well develop over time into separate species that, because of their common ancestry, had a strong "family resemblance." Given *enough* time, a single ancestral species might give rise to a large group of species marked by certain traits inherited from the ancestral species that they shared. He began to search for evidence—by reading, by correspondence, and by informal conversation with other scientists—that would shed further light on the subject.

Simultaneously, he began to look for reasons why a single species would split into two distinct but closely related species. Geography and geology, in the broad sense of environmental conditions, seemed likely to be involved. So, too, did the mechanics of sexual reproduction. Darwin reasoned that if organisms reproduced asexually, by budding or fission, offspring would always be a perfect duplicate of their (single) parent. They *would*, in other words, reproduce "after their kind" indefinitely, just as special creation predicted. Sexual reproduction, in which the hereditary traits of both parents were somehow combined into each one of their offspring, offered the promise of variety. Exploring this idea drew Darwin into research on the nature of sex, fecundity, and patterns that often occurred among the offspring of animals. Looking at it from another angle raised the issue of hereditary variation: where it came from, what its limits were, and how it tended to be distributed within a given species, population, or generation.

Other streams of information also fed into Darwin's attempts to come to grips with the origin of species. His own geological work on the uplift of land surfaces and the formation of islands suggested ways in which two populations from the same species might become isolated from one another. Watching an orangutan at the London Zoo interact with its keeper "like a naughty child" called to mind his experiences with "savage" humans in Tierra del Fuego, Tahiti, and Australia and set him wondering about the thin line between humans and apes. His father's admonitions about not postponing marriage, lest his children be born unhealthy or deformed as a result, fed into his theories about sexual reproduction as well as his anxieties about finding a wife. Interest in the distribution of traits among human populations led him to the work of Adolphe Quételet, who had coined the concept of the "statistically average" individual a few years earlier.

Jemmy Button

When *H.M.S. Beagle* moored at Tierra del Fuego in 1830, a group of Fuegians stole one of her boats, and four young hostages—ranging in age from nine to twenty-four—were taken to guarantee its return. Robert FitzRoy, then in command, had decided to take the hostages home to England so that they could be civilized, educated, Christianized, and eventually returned to spread their newly acquired culture to the other members of their tribe. The four were given new names by the *Beagle* crew—Fuegia Basket, Jemmy Button, York Minster, and Boat Memory—and treated like honored guests on the trip to England. Boat Memory died of smallpox soon after arriving but the other three were enrolled in school at Plymouth, given European-style clothes, and escorted by FitzRoy to social events where they were greeted as celebrities. When the decision was made to return them a year later, FitzRoy was prepared to book passage on a merchant ship, but his assignment to command the *Beagle*'s second surveying expedition made the expense unnecessary. The three Fuegians boarded the *Beagle* loaded with well-intended gifts from their admirers: dishes, glassware, silverware, and of course their clothes. FitzRoy returned them to their village in 1832, and continued on his surveying mission.

When *Beagle* returned a year later, FitzRoy and Darwin were appalled to discover that Jemmy Button and his companions had shed all traces of their stay in England. Their parting gifts were discarded, or dispersed (by sharing or theft) among the members of the village. Jemmy had thrown away the clothes he had worn with pride in England and was back to wearing a loincloth. His hair, which in England had been styled according to the latest fashion, was back to being stringy and unwashed. When Darwin and FitzRoy talked to him, he reported that—though he bore them no ill will—he was quite happy in his current state and had no desire to return to England or live like an Englishman in his native land. FitzRoy, who had a modest financial investment and a deep emotional one in the project, was devastated. Civilization, he concluded gloomily, was merely a veneer covering humans' animal-like true selves.

Darwin, who saw all the indigenous peoples he met in the Pacific through the parochial eyes of an upper-class Englishman, drew similar conclusions. He was appalled by the conditions under which the Fuegians lived, and doubly appalled that Jemmy Button and his companions embraced them even after tasting civilized life. He recorded, in his journal, sentiments close to FitzRoy's: that humans were animals at heart. That idea, reinforced by encounters with natives of Tahiti and Australia, took on a life of its own after Darwin returned to England. His first encounter with a live orangutan left him astonished at the humanness of the ape's emotions and expression of them. Jemmy Button and Jenny the orangutan were, he privately concluded, evidence that the gulf dividing humans and animals was neither wide nor deep.

Lyell's ideas loomed especially large. His doctrine that the observable present was the key to understanding the unobservable past steered Darwin into conversations with animal breeders: farmer, pigeon fanciers, dog enthusiasts, and the like. Such people would, Darwin reasoned, be more familiar than anyone with how hereditary variation expresses itself today. Lyell's belief in the cumulative power of small changes sharpened Darwin's determination to understand how such changes might accumulate within a species over the course of generations. Lyell's attack on Lamarck—still present, in all its ferocity, in the new edition of *Principles of Geology* that Darwin picked up and read in 1837—provided a useful catalog of pitfalls that, Darwin knew, his own theory would have to avoid.

The catalyst that brought these disparate threads together was the sixth (1826) edition of Thomas Malthus's *Essay on the Principle of Population*, which Darwin read in October 1838. From it he took the idea of a perpetual "struggle for existence" in which individuals competed for the resources they needed to survive. Malthus was writing about human societies, but Darwin applied the idea to the natural world, where he had seen ample evidence for such competition. Once he did so, pieces of his theory began to fall into place. Any given generation of a species would have some individuals that triumphed in the struggle for existence and others that failed. The individuals in the first group would flourish, and have ample opportunity to breed. Any inherited trait that contributed to their success would be passed on to their numerous offspring, and therefore be preserved. The individuals in the second group would die out quickly, with little or no opportunity to breed. Any inherited trait that handicapped them would tend to die with them and therefore be suppressed. Over successive generations, individuals with the beneficial trait would make up an ever-larger percentage of the population and individuals with the detrimental trait would make up an ever-diminishing percentage. All other things being equal, the species as a whole would grow better-adapted over time. The power of this kind of "natural selection" lies in the fact that it works comprehensively and continuously, on every trait in every individual member of a species.

Everything about Darwin's background primed him to accept natural selection as *the* mechanism by which evolution operated. He was a citizen of a society where unregulated capitalism and ferocious competition in the marketplace was the norm. He had grown up around the farms, dogs, and horses of the landed gentry, and was intimately familiar with "artificial selection" in the form of selective breeding. He had firsthand knowledge of fierce, life-or-death competition both in nature and in human society, including the triumphant conquest of non-European peoples by European ones that made the British Empire possible. Natural selection was, he quickly recognized, a process that operated gradually and in small increments, and so appealed to his Lyellian sensibilities. It clearly continued to operate in the present as it had in the past, and artificial selection—which could be studied directly, and in depth—provided a useful, if less powerful, analog for it.

Natural selection became the linchpin that held the other elements of the theory together. It explained why variations within a species would be preserved from one generation to the next, and others would disappear. It accounted for adaptation, and showed why populations of a single species would, if isolated in different settings, gradually diverge from one another. It did not make the theory complete, but it locked what had been a series of loosely connected ideas into a solid framework that Darwin could build on. Over the next three and a half years, through the early years of his marriage and the birth of his first two children, he did just that. He continued to read, to correspond, and to talk to anyone— now, more than ever, including farmers, animal breeders, and other experts on artificial selection—he thought might help him. He investigated embryology, the anatomical similarities between hands and flippers, and seemingly pointless organs such as male nipples. He considered what was known about variation in nature and how it differed from variation in the barnyard. He examined potential difficulties with the theory: Why were there no transitional forms in the fossil record? Where did complex organs come from? How could complex social behaviors evolve? All of it, in one form or another, went into notebooks or onto scraps of paper.

Beginning in late May and extending into early June of 1842, he wrote out what he called a sketch of his theory in longhand. It was a device for organizing his thoughts and for fitting together the ideas and evidence that he had been juggling in his head and his notebooks for nearly six years. When he finished, it was thirty-five pages long and extremely rough. He returned to it a year and a half later, in the winter of 1843, and began revising and expanding it into something suitable for publication. The result was an essay of 230 handwritten pages, filled with well-organized arguments and supporting evidence that he completed in July 1844. The 1844 essay was a remarkable document: the first detailed statement of Darwin's ideas on the origin of species. It was good enough to be published, but for reasons both professional and personal Darwin chose not to do so.

The professional reasons for his reluctance were simple: He felt that he did not yet know enough. There were, he felt, too many areas of the theory where the details were not adequately worked out, too many potential objections that he had not thoroughly thought through, and too many relevant fields and subfields of science with which he had only a passing familiarity. Three years earlier, Darwin had weathered a storm of criticism from his colleagues over a paper he had written about a geological feature called the "parallel roads of Glen Roy." The essence of the criticism was that his paper had been too much speculation supported by too little evidence, and he had no desire to face that accusation a second time. His theories about species were considerably further from contemporary orthodoxy than his theories about Glen Roy, and he wanted them to be backed up by massive quantities of unassailable evidence.

The personal reasons for Darwin's reluctance were more complex. They were bound up with his feelings for Emma, his aversion to controversy, his ideas about religion, and his anxiety over how his theory would be perceived.

Darwin had known, from the time he started working on the problem, that attributing the origin of species solely to natural processes would have metaphysical and religious implications. The combined effect of the seventeenth-century Scientific Revolution and the eighteenth-century Enlightenment had been to slowly remove the hand of God from the day-to-day workings of the natural world. Scientists had, by the 1840s, grown used to seeing the universe as a machine that—once created and set in motion—would run forever with no need for divine intervention to keep it in balance. It was a short step (but an optional one) from there to deism: the belief that God's non-intervention since the Creation extended to human affairs. Deism, by definition, rejected the basic elements of organized religion, and many devout Christians saw it as the top of a short, slippery slope that ended in outright atheism. Any suggestion that new species were the product of natural forces rather than God would, Darwin recognized, be greeted as a firm shove toward the top of that slope.

The idea of a universe where nature, not God, created new species troubled Darwin little enough. His religious upbringing had been unconventional and his commitment to the Church of England more practical than principled. He was a thoroughgoing deist by the mid-1840s, and the death of his daughter Annie six years later would blot out his faith in a benevolent God entirely. Emma, however, had a more conventional religious life. She was a devout (though not strident) Anglican, and the gap between her commitment to organized religion and his lack of it had been a source of anxiety for both throughout their marriage. Darwin had little desire to publish a theory that would underscore the differences between them, and still less to stir up a controversy that would engulf him and, by extension, her. He imagined himself denounced in print and from pulpits, and Emma snubbed and socially isolated. And so he waited.

The appearance, in October 1844, of an anonymous book titled *Vestiges of the Natural History of Creation* confirmed all of Darwin's worst fears. It anticipated many of Darwin's arguments—the power of small changes accumulating over time, for example—and drew its supporting evidence, as he did, from a wide variety of scientific disciplines. Far from skirting the most controversial implications of evolution it confronted them fearlessly. Darwin, in his 1844 essay, had sidestepped the question of human origins and glossed over the religious implications of evolution. The unknown author of *Vestiges*, later revealed to be Edinburgh publisher Robert Chambers, forthrightly declared humans to be a product of evolution, and reassured readers that belief in evolution was fully compatible with belief in God.

Vestiges fascinated the public, but appalled both the clergy (who saw it as a rejection of God, the Bible, and the idea that humans were anything but brutes) and leading scientists (who declared its reasoning slipshod and its grasp of scientific details feeble). Among dozens of scathing reviews of the book, the one written by Adam Sedgwick for the *Edinburgh Review* stood out. Sedgwick, who was both a scientist and a clergyman, produced a thunderous denunciation of the book, the author, and everything either of them stood for. His impassioned language, which sailed past dislike into outright hatred, startled Sedgwick's friends and appalled Darwin, who could now imagine his old teacher leveling a similar broadside against his own work.

After—and, to a large degree, because of—*Vestiges*, Darwin turned most of his attention to other projects for a decade. He did not, however, give up the species question entirely. Editing a second edition of his *Journal of Researches* for John Murray, he subtly revised his descriptions of the Galápagos finches and other subjects to hint at the possibility of evolution. Having become friendly with Joseph Dalton Hooker, a young botanist with a growing reputation, he asked him in 1847 to read and comment on the 1844 essay. His major project during this period was an eight-year (1846–1854) study of barnacles that resulted in a definitive two-volume book on the subject. The barnacle project gave Darwin, a confirmed generalist, something that he had never had: an intimate, comprehensive understanding of a particular class of animals. It also showed him that every part of a barnacle's anatomy exhibited a wide range of variation, and that the variations graded into one another so subtly that defining barnacle species was difficult. The implications for his ideas about evolution were clear, and drove him back to the project.

THE BOOK OF THE CENTURY, 1854–1859

Preparation of what Darwin came to think of as his "big book," tentatively titled *Natural Selection*, began in 1854 and lasted until mid-1858. He intended it to be his definitive statement on the subject, bringing together a fully developed version of his argument and all the supporting evidence that any reasonable scientific audience could want. He worked using the same methods he had used when he was first developing the theory in the late 1830s and early 1840s: gathering data through extensive reading, correspondence, and conversations with scientific colleagues. Darwin also began a series of experiments designed to test key points of his theory. He planted seeds, for example, to see if they would germinate after being soaked in salt water or passing through the digestive system of a bird or fish—part of a larger investigation of how plants might be distributed to offshore islands.

Now firmly rooted in Down, a respected member of the local community, he went to London only rarely. Instead, members of his "inner circle" increasingly came to the country to see him, or carried on their discussions by letter if the distance was too great. The principal members of the inner circle included Hooker, zoologist and anatomist Thomas Henry Huxley, polymath John Tyndall, American botanist Asa Gray, and Charles Lyell, who had cautiously embraced evolution despite concerns about its implications for human dignity. All, except for Lyell, were ambitious young scientists on the way up, who exhibited varying degrees of impatience with the tradition-bound minds of the scientific establishment. They provided Darwin with valuable observations—Huxley analyzed the anatomy of gorillas and found them equally similar to baboons and humans—but also with insightful criticism of his increasingly polished ideas. Responding to an 1857 letter outlining natural selection, for example, Gray cautioned Darwin not to turn a blind natural force into a personified "hand of Nature." It was an astute point, and represented a level of fine-textured analysis that Darwin's ideas and writings had rarely received before.

Darwin's inner circle also formed a buffer that protected Darwin not only from the outside world but also from its own, sometimes faulty professional instincts. This last function proved especially vital when Darwin—hard at work on *Natural Selection*—but beset by personal crises including the death of his infant son from scarlet fever, received a package from Alfred Russel Wallace, a virtually unknown biologist working in Borneo. The package contained twenty pages of manuscript that Wallace asked him to send on to Lyell. Those twenty pages, it seemed to Darwin, mirrored the basics of his own theory almost perfectly. Darwin's first instinct was to let Wallace's paper be published alone, to sacrifice his own claims to priority rather than to risk even a hint that he had tried to steal Wallace's thunder. It was Hooker and Lyell who persuaded him that Wallace's paper should be presented alongside extracts from Darwin's 1844 essay and 1857 letter to Gray, and they who got both papers on the agenda of the July 1, 1858, meeting of the Linnean Society.

Darwin abandoned *Natural Selection* in mid-1858 and began producing what he thought of as an "abstract" of it. Composed in bursts of furious activity separated by debilitating bouts of illness, it was a compromise between the compact, accessible essay he had written in 1844 and the all-inclusive, massively documented "big book" that he hoped to write. It is sign Darwin's skill as a writer, and of just how long he had been immersed in the material, that he could compose substantial parts of it from memory. Hooker, who had first seen Darwin's arguments more than a decade earlier, read and corrected the chapters as Darwin finished them. Lyell arranged for John Murray to publish the result. Murray himself made a small but significant contribution to the literary history of science by encouraging Darwin to prune his original title. *An Abstract of an Essay on the Origin of Species and Varieties through Natural Selection* thus became *On the Origin of Species by Means of Natural Selection*. When pre-publication interest in the book outstripped the planned print run of 500 copies, Murray raised the number to 1,250 copies. When the book officially appeared on November 24, 1859, every one of them was already spoken for by booksellers.

Origin of Species itself went from initial concept to finished book in fourteen months, but in a broader sense it was the product of nearly thirty years' work. The fact that it was conceived specifically *as* an abstract—a distillation of the arguments and evidence that would have gone into the never-written *Natural Selection*—gives it a forceful, streamlined quality rare among scientific works of Darwin's era. It is not just (as Darwin said of it) "one long argument," but an argument is developed in depth, in which each large point backed up by smaller ones and each smaller point backed up by examples. References to specific plants and animals follow one another on page after page, and the results of thirty years of reading, correspondence, and conversation are evident in the range of authorities Darwin cites. A single, randomly chosen page (p. 85) refers (in the context of a discussion of rump-striping in horses) to Belgian cart horses, English racing thoroughbreds, Welsh ponies, Arabians, and Kattywars from northern India, as well as to related species such as the ass, the zebra, and the now-extinct quagga. Seven individual informants are mentioned by name and substantially more are alluded to without names. Despite the quantity and diversity of example, *Origin* never reads like a simple compendium of natural history. Like the facts adduced in a lawyer's examination of a witness, Darwin's examples are always leading somewhere.

The larger structure of *Origin of Species* is reflected in the titles and section headings of its fourteen chapters. So, too, is the structure of Darwin's career. Contemporary readers, except for a few select friends like Lyell and Hooker, would have seen only the former, but to modern readers familiar with Darwin's life the latter is equally clear. One chapter after another shows Darwin's thirty years of research—observation, experiment, reading, correspondence, and conversation—paying off.

Chapters 1 through 4 lead the reader from variation in domestic animals through variation in nature to the existence of a Malthusian struggle for existence and, finally, to the concept of natural selection. They are the final, polished expression of ideas that began to coalesce in Darwin's notebooks in 1837, crystallized with his reading of Malthus in the fall of 1838, and became a coherent argument in the 1842 sketch. In a sense the first four chapters are a revised, expanded version of the 1844 essay, but they are reinforced, deepened, and fleshed out by work done in the fifteen years afterward. Chapter 1, on variation in domesticated species, rests on examples drawn from Darwin's research among pigeon fanciers, dog and horse breeders, and farmers—a major element of his post-1854 research for *Natural Selection*. Chapter 2, on variation in nature, and Chapter 5, on recurring patterns (Darwin calls them "laws") of variation, make extensive use of his work on barnacles. The core ideas of Chapters 3 and 4, on the struggle for existence and natural selection, date back to the late 1830s, but the examples used to illustrate them include botanical observations from Hooker and Gray, and anatomical material from Huxley, as well as more about Darwin's own beloved barnacles.

Chapter 6 deals with "difficulties on the theory" and so, in a sense, do Chapters 7 (on instinct), 8 (on interspecies cross-breeding), and 9 (on the gaps in the geologic record). The substantial amount of space devoted to raising, then meeting, possible objections to the theory is one of the things that distinguishes the *Origin* from earlier work such as the 1844 essay. It is difficult *not* to see this particular choice as Darwin's response to *Vestiges* and the ferocious reaction it drew from the scientific community. Chapter 9, "On the Imperfection of the Geologic Record," is particularly interesting. It addresses the most obvious objection to the theory—that the transitional species predicted by it are not visible in the fossil record—by arguing that the fossil record is too fragmentary to give a true picture of the history of life. Many species fossilize poorly or not at all, Darwin argues, and even for species well suited to it fossilization only occurs under precise conditions. Individual fossils are also subject to destruction by erosion, volcanism, or metamorphism, and may (because of changes in the land surface) become inaccessible to human observers. The conclusion that Darwin drew from this was that transitional species *should* be rare, since the fossil record is biased toward the common, the numerous, and the recent. It was, as Darwin acknowledged, a line of argument pioneered by Lyell in *Principles of Geology* decades earlier. Lyell, however, had used it to account for the absence of ancient, complex life, and so to advance his theory of non-progressionism.

Chapters 10 (the geologic record), 11 and 12 (geographic distribution), and 13 (taxonomy, anatomy, and embryology) lay out what Darwin saw as the best evidence for evolution by natural selection. The data he collected on the *Beagle* expedition are, naturally, prominent. The fossil animals of South America, described by Owen more than two decades before, have their day in Chapter 10. The Galápagos finches and mockingbirds that Gould helped Darwin to unravel are prominent in the chapters on geographic distribution. These chapters are also, however, showcases for those done by Hooker, Huxley, and Gray in the 1850s, and for more of the reams of data that Darwin had gleaned from scientific papers read and heard, from letters to naturalists on three continents, and from the accounts of travelers, soldiers, and colonial officials. A good deal of the material had been gathered in the 1850s; a few pieces of it may have entered his consciousness as an Edinburgh medical student in the 1820s.

Chapter 14, the conclusion, neatly wrapped up lines of argument and evidence. True to Darwin's preference for avoiding controversy, however, it remained silent on what the theory might say about the origins of human race, and on the broader philosophical and religious conclusions that might follow from the theory. On the issue of human origins, which had been central to his thinking since the *Beagle* called at Tierra del Fuego, Darwin contented himself with a cryptic half sentence. Through the theory, he wrote, "light will be thrown on the origins of man and his history." His only acknowledgment of the fact that evolution by natural selection turned species from the carefully crafted handiwork of a benevolent Creator into the products of natural forces was the final sentence of the book. In its acknowledgment of the continuity between past and present, it was an appropriately Lyellian benediction. "There is grandeur," he wrote, "in this view of life, with its several powers, having originally been breathed into a few forms or into one; and that, whilst this planet has gone cycling on according to the fixed law of gravity, from so simple a beginning endless forms most beautiful and most wonderful have been, and are being, evolved" (Darwin, 1966, p. 490).

THE UNFINISHED REVOLUTION, 1860–1882

Charles Darwin's career and the development of evolutionary theory were closely intertwined for thirty years, from the early 1830s to the early 1860s. After the publication of *On the Origin of Species*, however, the two quickly began to diverge. The ideas that Darwin presented in *Origin* led—as he had known they would—to intense discussion and debate involving not only biologists and geologists but engineers (Fleeming Jenkin), physicists (William Thomson, later Lord Kelvin), and members of the clergy (Bishop Samuel Wilberforce). A steady stream of reviews, responses, and commentaries followed in the wake of the *Origin*, and full-scale books dealing wholly or partly with the origin of species proliferated. Huxley, Wallace, Gray, and Lyell had all weighed in on the subject by 1870, and they were soon joined by Herbert Spencer (author of the term "survival of the fittest") and many others. Darwin took only a limited role in the flurry of activity his ideas had generated. He continued to work on evolution-related problems almost until his death in 1882, but kept his distance from the cutting edge of the field.

Darwin's publication record from 1860 to 1880, though it did not match the productivity of the miraculous decade following the *Beagle* expedition, was nonetheless impressive. It was all the more impressive for a man in his fifties and sixties who had suffered debilitating health problems for decades and made the sixteen-mile trip to London only occasionally. *On the Various Contrivances by which British and Foreign Orchids are Fertilized by Insects*, the first of his books on plants, appeared in 1862 and the second, a slim (118-page) volume on *The Movement and Habits of Climbing Plants* two years later. He continued his decades-long study of variation in the barnyard and the breeding shed in the two volumes of *The Variations of Animals and Plants under Domestication* (1868), and returned to ideas first raised by an encounter with Jenny the orangutan in *The Expression of the Emotions in Man and Animals* (1872). *The Descent of Man and Selection in Relation to Sex* appeared between *Variation* and *Expression* in 1871. *Insectivorous Plants* (1875) was followed in rapid succession by a new edition of *Orchids* (1876), *The Effects of Cross and Self-Fertilization in the Vegetable Kingdom* (1876), *The Different Forms of Flowers on Plants of the Same Species* (1877), and a new edition of *Climbing Plants* (1879). His last scientific book, *The Formation of Vegetable Mold, through the Actions of Worms, with Observations on Their Habits*, appeared in late 1881, less than a year before his death.

The titles of these books have often been used to paint a picture of the aging Darwin as a scientist past his prime and out of ideas, pottering in the greenhouse and writing about topics of little significance. A closer look suggests, however, what he did after the *Origin* was of a piece with what he did before. *Variation under Domestication, Cross and Self-Fertilization,* and *Expression of the Emotions* were all topics that figured prominently in his notebooks as early as the late 1830s and early 1840s. The first two received some attention in *Origin,* the third considerably less. *The Descent of Man,* with its dual focus on human origins and sexual selection, pairs two more topics that Darwin had treated at length in his research before the *Origin* but touched on only in passing in the book itself. The remaining botanical books—*Different Forms of Flowers, Climbing Plants, Orchids,* and *Insectivorous Plants*—cluster around a shared theme of natural selection's power to produce elegant adaptations in seemingly simple organisms. *The Formation of Vegetable Mold* is, beneath its yawn-inducing title, a case study in miniature of Lyell's (and Darwin's) belief in the cumulative power of small changes. It brought Darwin's career full circle, and its publication date, fifty years after Darwin picked up *Principles of Geology,* seems oddly fitting.

The history of evolutionary theory was, for many decades after *The Origin of Species,* more about departures from Darwin's ideas than about extensions of them. Darwin's work was brilliantly successful in establishing that evolution accounted for the origin of species, but not in establishing natural selection as the mechanism by which it operated. Darwin's inability to explain *how* new variations arose and were passed on, combined with the looming philosophical implications of "blind" natural selection, made it seem less than attractive even to Darwin's staunchest supporters. Lyell, always worried about the dignity of the human race, suggested that human intelligence and consciousness might be a product of something other than evolution. Asa Gray hypothesized that natural selection was guided, behind the scenes, by the hidden hand of God. Wallace—along with many other evolutionists—came to believe that *some* form of guidance was at work in evolution. T.H. Huxley fiercely defended Darwin in public, but harbored suspicions that "saltation" (periods of stasis, punctuated by rapid leaps forward) might be closer to the reality of evolution than the slow-and-steady grind of natural selection. Lamarck's model of linear, goal-directed evolution enjoyed a renaissance in the decades bracketing 1900. Particularly in the United States, it eclipsed Darwinian evolution for most of two generations.

None of this, ironically, diminished the already strong association of Darwin and evolution. When cartoonists wanted to reach for a human symbol of evolution, it was Darwin's familiar face and beard (often attached to an ape's body) that they used. When William Jennings Bryan railed against evolution at the Scopes Trial in 1925, it was Darwin's name that he invoked. It was not until the 1930s, however, that the merging of genetics and evolutionary theory in the "modern synthesis" resolved long-standing questions about heredity and variation, and returned natural selection to the prominent position that Darwin had envisioned for it. Only then did the dominant model of evolution actually, as well as symbolically, become Darwinian.

CONCLUSION

It is the central theme of the "Darwin legend" that Darwin's career consisted of flashes of brilliance punctuated by long digressions. Condensed, it might read as follows: Darwin, after an undistinguished childhood, boarded the *Beagle* as an orthodox creationist innocent of transmutationist ideas, but discovered (or, depending on the telling, "invented") evolution by natural selection on or just after the voyage. Having completed the theory, he then dropped the subject for twenty years until fear of being "scooped" by Wallace forced him to act. He wrote *On the Origin of Species* in another white-hot flash of inspiration, but soon fell back into pottering with uninteresting and unthreatening subjects, managing only one more significant work (*Descent of Man*) before he died. The reality of Darwin's life is, as I have tried to show, significantly more complex.

Darwin's upbringing was far from orthodox, his commitment to special creation modest, and his pre-*Beagle* exposure to transmutationist ideas considerable. His experiences on the *Beagle* were essential to the development of his theory, as was the "aha!" moment when he first read Malthus, but both sets of experiences took place in the context of his ongoing attempt to build a scientific career for himself. They happened where and when they did, and affected Darwin they way they did, partly because of the social, professional, and intellectual context within which he worked. During the five years between Darwin's opening of his first notebook on evolution (1837) and his first sketch of the theory (1842), he was immersed in a thriving scientific community and actively reaching out to its members for help in his research. The end product of those five years of work was a coherent theory but not (particularly given the reception of *Vestiges* two years later) a developed, publishable theory.

Darwin's actions 1844 to 1858 can be read as an attempt to avoid the subject of evolution, but also as his conscious attempt to close that gap. Even if he had been under pressure similar to that which he felt in 1858, he could not have written *On the Origin of Species* in 1845. He could have written *some* book-length explanation of his ideas (the 1844 essay shows that), but it would have been a very different book. The extra years gave Darwin the opportunity to gather data and refine ideas, to master barnacles and delve into the world of pigeon breeding, and to tap the expertise of Hooker, Huxley, Gray, and the rest of the "young guard." It also turned him from a man of thirty-six—recently married, newly settled in the country, struggling with the religious gulf between himself and his wife—to a long-married man of fifty, respected by neighbors and colleagues, who had buried his faith with his favorite daughter and infant son. The Darwin who wrote *On the Origin of Species* in 1859 was, quite simply, not the man who would have written it in 1845.

Darwin's career after *Origin* can, similarly, be read as an escape from evolutionary ideas or as an ongoing engagement with them. If his post-1859 books lack the brilliance and force of *Origin of Species* (even *Descent of Man*, the most famous of them, feels flat alongside Huxley's and others' writings on the same subject), they do not lack substance. They may be, in spirit, sections of the never-written "big book" on *Natural Selection*: things that Darwin wanted to discuss at length, but never had the opportunity to in 1858–1859 because of the pressure to get *Origin* into print. Taken on those terms, they can be seen as Darwin's attempt to bring to an orderly close the wealth of investigations he began in the decade after he returned from the *Beagle* expedition.

Regardless of how we choose to interpret the arc of Darwin's career, one thing is clear: It was a career made possible by the wealth of scientific opportunities available to a man of wealth and leisure—men like Charles Darwin—in nineteenth-century Britain. The Darwin legend, by painting him as the archetypal lone genius, gets him spectacularly wrong. He was a genius, to be sure, but even when physically isolated he was never professionally "alone." His great talent was his ability to pool his observations with the observations of hundreds of others and, peering into the resulting sea of data, discover patterns that none before him had seen.

FURTHER READING

Bowler, Peter J. *Evolution: The History of An Idea,* 3rd ed. (Berkeley: University of California Press, 2003).

Browne, Janet. *Charles Darwin: Voyaging* and *Charles Darwin: The Power of Place* (New York: Knopf, 1995 and 2002).

Darwin, Charles. *On The Origin of Species: A Facsimile of The First Edition.* Edited by Ernst Mayr (Cambridge, MA: Harvard University Press, 1966).

Desmond, Adrian, and Moore, James. *Darwin: The Life of a Tormented Evolutionist* (New York: Warner Books, 1992).

Keynes, Richard Darwin. *Fossils, Finches and Fuegians: Darwin's Adventures and Discoveries on the Beagle* (Oxford: Oxford University Press, 2006).

Stott, Rebecca. *Darwin and the Barnacle: The Story of One Tiny Creature and History's Greatest Scientific Breakthrough* (New York: Norton, 2003).

T.H. Huxley: The Evolution of the Bulldog

Lisa Nocks

Thomas Henry Huxley (1825–1895), another legendary figure in the history of evolution studies, held a key place in the London scientific community from the 1850s until the end of his life. He contributed to the transition of natural science from a gentleman's avocation to a profession, and to popularizing of scientific thought, preparing teachers in the applied sciences, and to incorporating science studies into the public school curricula. He had become a member of the British Association for the Advancement of Science (BAAS) in 1846, and by the time he returned from his stint as assistant surgeon on the Rattlesnake expedition (1846–1851), he had already published a half dozen scientific papers. Although it took him several years to begin earning a decent living by his vocation, Huxley was recognized for his contributions to science early on with a fellowship in the Royal Society (1851), which presented him with its gold medal in 1852, and memberships in a number of other professional organizations, including the Zoological and Geological societies (1856).

He began his teaching career in 1854 with a lectureship in natural history at the Government School of Mines (professor 1857). Among his early career appointments were naturalist to the Geological Survey, lecturer at St. Thomas Hospital, Fullerian professor of physiology at the Royal Institution (1855), and examiner in physiology and comparative anatomy for the University of London (1856). He soon became a popular science lecturer to the workingmen. In addition to numerous papers in the various society journals and in periodicals, his workingmen's

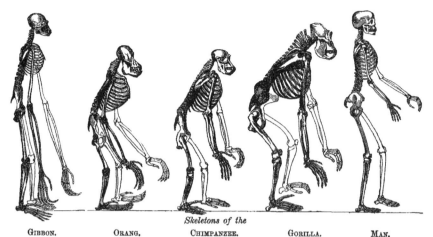

GIBBON.   ORANG.   CHIMPANZEE.   GORILLA.   MAN.

Skeletons of the

Photographically reduced from Diagrams of the natural size (except that of the Gibbon, which was twice as large as nature), drawn by Mr. Waterhouse Hawkins from specimens in the Museum of the Royal College of Surgeons.

Huxley's primate comparison illustration from *Evidences as to Man's Place in Nature.* The most well known, but misunderstood piece of artwork in the history of evolution studies. (Author's collection.)

lectures were published and distributed widely. In 1862 he was appointed Hunterian lecturer at the Royal College of Surgeons, where he served as until 1869. William James remarked in 1865 that before Huxley was forty years old, he had "made valuable contributions to almost every province of anatomical science" leaving "the mark of the strong grasp he takes on many other subjects" (James, p. 290).

Huxley's contributions to science and education have been a continual source of study since his *Life and Letters* were published by his son Leonard after his death. Most recently they have been detailed and analyzed in Adrian Desmond, *Huxley: From Devil's Disciple to Evolution's High Priest* (1997) and Paul White, *Thomas Huxley: Making the "Man of Science"* (2003). This chapter is meant to complement those excellent works by framing Huxley's iconic reputation within his pivotal works: his reviews of *On the Origin of Species* (1859), his *Man's Place in Nature* (1863), and his 1893 lecture *Evolution and Ethics* and its *Prolegomena*, published in 1894. It is through Huxley's own words that we may understand how recent claims by anti-evolutionists that Huxley rejected Darwin's theory are based on an incomplete reading of his work. Moreover, it is within this framework that we can discover how Huxley earned the (self-imposed) title for which he is today most often remembered: "Darwin's Bulldog."

Late in life, Charles Darwin called Huxley "the mainstay in England of the principle of the gradual evolution of organic beings" (Darwin in

Barlow, 1993, p. 106). It was Huxley (along with Asa Gray in America) who publicly promoted Darwinian theory. By the late 1850s Darwin already counted him (along with his friends, botanist Joseph Dalton Hooker and geologist Charles Lyell) among his "three judges"—the men whose scientific opinion Darwin valued most (*Correspondence of Charles Darwin* (*CCD*) v. 7, p. 398). Huxley's contribution to that court of opinion was an important one: Darwin, who earlier had reservations about putting the scrapper up for the Athenaeum Club, found that the pugnacious Huxley was the ideal proxy in the public arena. While Hooker discouraged Darwin from pursuing scientific debates in the press, Huxley eagerly took up perceived affronts to himself and Darwin in periodicals like *The Times*, to Darwin's delight. In a real sense Huxley was a proxy for Darwin whose botanic investigations, poor health, and reticence about joining in public debate kept him mainly at Down and the medicinal baths.

In their personal correspondence, Darwin playfully referred to Huxley as "my good and admirable agent for the promulgation of damnable heresies" (*CCD* v. 7, p. 434). The joke alluded to their common mission to extract theology from science, for as he had confided to Lyell, "I would give absolutely nothing for the theory of natural selection if it requires miraculous additions at any one stage of descent" (*CCD* v. 7, p. 345). He thought Huxley's positive review of *Origin* in *The Times* "nobly soars above religious prejudices" (*CCD* v. 7, p. 457).

THE ORIGIN OF SPECIES

It is important to understand that Darwin was not the first to suggest that life evolved on the earth; however, by the time he was working on his species book, there was no consensus about exactly how species developed. In the early 1800s Jean-Baptiste Lamarck had argued that variation occurred out of organisms' need and desire. One of the most circulated examples of Lamarck's theory is that the physical act of a giraffe stretching its neck to obtain food from higher branches results in the offspring's inheritance of the acquired characteristic—a longer neck. According to Lamarck, such a change makes the offspring better adapted to their environment. Huxley accepted neither Lamarck's view, nor the planned progress of Robert Chambers's *Vestiges of Creation*, and had already rejected the idea that new species were a result of "special creations," on the grounds that there was no evidence to support such a claim.

Huxley had earlier rejected the notion of an evolution based on a progressive "fish to Man" chain of being. When he lectured "On the Persistent Types of Animal Life" at the Royal Institution (3 June 1859), he hinted his support for Darwin, asserting that "the gradual modification of pre-existing species" was far more illustrative for explaining the facts of paleontology than assigning them to some creative power (Huxley 1859a, p. 153; *CCD* v. 7, p. 302).

Darwin's non-teleological model was compatible with what Huxley eventually termed his "agnosticism"—a method of inquiry about the nature of the universe that neither acknowledged nor denied the existence of God. He later wrote that what he and some of his colleagues were looking for was a system of evolution "which could be brought face to face with the facts and have their viability tested," and the Darwinian model "provided us with the working hypothesis we sought" (*The Life and Letters of Thomas Henry Huxley* (*HLL*) v. 1, p. 182). Consequently, the non-falsifiable beliefs of natural theologians like anatomist Richard Owen could not be the basis for evolution.

But he also perceived weakness in the views he knew Darwin was preparing to publish. Huxley had earlier gravitated toward another, spherical model of evolution, in which new forms emerged from a common archetype, though not from other variations, and was lingering on the point. Having been working with marine invertebrates and aquatic animals, he found the persistent characteristics of organisms like *mollusca* in the fossil beds irreconcilable with Darwin's idea about gradual transition.

In November, after reading an advance copy of *On the Origin of Species* (December 1859), he privately expressed to geologist Charles Lyell his continued reservations about transitions. He wrote to Darwin praising most of the book, but wondering, "Why, if continual physical conditions are of so little moment as you suppose, variation should occur at all." Huxley also cautioned: ". . . you have loaded yourself with an unnecessary difficulty in adopting *Natura non facit saltum* [nature doesn't make jumps] so unreservedly." Nevertheless he told Darwin that he would read the book over before making further criticism, and that Darwin had "earned the lasting gratitude of all thoughtful men . . . I am sharpening up my claws and beak in readiness" to defend *Origin* against the "curs who will bark and yelp" (*CCD* v. 7, p. 189).

THE DARWINIAN HYPOTHESIS

On 26 December 1859, Huxley anonymously posted "The Darwinian Hypothesis" in *The Times*. He opened his almost 5,000-word book review of *Origin* by quoting Darwin's own summary of the vehicle for transmutation: "Species originated by means of natural selection or through the preservation of the favoured races in the struggle for life"

(Huxley, *The Times*, 1859, p. 1). Huxley summarized the debate among naturalists over the definition of the term *species*, and asserted the logic of uniformitarianism to support evolution (the idea that change happens very gradually, and that natural forces like erosion and volcanic activity work today just as they always have). He then summarized Darwin's theory: that all life is a struggle, since all organisms, from the simplest to the most complex, must compete for living space and food; that all offspring vary slightly from their progenitors, and those *random variations* (we now call mutations) that make the offspring more apt to thrive and reproduce in that competitive environment survive, while those variations that do not eventually die out. Offspring that vary enough that they cannot reproduce with the former strains are new species. Other factors, like geographical divergence, play a part in evolution, but plan does not. According to Darwin, change occurs neither by supernatural means nor by the will of the organism. The fossil record—preserved evidence of long-dead organisms in the stratified layers of the earth—provides evidence of gradual change; and further excavations will yield more complete evidence of speciation.

Even in his essentially positive reviews of *Origin* however, Huxley expressed reservations about natural selection, not as impossible, but as a theory that had not yet been scientifically proven. He thought that although Darwin had been meticulous in basing his hypothesis of speciation on experiment and observation rather than blind faith, he should not have supported natural selection as the mechanism of speciation by the analogy of domestic breeding (in which certain pairs of animals or plants possessing desired characteristics present in one or both parents are bred to reproduce those characteristics. The offspring that most successfully inherit the desired traits will in turn be bred to reproduce them. For example, the offspring of a racehorse that exhibits strength of limb, or those of a vegetable that can withstand certain pests would be crossed with others of advantageous characteristics. Repeated selective breeding is meant to develop stock with only the most desired traits.) Huxley observed that human intervention was strongly at work in domestic breeding, but nature's laboratory was not so controlled as a stable or a garden. Furthermore, breeders' modifications were not exclusively operating to further survival by reproduction, but to satisfy certain desires of the breeder: better tasting meat, hardier plants, winning show animals. Finally, he did not think domestic breeding consistently produced sterile hybrids.

Notwithstanding these complaints, Huxley asserted that Darwin had provided the best hypothesis about evolution thus far articulated. He was not rejecting natural selection, but insisting on scientific proof to support the theory. In *Macmillan's Magazine*, he encouraged the public that if it was patient, it would see the paleontological proof of Darwin's hypothesis; and that "his fellow-workers will soon find out the weak points in his doctrines, and their extinction by some nearer approximation to the truth *will exemplify his own principle of natural selection*" (emphasis mine). It would thus be the "painstaking, truth-loving investigation of skilled naturalists" and not the "prejudices of the ignorant, or the uncharitableness of the bigoted, on either side of the controversy" that would draw out the mechanism of evolution from the fossil record (Huxley, *Macmillan's*, 1859, p. 148). He told Lyell privately how important it was to him that the public understood his appreciation for Darwin; that if it turned out Darwin was right about natural selection his contribution would far outweigh his predecessors, and Darwin "could take his place with such men as Harvey" (the seventeenth-century physician who had discovered and described the circulation of the blood). Even if he was wrong "his sobriety and accuracy of thought will put him on a far different level from Lamarck" (*HLL* v. 1, pp. 215–216).

Huxley's difficulties with Darwin's theory were gradually diffused as new data emerged. For example, he credited the discoveries of extinct species of the Tertiary epoch by paleontologists Marsh and Cope in America, and others elsewhere, to support Darwin's claim in *Origin* that further discoveries would add to the fossil record and bear his theory out. Huxley also credited Marsh's work on Cretaceous-era fossils of toothed birds of North America for providing evidence for Darwin's theory that reptiles and birds shared common ancestors. In 1868, Huxley delivered a paper at the Royal Institution on the anatomical evidence of transition from reptile to bird based on the 1862 discovery of fossil remains of *Archaeopteryx*, which was published in several science periodicals. Although Huxley's coming around to transmutation with transition was supported by paleontological evidence, it was in part motivated by his feeling that saltations smacked of special creation. Therefore, despite his reservations about Darwin's theory, he told Lyell, "follow [transmutation] out, and it will lead us somewhere; while the other notion is, like all the modifications of 'final causation' a barren virgin" (*HLL* v. 1, 25 June 1859). Huxley continued to disapprove of the domestic breeding analogy until it could be demonstrated that it would give rise to varieties that were infertile with each other. Still, he saw in Darwin's careful observation, experiment, and exchange of data the proper basis for science, and continued promoting the Darwinian hypothesis in his public lectures.

By his own account, Huxley had begun to draw a relationship between man and other animals in the 1850s. (It is important to note that neither Darwin nor Huxley ever said that man descended directly from an ape, but that all primates share a common ancient ancestor.) In 1857, Huxley criticized a paper delivered to the Linnean Society that classified humans as distinct from and above apes and other mammals by proposing that certain anatomical features of the human brain were peculiar to genus *Homo*. He continued writing and lecturing on a close biological relationship between man and other primates.

MAN'S PLACE IN NATURE

In 1861, after accepting an invitation to speak at the Philosophical Society in Edinburgh, Scotland, T.H. Huxley told his wife Henrietta, "they know my views, so if they do not like what I shall have to tell them, it is their own fault" (*HLL* v. 1, p. 207). His views certainly were well known: He had already established himself as a popular lecturer to the workingmen; and was publicly connecting his position on primate classification with Darwin's theory of evolution in essays including "On the Zoological Relations of Man with the Lower Animals" (*Natural History Review*, January 1861), and "On the Brain of Ateles Paniscus" (*Zoological Society Proceedings*, June 1861). A long-running argument with anatomist Richard Owen over the comparative anatomy of human and simian brains that had become known as the "Hippocampus Debate" had made its way from the scientific societies to the *Athenaeum*, where Owen had to defend himself against accusations from the Darwin circle of nonprofessionalism in not citing their contributions to his research, and supporting Huxley's claim that Owen's own research was faulty. Owen had placed humans into a suborder he called *Archencephalus*, on account of three peculiarities of the human brain, though some argued that his blindness to the morphological facts was more a case of his need to maintain for humans a special relationship with God. Their public sparring over the existence of a hippocampus in primate brains had even been spoofed in *Punch*. The liberal Reverend Charles Kingsley, who maintained a cordial correspondence with Huxley for years, had also spoofed the feud in "Speech of Lord Dundreary in Section D, on Friday last, On the Great Hippocampus Question" (1861). He picked up the topic again in his children's book, *Water Babies* (1863), depicting the religious anxiety that underlay public debate over human origins:

> The professor . . . had even got up once at the British Association and declared that apes have hippopotamus majors in their brains just as men have Which was a shocking thing to say, for if it were so, what would become of the faith, hope and charity of immortal millions? . . . if a hippopotamus major is ever discovered in on single ape's brain, nothing will save your great- great- great- great- great- great- great- great- great- great- great-greater-greatest grandmother from having been an ape too. (Kingsley, pp. 153–154)

However, when Huxley told his audience in "saintly Edinburgh" the following January (1862) that he "entertained no doubt of the origin of Man from the same stock as apes," it wasn't Owen he was baiting (*HLL*

v. 1, p. 210). As he expected, the evangelical *Witness* called his lectures on "The Relation of Man to the Lower Animals" a "blasphemous contradiction to Bible narrative and doctrine" and "the most debasing theory that has ever been propounded before a civilized audience" (*Witness*, January 14, 1862, np; January 18, 1862, np). Huxley glibly credited that reviewer with helping to "disseminate my views through regions which they might not otherwise reach." Gloating over his audience's "unmitigated applause," he wrote that he would "make something more" of the lectures in London, and then publish them. (*HLL* v. 1, p. 209). Huxley repeated them to the working men that fall, then incorporated them into *Evidences as to Man's Place in Nature* (MPN).

It is evident that he saw himself embroiled in a contest for the public's hearts and minds: Equating secular science with patriotism, he had once told listeners that "free discussion is the life of truth, and of true unity in a nation." Whether England would play its part "depends upon how you, the public, deal with science" (*Proceedings*, 1860, pp. 195–200). He had reminded Hooker in 1860 that that soon they would be "the representatives of our respective lines in this country" with "certain duties to perform to ourselves, to the outside world, and to science" (*HLL*, v. 1, p. 239).

Although not an atheist, Huxley was among those who held the view that religious dogma had no place in the pursuit of science. He had once told the reverend Charles Kingsley that Carlyle's *Sartor Resartus* "led me to know that a deep sense of religion was compatible with the entire absence of theology" and that if the Church of England were to survive it would be by the work of "men like yourself who see your way to the combination of the practice of the Church with the spirit of science" (*HLL*, v. 1, pp. 237–238). At the heart of Huxley's work was the promotion of the view that human beings developed through the evolutionary process with other forms of life. He quipped to his wife Henrietta about the success of his lectures on evolution, "My workingmen stick by me wonderfully, the house being fuller than ever last night. By next Friday evening they will all be convinced that they are monkeys" (*HLL* v. 1, p. 205). By the time he began preparing the manuscript for *MPN*, he was interpreting the positive reactions to his lectures as proof of "the general disintegration of old prejudices which is going on" (*HLL* v. 1, p. 210).

Despite already being overextended with personal and professional obligations, Huxley took on the editing of material that was even by his own admission a case of "reslaying the slain." We can attribute his decision in part to a profit motive, since any economic security he might have achieved by the early 1860s was compromised by the continual demands placed on him by his family. Besides his wife and children, he was caring for an alcoholic sister, and sending money to another in the United States. When his beloved brother George died, Huxley sold his own Royal Society medal for £50 to clear his brother's debts. Why

should he not make a few quid on material that others had already prof-
ited from? The publisher Hardwick had made his share on his transcrip-
tions of the lectures. With Huxley's permission, he had quickly advertised
and sold 2,000 copies. Huxley took Lyell's advice to "lose no time in
considering what steps to take to rescue the copyright of the third thousand" (*HLL* v. 1, pp. 223–224). The initial sales of *MPN* were not as
impressive as those of Darwin's *Origin*, whose first edition of 1,250 copies sold out in one day. However, *MPN* made a very respectable debut.
Williams and Norgate released it on 20 February, and despite its place-
ment on a specialist list, the first thousand copies were gone in weeks,
and Mudie's lending library could not keep it on the shelves. By the end
of February, the publishers were nagging him for corrections for the next
printing. A second thousand copies were put out in March, followed by
German, Italian, and Russian translations and an American edition. The
response to *MPN* is not so surprising, considering its populist style,
Huxley's renown as a lecturer, and that its subject fed the public's hunger
for controversy.

He was motivated by more than the potential for profit, however. A
few years earlier he had encouraged the German embryologist Ernst
Haeckel on the publication of Haeckel's *Morphologie* (1867), "I am
much inclined to think that it is a good thing for a man, once at any rate
in his life, to perform a public war-dance against all sorts of humbug and
imposture" (*HLL* v. 1, p. 309). It was the war dance that seemed to drive
Huxley. His objective was to remove Theology once and for all from
scientific theory, and to make his views accessible to a wider audience.

PRIVATE RECEPTION

Huxley sandwiched his argument for a non-miraculous human origin
between a descriptive chapter on primates and one on the recently discovered skeleton fragments of ancient humans. Though weary of Huxley's
insistence on including his reservations about natural selection, Darwin
told him he "never in my life read anything grander" than the conclusion to the second chapter—a rather poetic comment Huxley included to
assure his readers that an intellectual, "psychical" gulf still existed between Man and brute. It was such a piece of eloquence, Darwin thought,
that Huxley ought to have saved the best for last and used it for the
conclusion. In his praise of *MPN*, Darwin confided his pleasure with
Huxley's attack on Owen's "continuous operation of ordained becoming." He called Huxley's remarks "a delicious sneer, as good as dessert"
(*CCD* v. 11, p. 180, and n. 3, p. 181).

Reading "The Monkey Book" as Darwin called *MPN*, had helped
to neutralize his disappointment at Lyell's equivocation on human

classification in his long-awaited *Antiquity of Man* (February 1863). Darwin had told Hooker that he once thought Lyell could have done more "to convert the public than any of us." Now Darwin "wished to heaven he had not said a word on the subject" (*CCD* v. 11, p. 174). From Harvard, botanist and Fisher professor of natural history Asa Gray comforted Darwin, "you think Lyell [in *The Antiquity of Man*] too non-committal and timid. Well Huxley makes up for it, I should think!" (*CCD* v. 11, p. 254, and n. 10).

For his part, Hooker complained that *MPN* was a "coarse little book," perhaps because of its frontispiece, a succession of primate skeletons drawn by Waterhouse Hawkins, which someone had said were "not fit for a gentleman's table" (*CCD* v. 11, p. 179). Still, both men praised Huxley's articulate writing. Darwin extolled its "clearness and condensed vigour" of thought, while Hooker concluded, "Lyell never rises to the magnificence of Huxley's language nor to the sublimity of some of the passages . . . which one can read 1000 times with fresh delight" (*CCD* v. 11, p. 228, and n. 18). Botanist and president of the Linnean Society George Bentham admired Huxley's "raciness of style engrafted upon solidity of thought and correctness of views" (*CCD* v. 11, p. 427, and n. 6). German naturalist Carl Vogt called it a "beautiful little book," and incorporated Huxley's observations into his own work, *Lectures on Man: His Place in Creation and in the History of the Earth* (1864). In America, Asa Gray, who had not yet read *MPN*, anticipated that Huxley would present his views on Man "in a way that would frighten us off (*CCD* v. 11, p. 336). Still, he conveyed the news to Darwin that Jeffries Wyman, Hersey professor of anatomy at Harvard, was delighted with the book.

THE PRAISE

In 1861, despite being professionally overextended, Huxley had taken on the chief editorship of the *Natural History Review* (1861–1864) originally a failing, largely anti-Darwinian quarterly based in Edinburgh. He moved it to London, and staffed it with pro-Darwinian editors, hoping to create a kind of headquarters for scientific naturalism. The *Review* lasted only four years, in part because, as Lyell had predicted, Huxley ended up taking on most of the work himself. Nevertheless, while it existed, it was a tribute to the new theory of evolution. It is not surprising then, that its coverage of *MPN* was essentially an advertisement for the book. It comprised long extracts that summarized his main arguments: First, based on morphological comparisons of primate skeletons with human ones, Linnaeus's classification of *Homo* as a genus of the order *Primate* was valid. Second, despite there currently being no evidence for natural selection, Darwinian evolution was "the most powerful instrument of investigation which has been presented to naturalists since the invention of the natural system of classification and the commencement of the systematic study of embryology." Finally, the fossil record would eventually turn up a more precise age for the earliest humans; nevertheless recent discoveries showed that Man was ancient and derived from a common stock with other primates (*Natural History Review*, April 1863, pp. 381–384).

Despite Huxley's distaste for atheists, a number of them embraced *MPN* without reservation. Engels let Marx know that it was "very good" (Desmond, *Huxley*, p. 313). The free thought *National Reformer* decided Huxley was as much an agent for dissent as he was for Darwin. Its reviewer acknowledged Huxley's contribution to "a war to the death . . . with tyrants of the mind" (*National Reformer*, March 28, 1863, p. 5). This reviewer took the occasion to remind readers of the advantage they had over Christians: "He is free to think and to investigate and fearless as to the consequences" with "no God to punish him, no devil to frighten him . . . no priest to censure him." He concluded by linking Huxley's insistence on evidence as a basis for scientific theory with free thinkers' reverence for "what is shown to be true" (*National Reformer*, 4 April 1863, p. 5).

The Reader, a liberal weekly run by Christian socialist Thomas Hughes (author of *Tom Brown's School Days*) ran a review by Frederick Dyster, a well-to-do local doctor, and a friend of Huxley. Dyster, who supported labor unions and worker education, and treated his poor patients for free, had co-identified a species of tube worm with Huxley, which they named *Protula dysteri*. Expectedly, Dyster called *MPN* "an admirable book" and thought it curious that anyone should prefer, "when seeking his origin, to suppose that he was modified mud rather than modified monkey" (Dyster, *The Reader*, 7 March 1863, pp. 234–235). It is likely that Huxley primed Dyster for the piece, since so much of it was taken up with a rehash of the "Hippocampus Debate," painting Huxley in a more favorable light than Owen, and using language Huxley had used himself. Dyster rehashed the rehash: During the late 1850s, Owen had placed humans into a suborder he called *Archencephalus*, based on his observation that three "peculiarities" of the brain of the genus *Homo* were clearly absent in other primates: "the posterior development of the lobes of the cerebrum [the two outer hemispheres of the brain that control conscious thought, intelligence, and memory as well as some involuntary regulation of motor patterns] was so marked in the genus *Homo*, that anatomists have assigned the character of a third lobe" which extends back, covering the olfactory lobes and the cerebellum. Further Owen argued, humans had a quite pronounced posterior cornu (rear horn) of the lateral ventricle (a chamber that carries spinal fluid to the brain); and finally that modern humans possessed a hippocampus (the seahorse-shaped structure that today is thought to be important in learning and in the conversion from short to long-term memories). Dyster related how other reputable anatomists had already established that these features were in fact extant in other primates. Dyster also pointed out that Huxley did not "insinuate that we are lost if an ape has a hippocampus minor . . . no one is more strongly convinced than I am" Huxley wrote "of the gulf between civilized man and the brute" (Dyster, p. 235).

THE CRITICS

The conservative *Athenaeum* called Huxley's view of the "intimate relations between Man and the rest of the living world . . . candid though heretical." Treating Huxley's and Lyell's books as if they were two volumes of the same work, the reviewer defined Lyell's object as "to remove Man remotely back in the scale of geological time," while Huxley's was "to degrade Man deeply in the scale of animal existence." Revealing its Tory sympathies, the reviewer wondered at the frontispiece with its "grim procession of skeletons . . . where is our pride of ancestry; our vaunted nobility of descent?" (*Athenaeum*, 28 February 1863, pp. 287–288).

DEFENDING THE ALMIGHTY

Some scholars have argued that Huxley's book was criticized on scientific, rather than theological grounds. However the scientific criticism that Huxley met in those reviews was very clearly underwritten by theological assumptions. No one would have been surprised to read accusations of materialism in conservative or sectarian papers like *The Telegraph*, which linked Huxley's book with the infamous *Pentateuch*, in which Bishop Colenso suggested that perhaps the work was not written by Moses himself. The *Quaker's Friend* thought that Huxley's views attacked "the sanctity . . . that enshrouds our being" (Desmond, *Huxley*, p. 314). However, lengthier reviews by scientists including the British comparative anatomist, anthropologist, and paleontologist Charles Carter Blake and American geologist and zoologist James Dwight Dana, then professor of natural history at Yale, were also essentially defenses of man as the Almighty's special creation.

The *New Englander and Yale Review*, which tended toward an anti-Darwinian if not anti-evolution view, criticized the theory in *MPN* as having been defended "with no special acuteness or ability." The anonymous reviewer (Dana) did not "observe a single point that is in advance of what Mr. Darwin has written" or find that Huxley answered any of the original objections raised to Darwin's theory. He thought the text showed "intellectual narrowness, moral degradation, and shallow science" on account of Huxley's refusal to see that after all, Man is designed for spiritual service (*New Englander and Yale Review*, July 1863, p. 592).

Blake attacked Huxley in both the *Anthropological Review*, which he had recently founded with the Anthropological Society's president, James Hunt, and the conservative *Edinburgh Review*, where he claimed that Huxley's conclusions on the age and classification of humans were based on "fallacies," misrepresentations," and "evasions" (Blake, *Edinburgh Review*, April 1863, pp. 541–569). For instance, Blake called Huxley's conclusion that "man is in all cases much nearer to the gorilla than the gorilla to the lowest Quadrumana" an "entirely illogical fallacy," essentially because "no intervening link has as yet been discovered between gorilla and man" (Blake, p. 549). (The search for the "missing link" was a key component of nineteenth-century evolution and human-origin research, integrated into both scientific essays and popular culture.) Blake also claimed Huxley had disregarded obvious morphological evidence that the hind extremities of apes were hands and not feet. In fact, Huxley had covered this in some detail in *MPN*, although coming to a different conclusion than Blake.

Blake had begun by asserting that "physical facts . . . must form the true basis of any accurate generalization," and later claimed "no intention of opposing Mr. Huxley's reasoning on theological grounds." Yet his argument reveals otherwise. After carefully picking apart Huxley's descriptions of the structural similarities between the various apes and of humans, he concluded that "all the powers that make us MAN—are visibly *independent of that mere structural organization* in which, as we have seen, many of the animals surpass us" (emphasis in original). Between "an animal gifted with the nicest sensuous facilities" and a man "deprived or destitute of all his senses and animal powers" Blake argued, the man would still possess—in addition to the *gift* of articulate speech, and the intellectual facility for numbers and generalization—"the power of conceiving the relation of Man to his Creator,—the power of foreseeing an immortal destiny" (Blake, p. 567).

Blake "suspected" that Huxley's theory was "indistinguishable from that of absolute materialism and even tends to atheism." He countered Huxley's statement that "the whole analogy of operations furnishes so complete and crushing an argument against any but what are determined secondary causes" by asserting that secondary causes "are only the means by which a primary cause operates." Quoting the canon of St. Paul's, Blake continued, "the weakest and most absurd arguments ever used against religion have been the attempts to compare brutes to men" as

Huxley had done. Materialistic and atheistic ideas, Blake thought "make the universe itself quite unintelligible." Blake was "reluctant to suppose" that "contemporary writers actually elected to support such opinions of their scientific investigations" and suggested that Huxley had done both society and his profession a disservice (Blake, p. 569).

Dana concurred with Blake that Huxley had erred in his assertion there was enough structural commonality between the bones and musculature of apes and humans to show a common origin. Like Blake (and Owen) Dana contended that all four ape limbs were used for the same purpose—as hands to grasp, climb, and cling. Dana argued that "In Man the forelimbs are withdrawn completely from the locomotive series and transferred to the cephalic." The anterior portion of the body [in man] is "turned over to the service of the head and this . . . places Man alone" (Dana, 1863a, p. 452). However, as with Blake's, Dana's scientific argument was at heart a theological one. He wrote that the connection of the forelimbs to the erect stance and to the head indicate "Man's higher nature" and that the "raising of the forelimbs from the ground for aesthetic, intellectual, and spiritual service" was "in direct harmony with such spiritual service." For Dana, no other animal could attain this spiritual character of humanity through any developmental process. He described "something" in Man

> wholly distinct from . . . a physical or intellectual nature in the mere animal, is a spiritual one—that through which, man bears God's image. It is the *spirit* in man which suggests a sense of dependence on a Power above; which makes man a moral being and . . . which prompts him to approach the Spirit on high with words and rites of devotion. For only spirit can commune with spirit, or comprehend the revelations of a spiritual being . . . these high characteristics of man place a long interval between him and the brute. (Dana 1863b, pp. 283–284)

Dana concluded that through cephalization "the whole outer being is made to show forth the divine feature of the inner being" (Dana, 1863b, p. 287).

An anonymous critic for the *North American Review* who thought it unlikely that Darwinian evolution would remain the dominant theory confidently concluded that Huxley's position on Man "cannot cast a ray of doubt on our divine sonship and immortal birthright on the spiritual side, nor can it falsify the charter of our heavenly citizenship in the revelation that bears the incontestable signature of the Almighty" (*NAR*, 1863, p. 293).

CONTEXT OF THE CRITIQUES

The critics were responding to what they understood to be the larger aim of Huxley and his circle: By extending Darwin's "long argument" to include humans, Huxley was contributing to the cause of the secularization of science—itself part of a larger battle to secularize all British institutions. The contemporary reception of *MPN* is a case study in how that battle was fought out in the press during the 1860s. It would be inaccurate, however, to characterize that battle as strictly between belief and unbelief. At its heart was the concept of free inquiry, which had been promoted for decades by diverse groups, among them working-class radicals, medical students, free thought publishers, middle-class reformers like J.S. Mill, liberal clerics, and anti-clerical scientists. Those fighting for political, economic, and social reform represented diverse philosophical views including atheist, deist, and dissenting Christians (Broad Churchmen to Unitarians and Christian Socialists.) Still, science, and in particular evolutionary theory, had already achieved metaphorical status by the time Huxley published *MPN*. For conservative religionists, Darwinian evolution stood for materialism, and Huxley was rubbing their noses in it. In Australia, an anti-Darwinian reviewer thought *MPN* "might have been written by the devil" (Desmond, *Huxley*, p. 317 and n. 12). After all, who but Satan would stand against God?

To many of the elite, the book was undermining their hegemony. Anglican Tories had earlier shuddered at the radical secularist interpretation of Lamarck's evolutionary theory—promoted earlier in free-thought pamphlets and periodicals like G.J. Holyoake's *Reasoner* (f. 1846) and lately in Charles Bradlaugh's *National Reformer* (f. 1860)— that the lowly might rise above their station in life by their own will and by environmental influences. Earlier efforts of the Society for the Distribution of Useful Knowledge (SDUK) to promote a science based on natural theology might have kept order among the lower classes. But they had been corrupted by the "French school" of mechanistic thinking. These days freethinkers were threatening the traditional hierarchical social order by embracing Darwin's secular theory of struggle to justify their own political lobbying efforts. Now Huxley had picked up on Darwin's vague reference to Man at the end of *Origin*, and removed humanity from its special position in God's universe, but also muddied the line between the rulers and the ruled.

MPN was only one of a number of contemporary scientific works generating public curiosity and controversy. In the late 1850s, discoveries of stone tools found alongside fossilized mammal bones in France and England, and a skull fragment discovered in the Neander Valley in Germany were used to support hypotheses about the antiquity and classification of the human race. For instance, Lyell had used these findings as a basis for his theory of an old earth in *The Antiquity of Man* (1863). At the same time, anatomical comparisons between humans and the African mountain gorilla, first discovered by Europeans in the late 1840s, and descriptions of the more recently discovered West African lowland gorilla by Paul du Chaillu in his *Explorations and Adventures in Equatorial*

Africa (1861) were incorporated into ongoing discussions of human ancestry. (Attacks on the veracity of du Chaillu's account were published widely in periodicals, and satirized in *Punch*—e.g., "Lion of the Season", 25 May 1861, p. 213.) Owen was one of the first to examine gorilla skull specimens sent by missionaries to London. Reports were often followed by satirical editorial pieces that sometimes incorporated the news about gorillas with the feud between Huxley and Owen.

The anatomical points covered in *MPN* were also being put in service to opposing racial views. The book had been published at the height of the American Civil War, a conflict that aroused heated debate in Britain because of the potential involvement of Britain in the war, and also because of economic troubles related to British dependency on American cotton. Each side accused the other of using their scientific views to promote their respective positions on the war. Slavery (abolished in British colonies by 1833) was an important issue to many, including Darwin and Huxley, who had been vocal in their opposition. Darwin had long been an abolitionist, canceling his subscription to *The Times* when the paper came out in favor of the South. The Ladies Emancipation Society was quick to reprint and promote Huxley's statements about the immorality of slavery in their 1864 proceedings. On the other hand, Hunt and Blake of the Anthropological Society were polygenists who used their scientific view of separate origins for different races to justify slavery in America. In the spring of 1864, an ongoing debate over the comparative anatomy of Europeans and Africans had spilled over into the periodicals, especially *The Reader*, where for example, the polygenist James Hunt had published "The Negro's Place in Nature" (19 March 1864, pp. 334–335). Colleagues had hissed Hunt when he presented his views at the Society. Now they were talking behind his back: Anatomist George Rolleston wondered if the "low-bred, ill-instructed impostor" wasn't being "paid by the confederates to lie as he does." In fact, the Anthropological Society was receiving financial support from a slush fund in the Richmond government through a confederate on their board. Huxley declined an honorary membership in the Anthropological Society after Blake's scathing critique of *MPN* *was* published in the inaugural issue the *Review*. Later, Hunt acknowledged that "if the society as a body have shown unanimity of sentiment on any one point, it has been, I believe against the Darwinian theory of origin as propounded by professor Huxley" (*Anthropological Review* 6, 1868, p. 78).

Farcical letters to the editor of *Punch* occasionally focused on evolution. A letter published two months after *MPN* was released asserted,

> Look you Mr. Punch, I firmly believe in the antiquity of my race, which is as great as that of any family in Wales. But I cannot and do not want to trace up our lineage to the monad of a million years ago through the

gorilla, and jackanapes, and for ought I know, the slug!" (*Punch*, 21 March 1863, p. 122)

The magazine, which regularly used its pages for slurs against the British working class and the Irish, used the evolution debate as fodder. Public notices of the Neandertal skeleton and discoveries of a new species of gorilla inspired the suggestion that the "howling yahoo common in Ireland" might be the missing link being searched for ("Men and Monkeys," *Punch*, 10 January 1863, p. 19). Some scholars have warned against characterizing the reception of new science as a religion/science debate. However, we must be careful not to judge Victorians on our own terms. The society in which Darwinian evolution emerged was being asked to swallow quite a bit in the space of a few decades—from universal suffrage, to rejection of upper-class hegemony within the professions, to the replacement of gentleman vicars with professional men of science. The suggestion that perhaps there was no divine plan in nature increased the burden.

RELIGION AND SCIENCE

Late in life, Huxley recalled that at the time he published *MPN* it had been only decades "since my kind friend, Sir William Lawrence, one of the best men whom I have ever known, had been well-nigh ostracized for his book On Man, which might now be read in a Sunday school without surprising anybody" (Huxley, *MPN*, Preface). In 1816, the Tory *Quarterly Review* had viciously attacked surgeon William Lawrence's "questionable" assertions that it was unnecessary to look for some additional vital force to explain life. After the Court of Chancery ruled the material in his *Introduction to Comparative Anatomy and Physiology* blasphemous, Lawrence withdrew the lectures from sale. He was forced to resign from the College of Surgeons and to recant his views to retain his position at other hospitals. (Ironically Lawrence could not control the distribution of the pirated editions by radical publishers, since he no longer held the copyright.) On account of these editions, by Huxley's youth, the book had become an underground classic. At the time Huxley recalled this story, Lawrence had not only been redeemed, but knighted, and had served as the Queen's surgeon. It was in part fear of this sort of pressure earlier placed on Lawrence that had caused Darwin to hesitate publishing his evolution theory. Although the level of anxiety that sent him to the medicinal baths shortly before the release of *Origin* was unwarranted, in his case, liberal clergy did not fare as well.

For hundreds of years, Anglicanism had been the foundation of government and the professions, which were controlled through the church's members in Parliament, bishoprics, school boards, the courts, medical school and hospital councils, and the learned societies. Before mid-century, dissenters could not matriculate at Oxford or graduate from Cambridge, and neither school held honors exams in science. Questions of science had been weighed against matters of faith, and until well after mid-century, faith still held sway. Powerful reaction in the press over liberal interpretations of the bible in the "heretical" *Essays and Reviews* (1860)—which sold out eight editions in just over a year—generated two convictions for heresy (later overturned) and public petitions for the excommunication of the "Seven Against Christ." The 6 April *Athenaeum* ran a long letter against the book, drawing Darwin into the controversy with an addendum criticizing his theory of speciation. Twenty-five bishops wrote to *The Times* suggesting that the authors ought to be indicted for heresy. Although Reed Arthur Stanley, canon of Christ Church, enthusiastically supported the book, classifying the Bible with other historical works, at least 10,906 clergymen signed a declaration protesting the spread of historically based religion. Other religious controversies fed into the animosity. Just after the release of *MPN*, conservative Anglican reaction to William Colenso's *The Pentateuch* (1863) led to the dismissal of its author from his position as bishop of Natal in southern Africa. Among the claims that influenced their decision was Colenso's view that Moses may not have written those books of the Bible traditionally attributed to him. The liberal views of the natural world expressed by the authors of *Essays and Reviews* were linked in the press with the Darwin-Huxley circle.

Traditionally, scientific organizations such as the Royal Society (f. 1660) and the more inclusive BAAS (f. 1831) were the outlet for new descriptions, classifications, and theories about nature. Many of the men who presided over them were gentlemen naturalists or "squarsons"— squires who were also local parsons. In return for professing to the thirty-nine articles of the Anglican faith, sons of landed gentry were primed at Cambridge or Oxford as vicars, assigned a parish, and thus were free to pursue avocations such as horticulture, pigeon or horse breeding, beekeeping, rock climbing, or butterfly hunting. Charles Darwin himself had escaped the clerical orders by convincing his father to let him join the *Beagle* expedition—a voyage that deeply influenced his life and the future of natural science. "Amateur naturalists" had made a significant contribution to the geological and paleontological record and to zoological and horticultural studies by the middle of the nineteenth century. Darwin had drawn heavily on their fieldwork for *Origin* (1859) and later for *The Descent of Man* (1871) despite his rejection of the religious grounding of their conclusions. Thus, until later in the century, scientific findings were underwritten by the articles of the faith.

Belief in both a First Cause and humanity's moral and intellectual superiority were considered the fundamental principles on which not only faith but science rested, even for those who had let go of the idea of a six-day creation. Reverend William Paley's *Natural Theology: or Evidences of the Existence and Attributes of the Deity Collected from the Appearances of Nature* (1802) was required reading at "Oxbridge" during the first half of the century. Paley asserted that the blessing of natural science was that it provided clear evidence of God's handiwork. (Others have noted that the hierarchy of heaven was a rationalization for gentlemen's hegemony over politics, business, and social life.) Even more liberal clerics described Darwin's theory as demonstrating the work of God. It is not surprising then that before mid-century, the French anatomist Georges Cuvier's premise of "perfect adaptation"—each organism constructed to fit perfectly into its environment—could easily be reconciled with natural theology's conviction that nature was the result of a divine plan.

In Britain these views eventually gave way to a secular view of scientific research that was embraced not only by atheists, but believers of varying religious persuasions who were interested in defining a unified law of nature based on recent paleontological, geological, botanical, and zoological findings. Regardless of their personal religious views those in and outside of science saw a practical purpose for separating the workings of the natural world from metaphysical ones. Natural research was no longer seen as merely a matter of describing God's handiwork, but in locating and extracting the natural resources for Britain's growing industrial economy.

The expansion and professionalization of science was linked to a move towards institutional secularism, both of which accelerated after the end of the Regency period. As the growth of cities and factory towns increased the demand for physicians and surgeons, a large number of students were coming to medical schools in London. Many bypassed the Anglican universities to avoid the religious tests required for matriculation. They populated new secular medical schools, including the one at the Benthemite University of London (f. 1826), which had become a degree-granting institution. They were guided by more liberal-minded teachers who had migrated from Edinburgh—among them Darwin's former mentor, the atheist Robert Grant, who chaired the University of London's Zoology department. Between the 1820s and the 1830s, Reverend Paley's theory of design was largely described by these men "as a crude sign of the clerical hold over British science" that they gradually replaced with the concept of a unity of plan based on creative natural law (Desmond, *Politics of Evolution*, p. 379). It was the radical element of medical schools and hospitals who began lobbying for academic and administrative reform, in particular an end to privileged apprenticeships, board appointments, and religious tests.

The activities of medical reformers coincided with efforts to attract government support for the advancement of science. Mathematician Charles Babbage argued in *Reflections on the Decline of Science and Some of Its Causes* (1830) that only government intervention could restore British science to a level competitive with Europe. The government responded to this "declinist controversy" with pensions and awards for scientists. By the late 1840s, the lobbying efforts of the BAAS resulted in an annual £1000 parliamentary grant in aid for scientific research for the reformed Royal Society. Efforts to make Britain a modern, scientific nation were joined by students and teachers in the applied sciences. Chemists like Lyon Playfair were working to expand science education and the role of scientists in government. Through his connections in the industrial centers and with the Prince Consort, Playfair later made contributions as a member of several royal investigatory commissions. Huxley sat on more than one of these.

In the 1850s, the doors of the scientific research institutions had been beaten down by the new wave of professionalizers who fought for a secular meritocracy. These were mainly (although not exclusively) young dissenters and agnostics from the lower middle classes who like Huxley, had apprenticed or trained outside of Oxbridge, and who had committed themselves to research and teaching in the sciences. This new wave of scientists gradually replaced the natural theologians holding positions in the scientific societies and institutions. Between 1831 and 1865 at least forty-one Anglican clergymen presided over the BAAS alone; between 1866 and 1900, that number had fallen to three. The new circle of London scientists that included Huxley, physicist John Tyndall, botanist Joseph Dalton Hooker, naturalist John Lubbock, and others, were influenced by the earlier pursuits by Lyon Playfair on behalf of the progress of science. The motto of the X-Club that Huxley founded with several of his colleagues in 1864 was "devotion to science pure and free, untrammeled by religious dogma."

After years of hammering away at the political, economic, and intellectual control that they saw as collusion between landed gentry and the Anglican church, reformers had gained much ground: Decades of debate and agitation—much of it through the periodical press—had gradually resulted in the repeal of the Corporation and Test Acts, the removal of tests for matriculation at Oxbridge, and the reform of medical and legal institutions. In the sciences, the Benthamite University of London had become an established and respected degree-granting institution, and free-thinking industrialists were opening red-bbrick schools in the provincial towns in order to train chemists, engineers, and managers to run their plants. Although it would be another twenty years before Oxford would grant graduate degrees to dissenters, honors exams in the natural sciences were instituted at Oxford in 1848 and at Cambridge in 1850. The Chartists' "revolution" may have failed, but after decades of upheaval, reform activity, and technological growth, Britain was emerging as a secular, industrial nation.

A decade after the release of *MPN*, Huxley expressed his belief that "we are in the midst of a gigantic movement, greater than that which preceded and produced the Reformation, and really only the continuation of that movement." He concluded that no reconciliation is possible "between free thought and traditional authority." In the long run, free thought would win out, "and organize itself into a coherent system embracing human life and the world as one harmonious whole." It might take a generation or more of hard work by those who "teach men to rest in no lie, and to rest in no verbal delusion." Huxley hoped that he had been and would continue to play a part in that movement (*HLL* v. 1, pp. 427–428).

A significant element of the Darwinian legacy has been the adaptation of the concept of the struggle for survival certain socioeconomic philosophies. Darwin's use of Malthus's population pressure theory as a basis for evolutionary struggle was appropriated by some theorists to inject their laissez-faire agendas with scientific credibility. By the end of the nineteenth century for instance, the idea of the "survival of the fittest" (a phrase Darwin uses in the sixth edition of his *Origin* only in referring to others' use of the term) had become a justification for eugenics. It is not surprising. One of the byproducts of the industrialization of Britain was a rising population of urban poor. Based on tradition and statistics (including those collected by Darwin's cousin, Francis Galton) that character traits were inherited, population control and denial of public assistance for the indigent were widely promoted—to the chagrin of more liberal social reformers, and philosophers who promoted the idea that sympathy was also a "natural" human characteristic. Britain had also had a strong lobby for education reform since before mid-century. Huxley, seeing it as a useful alternative to the problems of modernity, became one of its staunchest promoters. He sat on more than one commission of inquiry as to the state of science education, including the Gladstone-appointed Devonshire Commission. With John Tyndall, John Lubbock, and other scientists, he fought for the inclusion of science in school and university curricula at the same level as theology and the classics. In 1868, Huxley and Tyndall sent a report from the BAAS urging government funding for science education, and the provision for the training of science teachers. Huxley had been an inspector of schools and fought for free primary schooling; and helped to block passage of grants to religiously based schools, thus contributing to the secularization of education. A strong believer in the progress of scientific inquiry, he argued that science education would open the door to economic and social opportunities. He promoted inclusion of all social classes in contributing to it and to reaping its benefits.

Huxley's lectures and activities on behalf of education reform demonstrated his agreement with Darwin that the pleasure principle took a back seat to the "rearing of the greatest number of individuals in vigor and health"—that is, to the general good. Huxley's intention was to demonstrate humanity's cause for optimism about its biological truth: By synthesizing Darwin's explanation of the benefit of social instinct, and the "genealogy of morals" from *Descent*, Huxley defined the rationale he had used for his service to science, family, and community. However, inherent morality also implied that (as Francis Galton was arguing) negative traits might also be "born in the blood." In the face of the alcoholism and emotional problems extant in his own family as well as in the rising population living in poverty in an industrializing Britain, eugenics seemed just as bad as the Church's traditional teaching that one must live with one's lot in life. *Evolution and Ethics* (1893) and its *Prolegomena* (1894), Huxley made his final pronouncement on the question he had long struggled with: Was there any strategy for ethical progress of human society that was not dependent upon teleological argument and religious dogma, or what he called the "fanatical individualism," of the eugenic and social Darwinist trends of the late nineteenth century? Huxley concluded that humanity's future lay in its intellectual capability to rise above the ruthlessness of what he called the "cosmic process." *Evolution and Ethics* may thus be read as Huxley's way of reconciling the struggle for survival with the descent of cooperative social organization.

EVOLUTION AND ETHICS

In early June 1894, a year before Huxley's death, MacMillan released a pamphlet version of *Evolution and Ethics*, the Romanes lecture he had given at Oxford University the previous spring. By that time, what Huxley had to say was important enough that 2,700 of the 3,000 copies printed sold out in two weeks. The pamphlet included his *Prolegomena*, an explanatory forward in which he responded to the criticisms of his lecture.

The language of the *Prolegomena* follows closely Darwin's concluding remarks in *Origin*. It also takes much from the material on social instinct in Chapter 7 of *Origin* and Chapter 3 of *Descent*. *Ethics* presents what Huxley called the "cosmic process" of natural selection as antagonistic to ethical social behavior and suggests that there are more reasonable ways to deal with population pressure and individualist instinct. Huxley's own career had been a demonstration of what he believed those reasonable ways were.

Huxley began by recapitulating the "long argument" of *Origin*, focusing on the struggle for survival:

> The state of nature is far from possessing the attribute of permanence . . . one of the most characteristic features of this cosmic process is the struggle for existence . . . the result of which is the selection, that is to say, the survival of those forms which on the whole are best adapted to the conditions which at any period obtain; and which are, therefore, in that respect, and only in that respect, the fittest. (Huxley, *Paradis*, p. 62)

Echoing Chapter 7 of *Origin* Huxley uses the beehive scenario to show how social organization is not peculiar to humans:

> The struggle for existence is strictly limited . . . each performs the function assigned to it in the economy of the hive, and all contribute to the success of the whole cooperative society in its competition with rival collectors of nectar and pollen. (Huxley, *Paradis*, p. 82)

Darwin's analogy had been the logical choice to communicate his theory to gentlemen naturalists who bred horses, bees, and pigeons. By the end of the nineteenth century, Darwin's gardener analogies were so familiar that Huxley could use them in service to his own message—that the natural and artificial processes are antagonistic. Having long rejected Darwin's domestic breeding analogy for natural selection (what Huxley now terms "the cosmic process") he now adapted it for his own purpose. Referring to evolution as "the enemy of ethical nature" (Huxley, *Paradis*, p. 133) Huxley argues that ethical behavior is "opposed to that which leads to success in the cosmic struggle" (Huxley *Paradis*, pp. 139–140).

Darwin was right, Huxley affirmed. Life *is* a struggle, and natural selection, he finally acknowledged, is the mechanism of evolution. Moreover, the ancient philosophers had erred in supposing that life is cyclical or that it is determined. Evolution "excluded special creation and all other supernatural intervention" and is susceptible to chance (Huxley, *Paradis*, p. 64). Furthermore, human beings are not set apart, but are an integral component of the natural world:

> I do not know anyone who has taken more pains than I have, during the past thirty years, to insist upon the doctrine, so much reviled in the early part of that period, that man, physical, intellectual and moral, is as much a part of nature, as purely a product of the cosmic process, as the humblest weed. (Huxley *Paradis*, p. 64)

Huxley addresses the Malthusian idea that human beings are continually at odds between individual desires and the good of the community, and that this opposition to the general good leads to a creative spirit—the development of mind, a method to survive. Huxley agrees that human beings suffered this psychical conflict, but by their intellect could, if encouraged by education, lead it to a beneficial future based on ethical social behavior. Huxley saw the development of an ethical Britain as a model for the development of the larger human community. His involvement in education reform, the development of technicians for a world market, and even his mentoring of younger scientists who went back to America to spread the "gospel of Huxley" and professionalize science, were expressions of this philosophy.

As to the determinist idea that the struggle for existence tends to a final good, Huxley replies that "there would be something in the argument if in Chinese fashion, the present generation could pay its debts to its ancestors; otherwise it is not clear what compensation *Eohippus* gets for its sorrows in the fact that, some millions of years afterwards, one of his descendents winds the Derby" (Bibby, p. 130). This is not a surprising response. The idea that the suffering of the ancestor is recompensed by the improvement of the progeny smacked too much of organized religion's suffer now/pleasure in paradise principle.

Rather, intellect, developed by education, provided its own solution to the cosmic struggle, and it was directly opposed to both laissez-faire economics and eugenics:

> The practice of that which we call ethically best—what we call goodness or virtue—involves a course of conduct which in all respects, is opposed to that which leads to success in the cosmic struggle for existence. In place of ruthless self-assertion it demands self-restraint; in place of thrusting aside, or treading down, all competitors, it requires that the individual shall not merely respect, but shall help his fellows; its influence is directed, not so much to the survival of the fittest, but as to the fitting of as many as possible to survive. . . . It is from neglect of these plain considerations that the fanatical individualism of our time attempts to apply the analogy of cosmic nature to society. (Huxley, *Paradis*, p. 140)

The human advantage over other organisms was intellect. It provided human beings with the ability to reflect on, and analyze their behavior and the technologies they invent. It was thus possible that society could find an ethical way for "fitting the most to survive."

Huxley had by 1858 already accepted the connection between the evolution of animal perception and human reason. He had stated publicly his belief that "the mental and moral faculties are essentially and fundamentally the same in kind in animals and ourselves. I can draw no line of distinction between an instinctive and reasonable action" (Desmond, 1997, p. 241, n. 23). Darwin's suggestion that the moral sense evolves further in man because of intellectual development provided

Huxley with a rationale for his insistence that humans were capable of rising above cruder instincts. He often told his workingmen that science knowledge went hand in hand with moral behavior and civil order.

Some scholars have interpreted Huxley's conversion to Darwin's gradual transition as a necessary component of his desire for gradual social reform (rather than sociopolitical revolution). To a certain extent this is true: His eventual replacement of persistence with gradual transition in his evolution lectures coincided with the steps he had taken to make education possible for the largest number of people in England. Huxley's objective was to replace salvation in the next life with betterment and achievement in this one.

Darwin wrote that Huxley, who had contributed much to science, might have done even more if he had not spent so much time in public service, including education reform. It might be said that in fact Huxley's divided focus provided a very significant, influential contribution to science. His activities on behalf of workingmen, and the red-brick schools, and funding for public education had the effect of widening the pool of intellects in the field across class lines. It was, after all, diversity, Darwin had once observed, that made for stronger stock.

Huxley's early willingness to defend Darwin's theory against natural theologians thus seems to support the hypothesis that it served Huxley's need for an evolutionary model whose validity could be tested. However, there was a reason why Huxley required such a model: While he would later argue for the retention of Bible reading in school because of its "poetry" and occasional positive moral messages, he was convinced that Anglican orthodoxy and what he referred to as superstitious mysticism must be removed from the British institutions for the good of society. Huxley, like many of his contemporaries, believed that organized religion's tenet of living with one's lot in life was to blame for the continued ignorance and poverty at one end of the class structure, and patronage and exclusivity at the other. Darwin, who lived at the upper end of the class system, articulated the irony himself:

> That there is much suffering in the world no one disputes. Some have attempted to explain this in reference to man by imagining that it serves for his moral improvement. But the number of men in the world is as nothing compared with that of all other sentient beings, and these often suffer greatly without any moral improvement. (Darwin, *Autobiography*, p. 90)

For Huxley too, creationist and idealist views were too deterministic to support his plan to broaden educational and professional opportunities, and to reshape the gentlemanly occupation of science into a professional meritocracy. Huxley may have interpreted Darwin's suggestion that inbreeding could have detrimental effects on populations as an implication that the best societal outcomes are the result of selection from a larger pool of people. His endeavor to make education, and especially science education, available to all was based on his notion that the society could only benefit from drawing from a larger pool of minds.

Of less prosperous antecedents than Darwin, the young Huxley had also experienced the disparity between the opportunities available to the working and lower middle classes, and the way that position and patronage was taken for granted among the upper classes. For instance, Huxley's position as assistant surgeon aboard the *H.M.S. Rattlesnake* was far less comfortable than Darwin's privileged position as FitzRoy's "friend" on the *Beagle*. Although Huxley's family was better off than the East-Enders, he was forever struggling financially. He was thirty before he could afford to marry Henrietta, to whom he had been engaged by that time for five years. Huxley, who worked tirelessly for every penny he had, spent most of it supporting his family. He had once complained "to attempt to live by any scientific pursuit is a farce. . . . A man of science may earn great distinction—but not bread" (Desmond, p. 161). While Darwin had the social and financial security to pursue a life entirely devoted to science, Huxley needed to teach and publish to earn a living, and decided to use his position and eventually his influence to change a system that he believed to dismiss the potential contributions of a large segment of society.

Huxley had been early on exposed to the radical social views of dissenters, who blamed the resignation of the poor on the Church. Huxley's personal journal is evidence that he formed this view as a teenager, after being exposed to the radical thought of the dissenting Unitarians. Huxley was in the habit of copying excerpts from various books he had read that had particularly impressed him. In his journal, *Notes and Doings* from 1840 (when he was fifteen), under the heading "Truths" he transcribed, " 'I hate all people who want to form sects. It is not error but sectarian error. Nay, and even sectarian truth, which causes the unhappiness of mankind.'—Lessing." This was followed with: " 'One solitary philosopher may be great, virtuous and happy in the midst of poverty, but not a whole nation.'—Isaac Iselin" (*The Huxley File, Notebook 1840*).

Huxley's own professional and financial struggles, combined with his firsthand experience with the plight of London's poor had apparently awakened in him a desire to create opportunities for the advancement of the lower classes. Huxley wrote of his experiences as a doctor's apprentice in Rotherlithe, "To me this advocacy of the cause of the poor appealed very strongly. . . . I had had the opportunity of seeing for myself something of the way the poor live . . . among the rest, people who came to me for medical aid, and were really suffering from nothing but slow starvation" (*HLL* v.1, p. 16).

Huxley later asserted that the cause of many of society's ills was people's general ignorance of the way that nature worked; and he implied that organized religion was to blame:

> Not only does our present primary education carefully abstain from hinting to the workman that some of his greatest evils are traceable to mere physical agencies, which could be removed by energy, patience, and frugality; but it does worse—it renders him, so far as it can, deaf to those who could help him, and tries to substitute an Oriental submission to what is falsely declared to be the will of God, for his natural tendency to strive for a better condition. (Huxley, "Liberal Education," pp. 90–91)

He complained that the state of education for the middle class was such that supposedly educated sons who went into business had no practical education, what they needed to conduct the businesses that they were entering—from manufacturing trades to commerce—was a "liberal education" that included among other subjects, geography, physics, and mathematics. Those who entered the House of Commons, Huxley warned, had little common sense to offer their constituents. And if primary and secondary education were failing the nation, the universities were equally faulty. Those of his colleagues who had gotten a university education were exceptions. For the most part Huxley complained, universities were nothing more than "boarding schools for bigger (and richer) boys" and "learned men are not more numerous in them than out of them" (Ibid., p. 101). In one address he called for the institution of a "Science Sunday School" in every parish. He said, "I cannot but think that there is room for all of us to work in helping to bridge over the great abyss of ignorance which lies at our feet" (Huxley, "After Dinner Speech," p. 133). He saw education as a way of bringing economic stability to the wage earner, and to provide the best of them with social mobility. Consequently Huxley was an active proponent of the red-brick schools. He lectured and served as dean of the school in South Kensington. In a speech he made on the opening of the Josiah Mason College in industrial Birmingham, he said,

> If the institution opened today fulfills the intention of its founder . . . no child born in Birmingham, hence forward, if he have the capacity to profit by the opportunities offered to him, first in the primary and other schools, and afterwards in the scientific college, need fail to obtain, not merely instruction, but the culture most appropriate to the conditions of his life. (Huxley, "Science and Culture," *Science and Education*, 1880, p. 157)

For Huxley what was "appropriate" was that which had practical value. The Bible, he argued, was a "vast residual of moral beauty and grandeur" but science education made vigorous minds, and provided useful information (Huxley, "The School Boards," *Science and Education*, 1880, p. 398). At the Workingmen's Institute, he asserted both his empathy with the working classes and the advantage of science in their development by drawing an analogy between the making of a chair and the making of science. Craftsmen, he said, knew how to deal with tangible things, with the facts, as it were, of engineering a chair. Likewise, scientists dealt with evidence and facts, while others speculated "about the occult powers of the throne of St. Peter" (Huxley, "Technical Education," *Science and Education*, 1880, pp. 405–406).

At one point he was challenged by representatives of public schools to put his money where his mouth was. At the British Association meeting held in Nottingham in 1866, Dean Farrar, a classical master at Harrow, had criticized the tradition of teaching boys classical composition whether they were suited to it or not, and thought the course might be exchanged for courses in elementary science. Mr. J. Payne complained however, that there were not enough teachers, and that "if men of science were really in earnest they would condescend to teach in the schools." In response, Huxley published *Lessons on Elementary Physiology* (1866) and *Elementary Instruction in Biology* (1875). He served with Tyndall on a committee on education for the British Association, which adopted their report in 1867; and between 1868 and 1869 he and Tyndall drew up a plan for the teaching of science in the International College (on whose council both men served). Huxley later chaired the London School Board, where during his tenure he fought successfully for a ban on continued funding for church-run schools.

Huxley's career was thus a complex marriage of personal ambition and public service—something akin to Darwin's explanation for the motive for altruistic behavior in organisms: the consequence being the good of the community. It is fair to say that when Huxley was not lecturing on biology, he was lecturing against the Malthusian legacy. His own sense of ethics motivated Huxley to speak out against applying the biological truth of natural selection for behavior to economics and government. In effect, his *Evolution and Ethics* and its *Prolegomena* are a response to the growing popularity of biological determinism and eugenic thought at the end of the nineteenth century.

Huxley insisted that any line of thinking that was not supported by evidence of sensory experience was unjustified. He therefore needed a scientific basis for moral development which was not determinist; a justification for his belief that human intelligence was ubiquitous and not the province of the upper classes. Further, he needed a theory that would demonstrate that the survival of humanity required social cooperation, and individual behavior that benefited the community. He apparently found these in Darwin. Two concepts drove him, it seems. One was the influence of Hume, whose philosophy he had reviewed. Here he found a philosophical basis for sensory and material evidence as the test of valid scientific claims. His public and private comments tell us that he was motivated by his own life experience to change the attitudes of the "souls who cannot see greatness in their fellow because his father was a cobbler" (Huxley, *Zoological Relations*, 1861, p. 472)—or in Huxley's case, a schoolteacher.

Furthermore, Darwin's publication of *The Descent of Man*, which reinforced the connection between moral and intellectual development of humans and lower animals served Huxley's rational for educational reform, a project to which he was already firmly committed. (Typical of their symbiotic relationship, Darwin later acknowledged that *MPN* had given him the confidence to assemble the ideas that he published in *Descent*.)

Late in life, Huxley complained that

There are two very different questions which people fail to discriminate. One is whether evolution accounts for morality, the other whether the principle of evolution in general can be adopted as an ethical principle. The first of course, I advocate, and have constantly insisted upon. The second I deny, and reject all so-called evolutional ethics based upon it. (*HLL* v. 2, p. 360)

CONCLUSION

By working to make Darwin both palatable and penetrable to the press, to colleagues, and to the general public, T.H. Huxley had by the 1870s been influential in reducing the power of the church in educational institutions and in the halls of science. He had accrued position and honors for himself across Britain, and had fought against interpreting natural selection as a ationale for a struggle among the wealthy to retain their eminence and their capital. The poor could hope to move out of the West End, graduates of the red-brick colleges could become successful industrialists, and even the son of a teacher could hope to land a seat on the Privy Council (as Huxley had done in 1892). He continued to defend Darwin's theory to the end of his life. He of all people deserved to be awarded the Darwin Medal, which was conferred on him in 1894.

Despite his reputation for pugnacity, or perhaps because of it, Huxley led by example. He promoted his moral view by generously providing for his extended family on a middle-class income, sticking his professional neck out on issues he deemed important to the development of science, and working tirelessly for the expansion of scientific education and the professionalization of science. His positions on social and political matters were a result of careful consideration of how the outcomes would affect society as a whole, and thus fitted with his social interpretation of evolution theory.

Huxley was not a humble man. Undoubtedly he believed he deserved every honor bestowed on him. *Evolution and Ethics* smells in parts of covert self-praise. Nevertheless, he arguably saw the lecture as a final tribute to Darwin by his bulldog and also as a testament to the value of a scientific education, and to science as a meritocracy, welcoming entrants from outside Oxbridge. The "vigorous minds" science education developed were delivering people from the social ills associated with poverty and unemployment, so that they could contribute to scientific knowledge, industry, the national economy, and perhaps find ways to alleviate the physical suffering and death that came from what Malthusian "natural checks" on the population like epidemic disease. At the end of his notable career, surely Huxley saw himself as proof of the value of "fitting the most to survive."

FURTHER READING

Desmond, Adrian. *The Politics of Evolution* (Chicago: Chicago University Press, 1989).

Desmond, Adrian. *Huxley: From Devil's Disciple to Evolution's High Priest* (Reading, MA: Addison-Wesley, 1997).

Huxley, Leonard. *The Life and Letters of Thomas Henry Huxley,* 2 vols. (New York: Appleton, 1901). (listed in text as *HLL)*

Huxley, Thomas Henry. *Evidences as to Man's Place in Nature* (London, 1863); rpt. in *The Works of Thomas Henry Huxley* (9 vols.), Westminster ed. (New York: Appleton, 1895).

Huxley, Thomas Henry. *Autobiography and Essays.* Ed. by Ada F. Snell. (Boston: Houghton-Mifflin, 1909).

Paradis, James, and Williams, George C. (eds.). *Evolution and Ethics* (Princeton, NJ: Princeton University Press, 1989).

Richards, Robert J. *Darwin and the Emergence of Evolutionary Theories of Mind and Behavior* (Chicago: University of Chicago Press, 1988).

White, Paul. *Thomas Huxley: Making the "Man of Science"* (Cambridge, UK: Cambridge University Press, 2003).

Gregor Mendel

Dawn Mooney Digrius

In November 1859, Charles Darwin's book, *On the Origin of Species by Means of Natural Selection, or the Preservation of Favoured Races in the Struggle for Life* was published by John Murray in London. When Darwin outlined his theory of evolution in the text, one thing not clearly understood was the manner in which characteristics were passed from one generation to another. In *Origin*, Darwin came to the conclusion that "the laws governing inheritance are quite unknown; no one can say why the same peculiarity in different individuals of the same species, and in individuals of different species, is sometimes inherited and sometimes not so" (Darwin, 2004 [1859], pp. 21–22). This lack of an explanation for inheritance in Darwin's time left a large gap in his explanation for the mechanism of evolution: natural selection. Unbeknownst to Darwin, an Austrian monk at a monastery in Brünn (then part of the Hapsburg Empire, now the Czech Republic), was experimenting with pea plants, nearly 30,000 of them, to understand how specific traits were passed from generation to generation. These experiments, conducted during the years 1856 to 1871, revealed that the knowledge of how heredity and variation worked could be applied to understandings of the past or the future of living things. This chapter will discuss the life and works of Johann Gregor Mendel (1822–1884) and his contributions to science and why he is an icon of evolutionary thought.

JOHANN GREGOR MENDEL

Johann Mendel (he assumed the name of Gregor later when he entered monastic life) was born on 22 July 1822, in the village of Heinzendorf,

Gregor Mendel, the first geneticist. (National Library of Medicine.)

situated in the foothills of the Moravian Mountains, to Anton and Rosine Schwirtlich Mendel. As a boy, he worked on the family farm until his father, all too brutally familiar with the plight of the peasant farmer, encouraged Johann to attend the village school. The school, under the direction of Countess Waldburg, taught natural history and natural science in addition to the ordinary components of elementary education. There were gardens next to the school grounds, and the children learned the essentials of fruit-growing and beekeeping along with their studies. The instructor, Thomas Makitta, recognized the potential in young Johann, and he was sent on to the higher school at Leipknik, roughly thirteen miles away from Heinzendorf. After completing his course of study there, on 15 December 1834, Johann was enrolled in Troppau High School. The headmaster at the school was Pater Ferdinand Schaumann, an Augustinian monk from the monastery at Altbrünn. Between 1834 and 1840, Mendel attended class regularly save for a short period of illness in 1838, and remained a distinguished pupil. In 1840, Johann was admitted to the second class for the study of humanities without examination, the highest class in the school. When he left school on 7 August 1840, a report noted that Johann Gregor Mendel had been one of the best students in the school, his marks being superior in almost all branches of study.

Unfettered Thoughts

Little documentation remains of Mendel's early years, save two poems composed during his school days. A snippet of one of the poems foreshadows the intellectual framework of the adult Mendel, who continued to think "unfettered thoughts" and searched for truth despite what official church sources held to be so. A portion of the poem is presented here:

As the Master willed, you shall dispel

The gloomy power of superstition

Which now oppresses the world.

The works of the greatest of men,

Which now, of use only to the few,

Crumble away into nothingness,

You will keep in the light and will preserve.

For in many a head still wrapped

Despite his academic promise, Mendel's prospects of continuing a life of study were low. His family did not have the resources to pay for a university education. His younger unmarried sister renounced her share of the family estate in order that Johann could take philosophy courses at the Olmütz Philosophical Institute. The institute was affiliated with the university there, and it offered compulsory coursework in philosophy, religion, elementary mathematics, Latin literature, and physics. History, natural history, and the science of education were optional courses of study. Professor Friedrich Franz was now a lecturer at Olmütz University, after teaching for nearly twenty years at the Philosophical Institute in Brünn. Franz, a physicist, would play an instrumental role in Mendel's life and career. During his teaching years at Brünn, Franz had resided at the Altbrünn monastery and was acquainted with many members of that religious community. He had recently been asked to select from among his pupils a few young men who might be candidates for admission to the monastery. On 14 July 1843, Professor Friedrich Franz sent a letter to the monastery at Brünn, stating,

> Up to now, two candidates have given me their names, but I can only recommend one of them. This is Johann Mendel, born at Heinzendorf in Silesia. During the two-year course in philosophy he has had, almost invariably, the most exceptional reports, and is a young man of very solid character. In my own branch he is almost the best. (Iltis, 1932, p. 42)

While this letter survives, there are no documents detailing Mendel's decision to enter the monastery. Mendel joined the monastery in Altbrünn in 1843. He assumed the name Gregor as his monastic appellation, and from this time forward would use it before his baptismal name.

Mendel found, as he arrived at Brünn, that the place he had come to held great potential to be one in which significant research could be conducted. The monastery was far more than a secluded place of theological contemplation, but a veritable center of academic study. Fellow monks included Fr. Aurelius Thaler (1796–1843), a mathematician who was also recognized for his botanical pursuits, and Fr. Matthaeus Klácel (1808–1882), a botanist with an interest in mineralogy and astronomy. Mendel first took on the studies necessary to complete his novitiate period, while at the same time devoted every waking moment to the study of the natural world. Klácel became Mendel's mentor, largely because he was more experienced as a botanist and had begun crossing plants. In October 1851, Abbot Napp of the Brünn monastery wrote to the Bishop's office recommending that Mendel be sent on to Vienna University to study the natural sciences because he had shown a remarkable intellectual capacity and industriousness as a student. Napp wrote:

> Your Excellency,
>   The views you have been good enough to express regarding Pater Gregor Mendel have decided me to send him to Vienna for higher scientific training. I shall not grudge any expense requisite for the furtherance of this training, and will only venture to beg Your Excellency to show him your gracious goodwill, of which I am sure he will prove himself worthy. (Iltis, 1932, p. 75)

It was at Vienna University in the years between 1851 and 1853 that Mendel learned the fundamental knowledge and skills which would make his later work on botany and heredity possible.

During his first term, Mendel attended lectures by Doppler on experimental physics. For the next several terms he continued his studies at the Physical Institute. The last terms at Vienna were spent studying zoology with Professor Kner, systematic botany under Eduard Fenzl (1808–1879), and the physiology and paleontology of plants under Franz Unger (1800–1880). Of particular interest to this story were the lectures on plant physiology given during Mendel's attendance at Vienna University by Professor Franz Unger and his work under Fenzl who taught Mendel how to use the microscope for botanical observation, in addition to the morphology and classification of Phanerogams (plants with visible seeds). Unger was, at the time, engaged in a debate over plant fertilization and had written on the subject of obtaining new horticultural varieties by using artificial fertilization. He had, sometime during the 1850s, shifted his point of view regarding the transmutation of species away from catastrophist tendencies of Frenchman Georges Cuvier to one more aligned with a pre-evolutionary frame. Unger changed his mind about how new species originated, invoking spontaneous generation in his early works and common descent later, but invariably he assumed that the process would follow a deterministic developmental law. Unger had the privilege of teaching Mendel botany at the Vienna University from the autumn of 1851 until the summer of 1853, during which time Unger's theory of universal common descent appeared. The public controversy it provoked in Vienna in 1852 (and again in 1856) perhaps impressed Mendel with the urgency of evolutionary questions.

Mendel returned to the monastery after the summer term of 1853. He was then appointed as a supply (substitute) teacher at Brünn Modern

School in May 1854, where he taught natural history and physics for the lower school. The collection of natural history specimens were also placed under his care. After two unsuccessful attempts at passing the teaching examination, Mendel began his experiments at the monastery. Iltis noted in his biography of Mendel that Inspector Nowotny, one of Mendel's colleagues at the Modern School, related that upon his return from Vienna, Mendel was much out of humor. It seemed that Mendel held a difference of opinion with the botany examiner and had maintained his own point of view. Nowotny believed that the dispute with the examiner drove Mendel away from teaching and into experimentation. His work certainly dates from very soon after the second defeat. Although there are no documents surviving to support this claim, it is clear that Mendel never made a third attempt at examining for a teacher's position at the school.

Mendel Distracted

Mendel became prelate of the monastery at Brünn during a period of major political change. In 1867, the Habsburg monarchy shifted from an autocratic state into a constitutional monarchy. The new governmental structure instituted a number of reforms, including its relation to the Church. In the spring of 1874, the German-Liberal Party in the Reichsrat introduced a bill to tax church property in order to "supply the financial needs of Catholic worship" (Iltis, 1932, p. 253). Mendel, saw the assessment as unconstitutional, for it went against his pledge to protect the monastery. He protested the payment, and when demands were made by the authorities for payment, he became increasingly more stubborn in his refusal to pay. By August 1876, two years after the law was enacted, Mendel argued for an explanation of the validity of the law, but to no avail. The state seized part of the monastery properties and enforced payment of the tax and so Mendel grew even more embittered in his belief that the government and his enemies (now growing daily due to his stubbornness) were doing him and the monastery a serious injustice. Mendel thus began charging the government 5 percent interest upon the amount of funds being withheld, considering the state a debtor of the monastery that eventually would have to repay the money with the accrued interest. The reason for his protests: the Catholic Church, he considered to be dependent upon its own resources and not a state church. Therefore, its financial situation should not be a state affair. Mendel spent nine years protesting the tax law, each protest meeting with little or no results or interest. As his health declined, Mendel grew bitter and mistrustful of everyone around him. He died before the matter was resolved.

IT BEGAN WITH PEA PLANTS

In 1856, shortly after his second attempt at passing the teaching examination, Mendel, now thirty-four years old, began his experiments crossing pea plants. In the introductory remarks to his classic monograph, Mendel wrote:

> If no one has hitherto succeeded in establishing a generally valid law as to the formation and development of hybrids, who can wonder that knows the magnitude of the task and the difficulties with which experiments of this kind have to contend. A final decision will only become possible when detailed experiments relating to the most diversified families of plants have been made . . . among the numerous experiments, now one has been carried out comprehensively enough or in such a way as to make it possible to determine the number of different forms under which the offspring of hybrids appear . . . or definitely to ascertain their statistical relations . . . Still, that would seem to be the only right way of ultimately achieving the solution of a problem which is of enormous importance in its bearing upon the evolutionary history of organic forms. (Mendel, 1965, p. 8)

The essay focused on the results of Mendel's experimentation with the genus *Pisum*, the edible pea. The plant necessary to conduct the experiments required several factors: forms distinguishable by sharply defined and constant characters; easily protected from fertilization with undesired pollen; fertile hybrids. *Pisum* had such characteristics—its hybrids are fertile, because of its reproductive nature dusting of pollen by another flower is unlikely, and despite difficulties in artificial pollination, it is, with practice, possible. Mendel chose thirty-four different varieties of pea, tested them, and finally chose twenty-two as candidates for his experiments. Most of them belonged to *Pisum sativum* (dry field pea), but he also used *Pisum quadratum* (which has wrinkled angular seeds), *P. saccharatum* (we know it as *Pisum sativum* L. *var. Saccharatum*—edible pod sugar snap peas), and *P. umbellatum* (common garden peas).

After choosing the varieties to work with, Mendel next chose the suite of characters he deemed appropriate for his observations, seven pairs of them. They were

1. shape of the ripe seed, which may be round or irregularly angular (wrinkled);
2. color of the cotyledon (A leaf of the embryo of a seed plant, which upon germination either remains in the seed or emerges, enlarges and becomes either yellow or green);
3. tint of the seed coat, which can be white, or grey, grayish brown, or tan with purple spots);
4. shape of the ripe pod, which may be curved (inflated) or deeply constricted between the seeds;

5. tint of unripe pods, which may be green or yellow;

6. position of the flower, which may be axial (main line of entire plant) or terminal (at the tip of a stem, twig, or branchlet); and

7. stature of plant, being either "tall" or "dwarf."

So, taking two plants with differences in one of the pairs of characteristics, Mendel crossed them, pollinating the first from the second and the second from the first. This resulted in a similar outcome; individuals always represented one another as regarding the characteristics under consideration in the first generation and did not seem to manifest an intermediate development of the character. In addition, Mendel found that within the paired characters in which the parents had differed, one character invariably manifested itself in the first generation, to the exclusion of the other. This he deemed the "dominant" characteristic. The character that seemed to disappear in the first generation of hybrid was then termed "recessive." Mendel used the term "uniform" in his eighth letter to Karl von Nägeli to describe the first generation when the parents belonged to a pure line, although he was careful to avoid arguing that there was some type of "law" regulating heredity.

Thus, although hesitant to argue for a law of uniformity, Mendel found that when the parent plants belonged to a pure line, there was a uniformity in the first generation of hybrids, and they were therefore designated as uniform by him. For example, in looking at the characteristics emergent in Mendel's experiments on seed shape, he found that in the first generation of hybrids, the color of the seed reflected yellow as dominant and green as recessive. Gliboff noted that with this paper "Mendel appears to have been addressing an evolutionary question: what developmental law governs the propagation of hybrids?" (Gliboff, 1999, p. 225). Mendel's research seemed, according to Gliboff, to be a daunting task, yet one worthwhile if it could answer important evolutionary questions such as what happened to the offspring of hybrids in the long run? Did they remain stable, take new forms, or revert to an ancestral type? What Mendel wished to find were laws governing the generational sequence of changes in living things.

Continuing his series of experiments, Mendel allowed his first-generation hybrids to self-fertilize. The ratio between the individuals who showed the

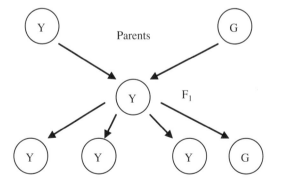

Diagram illustrating the concept of dominant and recessive traits in the first generation of hybrid, looking at color of seed. In the second generation (F_2), some of the individuals manifest the dominant, while some display the recessive as if never lost in the first generation.

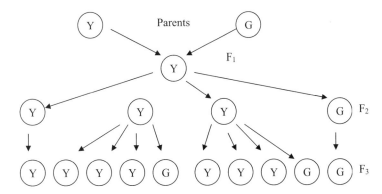

Illustration of the ratio between individuals which show the dominant characteristic and the individuals that show the recessive characteristic is 3:1, so that in every 4 plants in this generation (F_3) 3 manifest the dominant character and 1 manifests the recessive.

dominant characteristics and the recessive characteristics was 3:1; for every four plants in the generation three would manifest the dominant character and one would manifest the recessive. Mendel then noted that the proportions in which the descendents of the hybrids develop and split up in the first and second generations hold good for all subsequent offspring. He argued that "if *A* be taken as denoting one of the two constant characters, for instance the dominant, *a*, the recessive, and *Aa*, the hybrid form in which both are conjoined, the expression

$$A + 2Aa + a$$

shows the terms in the series for the progeny of the hybrids of two differentiating characters" (Iltis, 1965, p. 21). The offspring of each generation of hybrids separated in each generation into a ratio of 2:1:1 into constant and hybrid forms. Mendel then broke down the proportions into a chart that shows the relationship between the number of forms and the progeny of each generation. He assumed that only four seeds would be produced in each generation. The ratio chart, as noted in Fisher, shows the constant results of the hybrids produced through these crossing experiments.

Ratios

Generation	A	Aa	a	A		Aa		a
1	1	2	1	1	:	2	:	1
2	6	4	6	3	:	2	:	3
3	28	8	28	7	:	2	:	7
4	120	16	120	15	:	2	:	15
5	496	32	496	31	:	2	:	31
n				2^n-1	:	2	:	2^n-1

Mendel's idea was to put forward his theory in terms of mathematics, more like a physicist rather than a natural historian. His mathematical expressions of the combinations of traits in *Pisum,* like the chart above, helped establish the manner of planning genetic experiments and predicting the appearance of new combinations of hereditary characters. The explanation of the random combination of characters came to be known as Mendel's Law of Independent Assortment.

EXPERIMENTS IN PLANT HYBRIDIZATION

In February 1865, the same month that Darwin was reading W.C. Spooner's tract on blended characters, Mendel read his paper *Versuche über Pflanzen-Hybriden* before the Natural History Society of Brünn. In the introduction to his paper, Mendel found fault with previous hybridization work because it had failed to formulate a general law which could be applied to the development and evolution of hybrids. He wrote that the work of previous hybridizers was inadequate because it did not reveal the long-term pattern:

> None among the experiments has been carried out on such a scale or in such a manner as would make it possible to identify the number of different forms in which hybrids occur, so that one could assign these forms with certainty to particular generations and ascertain the relative numerical relationships. (1999, p. 226)

Mendel then described his experiments and findings. After determining that if several different characters are combined by cross-fertilization in a hybrid, the offspring form the basis of a combination series in which each pair of different characters are united. He noted that

> The uniformity of behavior shown by the whole of the characters submitted to experiment permits, and fully justifies, the acceptance of the principle that a similar relation exists in the other characters which appear less sharply defined in plants, and therefore could not be included in the separate experiments. (Mendel in Fisher, 1965, p. 29)

In all of his experiments, Mendel noticed that there appeared all the forms which his proposed theory demanded. This is significant, for Mendel concluded his paper by debating the research on *Pisum* that had been undertaken previously by Kölreuter and Gärtner. According to their research, hybrids present in outward appearance either an intermediate between the original species, or they closely resemble either and sometimes can hardly be told apart from it. However, Mendel pointed out that this is not the case with all hybrids. He argued that "sometimes the offspring have more nearly approached, some the one and some the other of the two original stocks, or they all incline more to one or the other side; while in other cases they remain perfectly like the hybrid and continue constant in their offspring" (Mendel in Fisher, 1965, p. 43). Thus, hybrids of varieties act like hybrids of species, but possess greater variability of form and a more pronounced proclivity to revert to their original type.

According to Mendel, the "nature of a hybrid consists in this, that it has resulted from the union of reproductive cells equipped with different hereditary factors" (Iltis, 1932, p. 147). The combination of these conflicting hereditary factors continues throughout the life of the plant. In his second letter to von Nägeli, Mendel wrote:

> The course of development consists simply in this, that in each successive generation the two primal characters issue distinct and unaltered out of the hybridized form, there being nothing whatever to show that either of them has inherited or taken anything over from the other. (Iltis, 1932, p. 148)

Thus, the doctrine of the purity of the gametes, the basic principle of Mendelism, was now fully realized. Mendel also dissented from the view that species diverged quickly into an endless collection of offspring who themselves change into variable forms over short periods of time. He argued "a number of facts combine to force upon us the view that our cultivated plants, with few exceptions, belong to various hybrid series, whose further development is modified and interrupted by frequent intercrossings" (Iltis, 1932, p. 159). This statement suggests an opposition on Mendel's part to Darwin's theory of deliberate selection as accounting for the origin of varieties in the world of cultivated plants. This was a theory that had gained wide acceptance at the time of Mendel's publication of his papers through the data put forward in the writings of Charles Darwin. But Mendel was disposed, based on his experiments, to insist upon the stability of species rather than their variability and upon the discontinuity of hereditary characters rather than the fluidity and continuity of hereditary characters. Iltis suggested in his book on Mendel that these opinions support the notion that Mendel was "somewhat skeptical as regards the theory of evolution in the form in which it had not long before been presented by Darwin" (Iltis, 1932, p. 160).

MENDEL AND EVOLUTION

The question of how Mendel understood evolution and of his attitude to Darwin's theory should be noted here. It has been argued that a number of different theories concerning transmutation were published around the middle of the nineteenth century, including those of philosophers and naturalists and theologians alike. Mendel was, as you may recall, familiar with the monk Klácel, who also called upon examples from the natural world as well as the natural sciences to support his philosophical view and stretched the concept of a gradual process of development put forward by the German philosopher Hegel, further. Mendel was also familiar with the work of Charles Lyell and in his examination essay in geology referred to Lyell's ideas of evolution in nature. Franz Unger's influence on Mendel while he was at the university was significantly directed toward evolution, for Unger did not accept the idea of the constancy of species and argued for research into plant hybridization, in order to reveal what principles might form the basis of the variability and development of characters.

According to Darwin, evolution "operates through the emergence of new forms, differing from the previous ones in minor respects. The force of natural selection favors the better adapted characters and organisms" (Orel, 1971, p. 70). Darwin, while aware of the role of heredity in variability, was not keen on the laws governing inheritance and admitted they were not fully known or understood. It was Mendel, of course, who was investigating the relationship of heredity in the origins of variety. His concept of discrete unit characters was, however, at odds with Darwin's conception of a steady flow of variations, according to which small changes were constantly occurring, which in turn gave rise to new species slowly over time.

Mendel did not accept the impact of the environment on the origins of hybrids and at the time of the 1865 paper did not mention either Unger's or Darwin's names explicitly in his works. He did, however, reference the views of other noted physiologists. Darwin would not be mentioned in Mendel's works until his 1870 paper on *Hieracium* in connection with von Nägeli's views on the transmutation of this species. Orel posits that the best way to tease out Mendel's attitude toward Darwin's theory was to examine the writings of Niessl, secretary of the Natural Science Society. Niessl frequently visited Mendel at the monastery at Brünn and openly argued that Mendel had been deeply interested in the idea of evolution, and far from being an adversary of natural selection, was an ally.

B.E. Bishop, in his article "Mendel's Opposition to Evolution and to Darwin," argued that as of 1996 only one writer, L.A. Callender (1988), had concluded that Mendel was opposed to evolution. However, upon closer examination, Mendel's *Pisum* paper reveals an anti-evolutionary stance and was written, according to Bishop "in contradiction" of Darwin's book *On the Origin of Species* (1996, p. 205). Bishop stated first that Mendel does not mention Darwin in his paper, although Darwin is mentioned in letters to von Nägeli. Mendel did own a copy of the second German edition of *On the Origin of Species*.

On the Origin of Species (1859), as Bishop posited, was "an onslaught on the doctrine of special creation" (1996, p. 207). Darwin's themes were found in abundance in Mendel's work on heredity. If this is the case, why was it that Bishop believed Mendel was, contrary to Niessl's statement, against Darwin's idea of natural selection? He felt that there was evidence, both from the historical background and in Mendel's paper itself that refuted Darwin's ideas and positioned Mendel as part of the theological mainstream. If Mendel was a staunch member of his Christian monastic community and an ardent defender of his faith then how could he be a supporter of Darwin's theory?

Returning back to the distribution of Mendel's paper, it was published in the Brünn Naturalist Society's *Proceedings* in 1866. Forty reprints were made and of those, only four have been located. One copy was sent to the German botanist Anton Kerner von Marilaun (1831–1898) and another was sent to the Swiss botanist Karl von Nägeli (1817–1891) who was then a professor of botany at Munich. Nägeli, you will recall, had corresponded with Mendel, read his paper but did not fully support its further distribution (1968, p. 78). A third copy fell into the hands of M.W. Beijerinck (1851–1931), professor of bacteriology at Delft. The final known reprint was located in the collection of Theodor Boveri (1862–1915), a German zoologist. In addition to the four reprints noted here, subscription issues were sent to 115 libraries and institutions, including the Royal Society of London.

As Vorzimmer and other historians of science have noted, it does not appear that Darwin was aware of Mendel's work. However, Darwin had read Heinrich Hoffmann (1819–1891), including his 1869 pamphlet "An Investigation into the Determination of the Quality of Species and Varieties." Hoffman summed up Mendel's paper, noting "Hybrids possess the inclination, in the following generation, to revert back to the parental [or specific] form" (1968, p. 81). Darwin's notations on Hoffman's pamphlet insinuate its importance and he cited the work in his book on self- and cross-fertilization in 1876. Despite the close proximity of Darwin to Mendel's work, through Darwin's interest in the subject and his read of Hoffman's pamphlet, they never communicated.

Part of the reason for this may have been Mendel's reverence and class deference to his more professional colleagues which may account for why he never sent a copy of his paper directly to Darwin. By sending his paper to Kerner and Nägeli (and possibly Hoffman), Mendel had done his duty in sharing his findings with scientific contemporaries, albeit at what he thought of as a lower level. It did not, however, wind up in the hands of any naturalist who had the training or subtlety of intellect to appreciate just how important and groundbreaking it was until almost thirty-five years later. If we are surprised that Mendel's paper remained unnoticed for thirty-five years, we shouldn't be. It was quite common for scientific ideas to go unnoticed at first. For example, over sixty years went by between when Camerarius (1694) discovered proof of sex in plants and when the great Swiss systematist Linnaeus wrote his essay on the sex of plants (1760). In the pre-Internet and mass publication era it was common—almost expected—that scientific ideas spread somewhat slowly or not at all.

THE REDISCOVERY OF MENDEL

It was not until 1900 that Mendel's experiments were brought to light. Three European botanists, Hugo de Vries (1848–1935), Carl Correns (1864–1933), and Erich von Tschermak (1871–1962), independently made the same observations on plant heredity and had come to the same conclusions as Mendel had so many years before. Tschermak was the grandson of Eduard Fenzl, who was teaching at the University of Vienna while Mendel was in attendance. The story of this simultaneous redis-covery of Mendel's classic paper after nearly three decades is well known, but as Olby noted in his study of the origins of Mendelism, "the circum-stances and events which led to this discovery are little known" (Olby, 1966, p. 124). How could such an important moment in the history of science, that being Mendel's contribution to evolution, been overlooked for so long? Part of the reason was Mendel's scientific contacts were concentrated in Central Europe and not Western Europe. Second, while Mendel did visit England, he did not meet with any British naturalists and more important, neither did Mendel meet Francis Galton, the pio-neer of statistical modeling and early promoter of eugenics. The history of genetics would, according to Olby, "surely have been different had these two original thinkers met" (Olby, 1966, p. 142).

An additional reason for the absence of Mendel's work in the larger scientific community at the time was that he published his papers locally since he had lectured locally. He may have published his work in a wider-reaching journal, such as the *Berichte der Deutschen Botanischen Gesell-schaft* if invited. However, he did not dare ask and when writing to Karl von Nägeli for encouragement Mendel received none.

Margaret Mann Lesley posited that the reason why Mendel did not receive any support from von Nägeli was "the fact that von Nägeli did not believe that garden varieties have any importance in evolution" (Mann Lesley, 1927, p. 370). Von Nägeli also believed that plants are "continually adapting themselves to their environment" so "he could not believe that Mendel's *Pisum* types would remain true" (Mann Lesley, 1927, p. 370). In addition, von Nägeli had a deep distrust of "amateurs" and thought that only a properly trained and professional physiologist had the capacity to study heredity with any chance of developing useful insights into its workings. Therefore, a monk at the monastery at Brünn (that is Mendel) would have no validity or position of authority worth respecting by which to discuss such matters. It is clear from his writings that Mendel's work had little or no impact on von Nägeli's research.

In the first letter (December 1866) that Mendel wrote to von Nägeli, he described why he had approached the distinguished naturalist from the start, noting,

> The acknowledged value of your contributions to the definition and clas-sification of wild-growing plant hybrids makes it a pleasant duty for me to submit this description of some investigations on artificial hybridization of plants to your friendly attention. (Mann Lesley, 1927, p. 371)

Mendel then commented on the work of Gärtner,

> The results to which Gartner arrived through his investigations are known to me; I have repeated his work and reexamined it in order where possible to see whether they agreed with the laws of development found for the plants which I have investigated. Try as I would, in spite of my trouble I could in no single case obtain a clear insight from it. (Mann Lesley, 1927, p. 371)

Von Nägeli did not answer Mendel's first letter for over two months. When he finally responded to Mendel, von Nägeli noted firstly that all amateur experimenters in heredity are hopelessly inferior to such researchers as Kölreuter and Gärtner in quality or in the ability to keep their work going for the many years needed to accomplish anything in the field. Von Nägeli had been carrying out experiments on *Hieracium* (hackweed) and encouraged Mendel instead to perform hybrid fertiliza-tion experiments with Hieracia, because its characteristics made it a prime subject to learn its hereditary behavior quickly allowing its inter-mediate forms to be more easily observed.

Mendel's last three letters to von Nägeli were largely concerned with his progress with *Hieracium*. This work was begun to work out the ori-gin of intermediate types in a botanical genus. Von Nägeli had found the genus difficult to classify because of intermediate types present in nature. Mendel wanted to discern whether they arise by hybridization and if their short duration was due to the fact that they were sterile or that they are more easily fertilized by pollen from their parents. Mendel recog-nized, in 1870, that his results with peas were vastly different than those he had obtained with *Hieracium*. Writing to von Nägeli Mendel noted "I can not on this occasion repress the remark that it is very impressive to observe that the hybrids of *Hieracium* in comparison with those of *Pisum* present an exactly opposite state of affairs. We have here clearly to do with separate phenomena which are a part of a higher general law" (Mann Lesley, 1927, p. 375). The Franco-Prussian War delayed correspondence between Mendel and von Nägeli from 1870 to 1873. His last letter to von Nägeli offered him any or all of his material for further investigation and "concluded with a statement of his general conclusions concerning the importance of hybridization to evolution" (Mann Lesley, 1927, p. 377). Interestingly, von Nägeli wrote twice to Mendel (in 1874 and 1875) after receiving his last letter, but received no response to either.

Thus, Mendel's contributions to the history of genetics during the 1850s and 1860s were kept isolated within a small group of naturalists in the region. The facts available in 1900, to those adept enough to appreciate them, were at the very least enough to show that Mendel's contribution was one of the great experimental advances in the history of the biological sciences, and more than enough to establish him as an icon of evolution.

HUGO DE VRIES

As the story goes, the manner in which Hugo de Vries (1848–1935) obtained a reprint in the early part of 1900 of Mendel's paper was through his friend Professor Beijerinck in Delft. De Vries, a Dutch hybridist, had begun experiments in hybridization as early as 1876. At first, he crossed varieties of maize (corn) with sugary and starchy, black and white grains. Due to poor weather in Amsterdam at the time, no harvest of the maize hybrids was made and he terminated this first set of experiments. However, in the late 1880s de Vries again turned to plant genetics and wrote his 1889 book *Intracellular Pangenesis*, in which he stressed the importance of treating the characters of organisms as separate entities in the study of both heredity and variation. De Vries took on hybridization experimentation soon after, crossing the plant white campion (*Silene alba* × *S. alba* var. *glabra*). The results did not lead de Vries to Mendel's explanation for inheritance, however.

In 1896 de Vries crossed *Papaver somniferum* var. "mephisto" (a poppy with a black heart-shaped spot at the base of each petal) with *P. somniferum* var. "Danebrog" (a poppy that has a white base on the petal) and saw segregation in the F_1 generation, with 158 black to 43 white. De Vries then self-pollinated the F_2 generation of plants, and collected seeds from ten of the recessive white type hybrids and thirteen of the dominant black type hybrids. In 1896, he sowed these seeds in separate plots, to find the recessive segregates did breed true in all cases while some of the dominant types did, and the others segregated in a ratio of 1095 black to 358 white. The significance of this round of experimentation was that it came close to the 3:1 ratio expressed in Mendel's work and also had shown the twofold constitution of the dominant F_2 segregates— part homozygous (possessing two identical forms of a particular gene, one inherited from each parent) and part heterozygous (possessing two different forms of a particular gene, one inherited from each parent).

Why was it then that despite de Vries's success in these experiments he did not publish the results in the years prior to and during 1896? Part of the reason for his apprehension may have been that success with a few species would not have convinced the world. Yet, by 1899 de Vries had found the segregational results of his experiments in over thirty different species and varieties. On 11 June 1899 the Royal Horticultural Society in London held a conference on hybridization and crossbreeding at which de Vries presented his results. The conference was significant in the history of genetics, for it brought together British, American, and continental European scientists with wide-ranging theoretical, practical, and commercial interests in botanical hybrids.

Resulting from this conference, de Vries published in 1900 a short notice for the *Comptes Rendus de l'Academie des Sciences* (Paris) titled, "Sur la loi de disjunction des hybrids." No mention is made of Mendel. After completing the French report and while placing the finishing touches on a report for the German journal *Berichte der Deutschen Botanischen Gesellschaft* titled "Das Spaltungsgesetz der Bastarde," de Vries supposedly received from Beijerinck a reprint of Mendel's 1865 paper with the comments: "I know that you are studying hybrids, so perhaps the enclosed reprint of the year 1865 by a certain Mendel which I happen to possess is still of some interest to you" (Olby, 1966, p. 127).

This may explain, according to Olby, why there is no mention of Mendel in the French publication and only a brief mention of him in the German report. The French version stated, "The totality of these experiments establishes the law of segregation of hybrids and confirms the principles that I have expressed concerning the specific characters considered as being distinct units" (Olby, 1966, p. 128).

De Vries wrote in the German version:

> From these and many other experiments I conclude that the law of segregation of hybrids in the plant kingdom, which Mendel established for peas, has a very general application and a fundamental significance for the study of the units out of which the specific characters are compounded. (de Vries, 1900, p. 90)

The German report, with its recognition of Mendel in the conclusion, brought forth the contribution that Mendel had made with his experiments on pea plants and that de Vries had made with his studies of a wider range of plant species.

One interesting observation that Olby made sheds light on an alternative view: that de Vries had been aware of Mendel's work far earlier (1896 or 1897) and that he had underestimated the importance of Mendel's contribution in light of his own. De Vries may have seen his work as superior to Mendel's and that because his work coincided with recent studies on plant cytology and Weismann's theory of the germ-plasm, it superseded Mendel's ideas it pushed his work off to the side. Whether de Vries did or did not obtain a copy of Mendel's 1865 paper prior to his published report in the *Comptes Rendus* is not as significant to the history of genetics as is the fact that by 1900 the scientific world was aware of the contributions that Mendel had made through de Vries's mention in the German report.

CARL CORRENS

Carl Correns (1864–1933) studied botany under Karl von Nägeli in Munich and completed his dissertation in 1889. As Rheinberger noted, "It has been widely accepted that he interpreted his crossing experiments with *Pisum*, performed in Tübingen (1896–1899), in a manner consistent with Mendel's rules, and that he reached his conclusions independently and before he learned of Mendel's paper" (Rheinberger, 1995, p. 613). In 1892, Correns received his *venia legendi* (Latin for permission for lecturing) in Tübingen and for almost ten years remained there as a Privatdozent. Correns was appointed extraordinary professor at Leipzig in 1902 and in 1909 became professor of botany and director of the Botanical Garden at Münster. Interestingly, Correns began his experiments on pea plants just after the death of von Nägeli, who, you may recall, had corresponded with Mendel during the years 1866 to 1873 but gave little support to Mendel's work.

Correns began breeding experiments soon after obtaining access to the botanical garden at Tübigen in 1892. As Rheinberger noted, at the start the crossing experiments seemed to make little impact on Correns's overall views, for they were concentrated mostly on problems of developmental physiology. However, this would change. Correns had also been studying more complicated cases of inheritance, specifically with maize (*Zea mays*) and peas (*Pisum*). De Vries sent a copy of his first paper to Correns because he was aware that Correns had been working on such experiments at Tübingen. Correns was stunned to receive a reprint of his report to the *Comptes Rendus* from de Vries. Yet Dodson saw Correns as not so much threatened by de Vries's paper, but confirming similar results he had seen in his own experiments. Whether Correns was disturbed or not by the reception of de Vries's paper it did spur him on to publish a short report in the *Berichte der Deutschen Botanischen Gesellschaft* in April 1900, revealing

The same thing happened to me which now seems to be happening to de Vries: I thought I had found something new. But then I convinced myself that the Abbot Gregor Mendel in Brünn had, during the sixties, not only obtained the same results through extensive experiments with peas, which lasted for many years, as did de Vries and I, but had also given the same explanation, as far as that was possible in 1866. (Dodson, 1955, p. 192)

Correns found through his experiments, like Mendel, that there were some instances where hybrids were intermediate between the parent types. Thus, Correns argued that de Vries's law of dominance was not universally applicable. In addition, Correns also confirmed the primacy of Mendel over de Vries with respect to such laws of inheritance. Correns would go on to publish the letters between Mendel and von Nägeli in 1905, quite possibly to further support the view that Mendel was the progenitor of such laws.

ERICH VON TSCHERMAK

Tschermak (1871–1962) was born in Vienna in 1871, the son of Gustav Tschermak, professor of mineralogy at the University of Vienna. His grandfather was Eduard Fenzl, who was professor of botany at Vienna and also director of the botanical garden. The connection between Fenzl and Mendel has already been noted earlier in this chapter, as Fenzl was at the University of Vienna while Mendel was attending courses there. Tschermak began studying agricultural science at the Hochschule für Bodenkultur (now the University of Agricultural Sciences in Vienna) in 1891 and simultaneously studying biology at the University of Vienna. Tschermak moved his studies from Vienna to the University of Halle in 1893, receiving his Ph.D. in applied botany in 1896.

Tschermak's initial work in plant breeding was undertaken at various breeding stations for vegetables and ornamental plants during the years 1896–1898. In 1898, disappointed by the loss of a position as university assistant in Vienna, Tshermak traveled to Gent and Paris, where he began crossing experiments with *Pisum vulgare* L. (common garden peas) and *Erysimum chéiri* (wallflowers). It was Darwin's work on the cross- and self-pollination of plants that Tschermak read in the library at Gent which spurred on his experimentation. Tschermak, limited by time, concentrated his studies on peas, a species easy to cross and of which the greenhouse at Gent was full.

After this period of experimentation in Gent, the seeds were sent off to Vienna and Tschermak traveled to the Netherlands to visit de Vries in Amsterdam. De Vries showed Tschermak his work on *Oenothera lamarckiana* but not on *Pisum*. Tschermak was overly reluctant to tell de Vries about the work he had done in Gent. If he had de Vries might have mentioned to him that he was closing in on the work of Gregor Mendel. Despite the lack of transparency, de Vries was delighted at the interest the young Austrian scientist showed in his work, so he put him on his mailing list for the exchange of scientific papers.

On 17 January 1900, Tschermak completed his doctoral thesis. The thesis contained the results of his experiments with crossbreeding. Tschermak observed the phenomenon of xenia in the seed pods of the same F_1 plants, that seed color and seed shape resemble the difference in parental characters at this early stage after hybridization. He also saw that parental characters appeared in a 1:1 segregation scheme when the hybrids were crossed again with their parental types.

Because Tschermak had been included in de Vries's mailing list, both Correns and de Vries sent copies of de Vries's first paper to Tschermak in April 1900. Like Correns, Tschermak lost no time preparing his work for publication soon after the arrival of de Vries's paper, sending off a paper to the *Zeitschrift für das landwirtschaftliche Versuchswesen in Österreich*. Concerned with the timeliness of presentation, Tschermak made sure that reprints of his paper were available in May 1900, prior to the publication of the journal. Soon after the rediscovery of Mendel's work, Tschermak promoted the idea that these fundamental principles of inheritance must be applied to achieve stable and uniform combinations of different characters of parental genotypes by crossing experiments. Tschermak advocated a system of combination breeding instead of individual selection of phenotypically equal plant types within populations, which was very common at the time to create improved varieties. In recognition of Tshermak's contributions to the history of genetics, the *Journal of Heredity* dedicated a translation of his rediscovery of the Mendelian laws to him on his 80th birthday in 1951.

WILLIAM BATESON

William Bateson was another central figure in the rediscovery of Mendel. Born to a master and scholar at St. Johns College, Cambridge, Bateson (1861–1926) showed interest in natural history early in his life and entered St. Johns College in 1879. In 1882 he won honors on his Natural Science Tripos for his work on the acorn worm (*Balanoglossus*). After two years working with W.K. Brooks at Johns Hopkins, Bateson published papers on the acorn worm and gained acclaim as a biologist. He showed that "Balanoglossus must be regarded as the humblest member of the group to which vertibrates belong, thus opening up a fresh view of the relation of this great group to the rest of the animal kingdom" (Punnett, 1952, p. 337). He was elected fellow of St. Johns College in 1885.

Bateson entered the Mendel rediscovery scene soon after he read the republished Mendel paper in 1900. By this time Bateson had a reputation as a critic of traditional biology. He demanded high scientific standards for his own work and that of others and due to his outspoken views, his ideas were not popular. Bateson's program of research included rigorous experimentation and the extensive collection of facts. In constructing the pedigree of the animal kingdom, naturalists had always taken it as a given that hereditary variations went in many different directions, and that evolution worked by natural selection which favored any variations that were favorable to the organism in its struggle for existence, at the same time rejecting any that were less so. Bateson's observation of discontinuity between species led him to believe that evolution does not take place through the selection of individuals possessing small, but valuable variations (Darwin's view). Instead, Bateson believed that evolution, particularly the origin of a new species, takes place by great leaps in variation (hence the term discontinuous). This was expressed in *Materials for the Study of Variation* (1894).

In this book, Bateson pointed out that the relatively new idea of variation must impact upon the notion of the nature of species formation. This was the fundamental problem of evolution that biologists had to deal with. Concerned that Mendel's view would be lost for a second time, Bateson formulated a vigorous defense of Mendel that initiated a bitter controversy. As Punnett noted, "the discontinuity in heredity revealed by Mendel was the logical sequence to the discontinuity in variation upon which he himself had so long insisted" (1952, p. 341). In fact, Punnett argues that the address Bateson gave to the Royal Horticultural Society before the hybrid conference in July 1899 was "vastly similar to the introduction to Mendel's own paper and was constructed so that it could be given its proper due" (1952, p. 341). Bateson recognized that Mendel's discovery was not just the recording of an interesting phenomenon which took place during the crossbreeding of plants, but that it was a fundamental law to be applied to all living things and that it explained a good deal about the passing on of hereditary traits.

BATESON'S REDISCOVERY OF MENDEL

In 1909, Bateson published *Mendel's Principles of Heredity*. The object, according to the author, was "to give a succinct account of the discoveries in regard to heredity made by the application of Mendel's method of research" (Bateson, 1930, p. v). The reason for Bateson's insistence on bringing Mendel's contributions to light was to undermine criticism of Mendel's work that had begun to appear. He wanted to protect Mendel's conclusions and the inferences drawn from it, by strengthening the theory with more evidence. Bateson promoted the notion that the year 1900 sparked a new era and his interpretation of Mendelism transformed his career. With L. Cuénot, Bateson proved that Mendelian behavior was the same in animals as it was in plant life, thus making it a universal mechanism of heredity.

Bateson made it a major part of his work to support and bolster Mendelian hereditary ideas from attack and to work them into the general understanding of how evolutionary change occurred. The biometricians (those who saw heredity working with an almost mathematical regularity) were firmly opposed to the idea of discontinuity in variation, and had adopted Galton's Law of Ancestral Heredity. Because they had, Mendel's conclusions were incompatible. Bateson became president of the Zoological Section of the British Association and now felt primed for attack. At the Cambridge meeting in 1904, both biometricians and Mendelians attended and it was before a large audience that Bateson gave a dynamic presentation supporting Mendel. While Bateson was an affirmed evolutionist, he was not a proponent of natural selection. He is, however, credited for coining the term "genetics."

THE REDISCOVERY OF MENDEL?

Ronald Fisher, in his piece "Has Mendel's Work Been Rediscovered?" (1936), argued that a careful scrutiny of the mythology surrounding the "sensational" rediscovery of Mendel's work thirty-four years after its publication and sixteen years after his death is needed. The tale of this event has become monumental in the history of biology (virtually an icon on its own), and Fisher cites Bateson as being the culprit. He posited that Bateson had a powerful interest in the rediscovery of Mendel and that he was building a career on it, and was not just a disinterested bystander in the discussion. Fisher saw Bateson as biased toward Mendel because (1) Bateson believed that Darwin's influence was responsible for the neglect of Mendel's work, and (2) he believed that Mendel was hostile to Darwin's theories. Fisher used a passage from Bateson's introduction in *Mendel's Principles of Heredity* (1909) to illustrate how Bateson saw the overshadowing of Mendel by Darwin:

> While the experimental study of the species problem was in full activity the Darwinian writings appeared. Evolution, from being an unsupported hypothesis, was at length shown to be so plainly deducible from ordinary experience that the reality of the process was not longer doubtful. With the triumph of the evolutionary idea curiosity as to the significance of specific differences was satisfied. (Bateson, 1909, p. 2)

Building upon this, Fisher then noted that Bateson's desire to use the work of Mendel in opposition to natural selection showed itself in his misrepresentation of Mendel's own views. Fisher next revealed Bateson's biases toward Darwin with Bateson's own extrapolation of Mendel's opinion of Darwin. Bateson suggested that "with the views of Darwin which were at that time coming into prominence Mendel did not find himself in full agreement, and he embarked on his experiments with peas" (Fisher, 1965, p. 62). The suggestion by Bateson that Mendel began his experimentation on *Pisum* because of a disagreement with Darwin's views remains unfounded.

R.A. Fisher was interested in biometrics and could be part of the reason for his dissatisfaction with Bateson's polemical view of Mendel. In 1933, Fisher was appointed Galton Professor of eugenics at University College, London where he remained for ten years. Where Fisher's career intersects with the discussion of Mendel's rediscovery is that in 1900, particulate inheritance seemed to cast doubt not only on Karl Pearson's extensive work on blending inheritance, but also on Darwinian natural selection as well (some argued that Mendelian genetics was incompatible with Darwin). Fisher did not agree and in 1918 published "The Correlation Between Relatives on the Supposition of Mendelian Inheritance," a paper that brought a host of powerful new mathematical tools into the discussion of evolution. Fisher's researches would initiate the modern discipline of biometric genetics.

Despite Fisher's disagreement with Bateson's agenda regarding the rediscovery of Mendel, it is important to recognize Bateson's contribution to the process. Bateson noted that Mendel's experiments were first undertaken in the hope that they would make the problem of speciation easier to explain. Mendel's discovery of segregation was one of the essential discoveries of the field because it was seen that segregation helped determine the rate at which differences in heredity were transferred from one generation to another. As Bateson concluded, the "consequence of the application of Mendel's principles [shaped a] vast medley of seemingly capricious facts" recorded on heredity and variation and brought them into "an orderly and consistent whole" (1930, p. 17). It was Bateson who elevated Mendel's work above Darwin's for evolutionists or even sociologists who wanted to achieve wider views and look in new directions for answers to problems of evolution. R.C. Punnett, for example, believed that "not until we had from Bateson the enunciation of discontinuity in variation, and from Mendel the revelation of discontinuity in heredity, could there come into being that irresistible body of facts which must henceforth play so great a part in the destiny of mankind" (1952, p. 347).

WHERE MENDEL MEETS DARWIN: THE MODERN SYNTHESIS

As Theodisius Dobzhansky once famously noted, "Nothing in biology makes sense except in light of evolution" (1973, p. 125). Similarly,

William B. Provine, in *The Evolutionary Synthesis* (1980) stated that "the evolutionary synthesis is unquestionably an event of first-rank importance in the history of biology." During the 1930s, the movement known as the modern synthesis or the unification of Mendelian genetics, systematics, paleontology, and ecology into a theory of evolution that made sense of all the new discoveries came into being. The insight that unified heredity, evolution, and development came at the beginning of the twentieth century. It derived from the discovery that the gene, localized to specific positions on the chromosome, was at once the unit of Mendelian heredity, the driving force for Darwinian evolution, and the control switch for development. It combined the theory of natural selection with the newly emergent science of genetics and helped explain how genes are transmitted from one generation to another.

Markert noted in his introduction to the centennial celebration of Mendel's work in 1965 that when a century ago Gregor Mendel read his paper "Experiments in Plant Hybridization" before the Brünn Society of Natural History there was no one in the audience who was prepared to accept or even understand his ideas on inheritance. This was because Mendel had used mathematics to explain himself and because his ideas were so new and radical. Unlike his contemporaries in 1865, Markert said, we today are "far better prepared to understand and to appreciate Mendel's analysis on inheritance" (Markert, 1968, p. 1). The reason why we are better prepared to understand Mendel's analysis is largely due to the development of evolutionary thought during the past century. This remarkable discovery can be traced directly to one person and to one institution: T.H. Morgan and Columbia University.

Thomas Hunt Morgan (1866–1945) began experimenting on *Drosophila melanogaster* (fruitfly) and challenged the assumption that germ cells are pure and uncrossed and was skeptical that species arose from natural selection. In his essay "The Rise of Genetics" (1932) Hunt explained that genetics had its beginnings in the experiments on pea plants that Mendel had conducted. Its rediscovery was not the simple result of his once-obscure treatise but because later researchers, like de Vries, Correns, and Tschermak, had independently come across phenomena identical to the ones Mendel had encountered.

In his article, "The Bearing of Mendelism on the Origin of Species" Morgan posited that "the evidence that all heritable variations may have had the same kind of origin rests on the following facts and argument: that mutant types that appear in our cultures follow Mendel's laws of inheritance and practically all the character differences of domesticated races also fall under these same laws" (Morgan, 1932, p. 239). How this fits into the discussion of Mendel and evolution appears when we look closely at Morgan's assessment that evolutionary biology would have to be concerned with the genes that remain unchanged because it would be these characters that clearly indicate the long line of hereditary connection between species: In other words they would help show a common ancestry. One can trace the unbroken chain of genetic material to identify a common ancestor, reflecting the branching tree illustrated by Darwin in the *Origin of Species*.

Morgan highlighted the importance of this in his article "The Rise of Genetics" (1932). Morgan saw the birth of genetics as dating from 1900, with the resurrection of Mendel's paper. He saw three lines of significance for genetics brought together in the mutation theory of de Vries, the rediscovery of Mendel, and the application of discoveries in cytology to the new theories. Morgan felt that the rapid developments and expansion of genetics after 1900 as "intimately connected with the application of the chromosome theory to the experimental work" being undertaken in genetics (Morgan, 1932, p. 264). For example, Morgan noted that if we accept that the method by which the number of chromosomes (the long strands of genes that exist inside each living cell in the form of the famous double helix) increased, by suppressing the cytoplasmic division of a cell when the chromosomes divide, the chromosomes do not unite, but instead the descendents of that cell will possess twice the original number of chromosomes. In theory, this might go on forever, with doubling of diploids leading to tetraploids; these crossed to diploids give triploids; double tetraploids crossed to tetraploids give hexaploids, and so on. Where this leads is an opportunity to solve certain genetic problems of theoretical significance, for without this knowledge, according to Morgan, some of the known genetic ratios would be difficult to interpret. With this knowledge, they conform to recognizable general principles.

During the 1930s, researchers like Oparin and Haldane postulated separately that the chemical structures that characterize living organisms could have arisen as a result of energies such as radiation, becoming part of the mixture of inorganic molecules in the primordial atmosphere. These experiments brought on an entirely a new era in evolutionary studies in which experiments were designed to work out the mechanisms of not just how one species evolved into another, but how life itself had come into being. In conjunction with these studies was the development of molecular genetics.

As Ernst Mayr noted, "what occurred during the period from 1936 to 1950, when the synthesis took place, was not a scientific revolution; rather it was a unification of a previously badly split field" (Mayr, 1998, p. 134). Many naturalists up to that time tended to separate characters into Mendelian (particulate) ones, which they considered to be of reduced evolutionarily unimportance, and gradual or blending ones, which, following Darwin's suggestion, were the genuine building blocks of evolution. The major achievement of the synthesis was the reconciliation of the Mendelians with those who worked on quantitative evidence of inheritance. We must be careful, however, not to assume that the synthesis occurred simply with the application of Mendelian inheritance to evolutionary biology.

Julian Huxley (1887–1975) wrote in his book *Evolution: The Modern Synthesis* (1943) "In Darwin's day biological inheritance meant the reappearance of similar characters in offspring and parent, and implied the physical transmission of some material basis for the characters" (p. 20). With the shift toward the employment of mathematics in the study of evolution at the beginning of the twentieth century, the biometricians adhered to hypothetical modes of inheritance and genetic variation and the Mendelians refused to acknowledge that a mathematically based theory of selection could be of value to the evolutionary biologist. It was the synthesis of these disparate views that led to what Huxley called the Modern Synthesis (Darwinism rethought and reconsidered by men ideologically and methodologically different from William Bateson).

The 1965 centennial of the publication of Mendel's *Versuche über Pflanzen-Hybriden* was cause for a multitude of publications celebrating his contributions to genetics and a major factor in reconciliation. Betty Smocovitis, in *Unifying Biology* (1996) posited that the mid-1960s saw an increase of interest in the history of biology and the history of science generally. This, she argued, opened the way to historical studies of the synthesis. When, in 1971, William B. Provine's *The Origins of Theoretical Population Genetics* was published, the "efficacy of natural selection under a range of different parameters" was noted and "made possible the synthesis of genetics and selection theory, which in turn led to the reconciliation between Mendelians and biometricians (biologists that study biological phenomena and observations by means of statistical analyses), who had opposed each other at the turn of the century" (Smocovitis, 1996, p. 28). It was this reconciliation that led to new frontiers in genetic research.

MENDELIAN CONTROVERSIES

In 2000, a century had passed since the rediscovery of Mendel's work and the birth of genetics. As Fairbanks and Rytting noted, "during that century, Mendel's name became indispensable to science" (2001, p. 737). It was largely due to the apparent lack of recognition given to Mendel during his lifetime that unfortunately little information on his scientific work was preserved for later generations to study. This lack of information plagued the appreciation of Mendel's work throughout the twentieth century. Soon after the rediscovery of Mendel's researches, some questioned the validity of his results. W.F.R. Weldon, whose criticism was the impetus for Bateson's propensity for defending Mendel, noted that the observed ratios of Mendel's were close to his own. Similarly R.A. Fisher, who published an examination of Mendel's paper in 1936, also uncovered that Mendel's results were "consistently so close to expected ratios that the validity of those results must be questioned" (2001, p. 738). Fisher's work spawned further critiques, yet none of them produced any kind of definitive solution. Despite the inability to solve the controversy over Mendelian ratios, other questions remained. Some authors question Mendel's description of his experiments. For example, Bateson, in a footnote to the Royal Horticultural Society's English translation of Mendel's paper, referred to Mendel's claim:

> This statement of Mendel's in light of present knowledge is open to some misconception. Though his work makes it evident that such varieties may exist, it is very unlikely that Mendel could have had seven pairs of varieties such that the members of each pair differed from each other in only one considerable character. (Bateson, 1930, p. 350)

So even Bateson, grand defender of Mendel at the beginning of the twentieth century had some reservations regarding the reality of the experimentation by Mendel on *Pisum*.

CONCLUSION

As this chapter has shown, the rediscovery of Gregor Mendel's 1865 paper represented a monumental shift in the history of biology. His work on *Pisum* became the stepping stone for the scientific study of inheritance: what we now call genetics. Without the discovery by Mendel of the deeper functioning of heredity, it is unlikely that evolutionists could have made as many strides in understanding the workings of biology as they did. Mendel's seemingly innocuous fiddling with pea plants, so completely overlooked in his time, came back to be recognized as one of the great breakthroughs of evolutionary biology. We are left today with the understanding that in order to understand the evolution of everything from dinosaurs to birds to humans themselves, we must start with the vision of a monk intently and quietly studying a garden of delicate and unassuming pea plants.

FURTHER READING

Caporale, Lynn. *Darwin in the Genome: Molecular Strategies in Biological Evolution* (Boston: McGraw-Hill, 2001).

Carlson, Elof Axel. *Mendel's Legacy: The Origin of Classical Genetics* (Cold Spring Harbor: Cold Spring Harbor Laboratory Press, 2004).

Corcos, Alain F., Floyd F. Monaghan, and Gregor Mendel. *Gregor Mendel's Experiments on Plant Hybrids: A Guided Study* (New Brunswick: Rutgers University Press, 1993).

Mayr, Ernst. *The Evolutionary Synthesis: Perspectives on the Unification of Biology*, with a new preface; reprint edition (Cambridge, MA: Harvard University Press, 1998).

Olby, Robert C. *The Origins of Mendelism*, 2nd ed. (Chicago: University of Chicago Press, 1985).

"Survival of the Fittest"

Peter Bowler

The "survival of the fittest" is a term coined in 1864 by the philosopher Herbert Spencer in his description of the evolutionary mechanism that Charles Darwin had described under the name "natural selection." Spencer's term seems to have caught the popular imagination and has remained in use as a catch phrase used to highlight the apparently ruthless and amoral character of Darwin's theory. Applied to human affairs it is a stock in trade of the critics of what has become known as Social Darwinism. In more recent times it has become the basis for a widely quoted critique of the selection theory based on the claim that it is no more than a tautology. Since the fittest are by definition those who survive, "the survival of the fittest" tells us nothing. This blatant misrepresentation of the theory was in part made possible by the fact that there has always been some ambiguity about the meaning of the terms involved. Spencer himself was deliberately vague about the precise definition of "fitness"—did it mean those best fitted (i.e., adapted) to the local environment, or was there an implication that the fittest were superior in some overall sense that would allow evolution to be seen as necessarily progressive? Darwin complicated matters by implying that it was the superior reproductive potential of the better-adapted that counted, not their mere survival. This chapter will look at the origins of the term and at the ways in which the process it symbolizes has been used in biology and in the human applications of evolution theory.

Surviving the Fittest

The "survival of the fittest" is a term that encapsulates the ruthless logic of the Darwinian theory of natural selection. To the theory's critics the phrase is a symbol of its origins in a materialistic value system promoting indifference to the fate of those who do not succeed in life's race. But how do we define the "fit" and what are the key benefits they gain from their fitness? These questions have proved surprisingly hard to answer, both in biology and in social theory.

The survival of the fittest is almost invariably understood in conjunction with another stock phrase linked to Darwinism: the "struggle for existence." In effect we are only interested in "the survival of the fittest in the struggle for existence"—although in principle one could imagine a rather pallid form of natural selection operating even in a world without struggle. Presumably the positively unfit would not survive, or at least would be too weak to breed, even if unlimited resources were available. But in such a world would the very fittest do any better than the average? Presumably not enough to count, in which case selection would play a negative role only. We only get the impression of a relentless process of selection when we include the element of struggle: An organism's degree of fitness matters because at every moment of its life it must compete against rivals who will, if they succeed, threaten its very survival. So closely is the survival of the fittest linked to the concept of struggle that we all too often forget that the relationship is not symmetrical. One cannot have a meaningful process of natural selection without struggle, but one can certainly imagine a world full of struggle in which there was no natural selection, for instance because there was not enough variation in fitness between individuals to make any difference. More important, a world dominated by struggle might allow evolution to take place in a way that did not depend on selection. Spencer himself discussed the survival of the fittest only after he had expounded what was, for him, a more important mechanism of change in which the pressure from the environment forced individuals to become fitter through their own efforts. And since he accepted the Lamarckian process of the inheritance of acquired characteristics, he thought that these self-improvements would be passed on to the next generation and thus become cumulative. As we shall see, much of what has passed for Social Darwinism is in fact social Lamarckism, driven by the struggle for existence but not depending on the actual elimination of the unfit.

Once we start to unpack the terms contained in the phrase "survival of the fittest" we begin to realize what a can of worms Spencer opened up. As Darwin explained, and Spencer accepted, the fittest don't really have to survive in order to make selection work, they just have to breed more effectively than their less fit counterparts. Differential reproduction will be just as effective as the actual weeding out of the unfit. There is competition to reproduce in nature as well as competition for survival, and Darwin's proposed mechanism of sexual selection drew on this point to explain how evolution could produce apparently unfit characters such as the peacock's tail. It was partly as a consequence of twentieth-century scientists taking a renewed interest in sexual selection that they fell into the trap of defining fitness in terms of reproductive success, opening the way for the claim that selection was a mere tautology.

We have already noted the ambiguity of the term "fitness" even when confined to the struggle for existence: Does it mean adaptation in the purely local sense, or will the fittest tend to be better in some overall sense, at least in the long run? And if so, how should one define this more general meaning of fitness if not in terms of mere survival? It must include the whole range of survival factors, including both access to resources and the avoidance of threats. If, for the sake of argument, we limit ourselves to adaptation and argue that we could measure the level of adaptive fitness, for instance in terms of the chances of a camouflaged moth escaping predation, how are we to understand the role played by individual differences? Are there only a few of the really fittest individuals who survive when the rest of the population perishes, or a few really unfit ones who die when everyone else survives? The intensity of the struggle is clearly a key factor. The model of variation we employ is also crucial and it was only in the twentieth century that biologists began to think clearly in terms of the differential survival and reproductive opportunities of individuals spread out along a continuous range of variation for each character.

Another potential minefield opens up when we ask about the different possible levels at which natural selection can operate. In the struggle for existence, who are the opponents? Darwin actually borrowed the phrase "struggle for existence" from Thomas Malthus, but although we remember Malthus mostly in the context of the British laissez-faire or individualist model of society, he actually used the term in a description of competition between rival groups. It was Darwin who applied it at the level of competition between individuals within the same population, thereby creating the basic concept of natural selection. As we have already noted, though, the idea of struggle could also be employed to sustain a Lamarckian theory of evolution in which all individuals strove to become fitter. Many of Darwin's contemporaries understood the struggle for existence as a struggle of the whole species against the limitations imposed by the environment, which sidelines the whole idea of any selective process within the population. But the environment does contain rival groups and rival species competing for the same resources, so one can imagine selection acting at various levels, not necessarily with consistent effects. Darwin developed his principle of divergence by realizing that selection could operate in a stable environment to divide a species up into daughter forms specializing for different ways of exploiting the limited resources available. Fitness could, in effect, be defined in terms of become increasingly different from potential rivals to avoid treading on each others' toes. Darwin also knew that species compete for territory and that this is a key factor which must be taken into account in explaining the geographical distribution of species. One species can invade another's territory with disastrous consequences, in effect completely transforming the pre-existing levels of fitness.

Each of these various levels of struggle has its equivalent in the human sphere, allowing Social Darwinism to be invoked at several levels: individual, tribal, national, and racial. But as Darwin was anxious to stress, the possibility of selection acting at the group level between rival tribes might have consequences that would seem in conflict with the effects of individual competition. If selection is driven solely by individual struggle one would expect it to generate purely selfish instincts, because anyone who behaved altruistically would place themselves at a disadvantage. But if we imagine rival groups seeking to occupy the same territory, it may well be that the group whose members cooperate most effectively will beat another composed of rugged individualists. Group selection will generate instincts which conflict with those produced by individual selection. What is fit for the individual may not be fit for the group. The possibility of group selection thus joins Darwin's model of sexual selection as a mechanism which can drive evolution in a direction that conflicts with the simplest definition of fitness.

To put some flesh on the bare bones of these distinctions I shall first look at Spencer's introduction of the term "survival of the fittest" and then at the different uses the concept of struggle had in Darwin's own thinking and in that of his followers both in the late nineteenth century and more recent times. I shall then move into the social sphere, showing how the rhetoric based on phrases such as "survival of the fittest" and "struggle for existence" were exploited to create the various models of Social Darwinism. In most people's eyes these models have been long discredited, but in conclusion we shall have to recognize that these catch phrases still have a power to affect our perception of the theory. By tagging Darwinism with phrases which evoke the murkier aspects of its public influence, modern opponents such as the creationists seek to discredit even the scientific applications of the theory.

SPENCER, DARWIN, AND WALLACE ON NATURAL SELECTION

Darwin published his *Origin of Species* in 1859, and Herbert Spencer immediately realized that here was a valid evolutionary mechanism which he would have to take into account, in addition to his long-standing acceptance of the Lamarckian theory of use-inheritance. In 1864 Spencer published his *Principles of Biology*, in the course of which he outlined his ideas on evolution and described the two mechanisms he now endorsed. Note that he described his Lamarckian approach first; this was the "direct" process by which organisms were brought into equilibrium with their environment, i.e., by each individual self-adapting to changed conditions and passing its improvements on to its offspring. Natural selection came second, as the "indirect" mechanism of adjustment. Here he introduced the phrase we are concerned with: "This survival of the fittest, which I have here sought to express in mechanical terms, is that which Mr Darwin has called 'natural selection, or the preservation of favoured races in the struggle for life' "(Spencer, 1864, v. 1, pp. 444–445). Note that Spencer immediately links the phrase with Darwin's vision of struggle, and although he does not mention Malthus he does acknowledge that what will happen is the "inevitably destruction" of the unfit. He also realizes that reproduction is as important as survival: "But this survival of the fittest, implies multiplication of the fittest" (Spencer, 1864, p. 444). At this point Spencer seems quite clear that fitness must be defined in terms of adaptation, as "fitness to the conditions of life." Elsewhere, though, he portrays evolution in more general terms, arguing that the constant change of the environment will promote a gradual differentiation of living forms and the creation of ever more heterogeneous (i.e., more complex) individual organisms. The scene was thus set for a generation of social writers to assume that increased fitness meant not better local adaptation but something more progressive. We should also note that for Spencer, natural selection had acted most effectively in the lower stages of evolution. As higher animals appeared they acquired increasing powers of self-adaptation, allowing the Lamarckian mechanism to become more important. This process had reached its culmination in human social progress.

What did Darwin and Wallace make of Spencer's phrase? As a matter of fact we can answer this question fairly directly, since they discussed the issue in an exchange of letters which nicely highlights the confusions implicit in its use. In a letter written to Darwin on 2 July 1866 Wallace said that he preferred the term "survival of the fittest" to "natural selection" because it would prevent critics assuming that there was an intelligent selector at work in nature (Marchant, 1916, pp. 140–143). His precise wording on the issue is significant and we shall return to it in a moment. But the letter was important enough to prompt Darwin to take the phrase seriously, and he responded three days later saying that he hadn't realized "the advantages of H. Spencer's excellent exposition of 'the survival of the fittest' " until reading Wallace's letter (Darwin, 1882, v. 3, pp. 45–46). The main disadvantage was that it could not be used as a verb. He informed Wallace that he intended to use it in his new work on variation and thought it was a pity he had now sent off the latest revision of the *Origin of Species*. He did indeed introduce the term into the *Variation of Animals and Plants under Domestication* (1868) using it in the first sentence of the chapter "Selection by Man" and in the first subtitle of the chapter "Natural Selection." The final edition of the *Origin* also uses the term in the title of Chapter 4, "Natural Selection; or the Survival of the Fittest."

So Darwin and Wallace evidently thought Spencer's term was a good shorthand way of describing natural selection—but what did they actually mean by it? The problem is that even in Wallace's original letter there are ambiguities in his descriptions of the process. In his first attempt he criticizes "natural selection" as a metaphor that is both indirect and incorrect, because nature "does not so much select special varieties as exterminate the most unfavourable ones." We shall return to the question of whether Wallace might actually have meant varieties in the sense of competing subspecies below. But even if he means individual variations, the implication here is that what happens isn't the survival of those of above average fitness, but only the elimination of those with actual disadvantages. Later in the letter he returns to the topic using words which certainly make it clear that he is thinking in terms of struggle between individuals, but which dramatically alter the understanding of how fitness matters. He claims that the struggle for existence leads to the constant destruction of by far the largest proportion of the population, now implying that it really is only those with positive advantages which make it through to breed. This is a far more relentless and by implication more effective method of selection than the mere elimination of the unfit. Here was a real difference between two conceptions of how fitness matters in the struggle, driven by different conceptions of how intensive the struggle is. A really fierce struggle kills off all but a few, those who survive having some variation which gives them a positive advantage over the norm. But there is a concept of struggle that is much less intensive, allowing most members of the population to survive and killing off only the unfit. In this case, the role of the better-adapted variants is to breed more effectively and thus increase the proportion of their character in the population indirectly—but Wallace doesn't seem to recognize this in the letter.

In the development of his theory Darwin had always been clear that the basic process of natural selection operated through competition between individuals within the same population. It is worth noting, though, that the term "struggle for existence" was used by Malthus to describe the competition between rival tribes of humans. Darwin borrowed the term, but realized that the key level of struggle was between individuals within a single population, not between groups. He realized that selection killed off any individual with even a slightly unfit variant character. For positive variations, he recognized that adaptive advantage not only helped to ensure survival but was also likely to confer better health and hence a better chance to breed. So nature both destroys the unfit and boosts the reproductive rate of the fittest. He came close to realizing that survival or elimination is merely the most extreme sanction, the final way of controlling breeding ability. Nevertheless, Darwin's early version of his theory imposed restrictions on how the process was supposed to work, as Dov Ospovat pointed out (1981). When he first conceived the theory he seems to have thought that struggle only came into play when the environment changed. By the 1850s he had realized that the pressure of population means that there can never be a permanent relaxation of the struggle for existence. Darwin also thought that a changed environment stimulated variation, but only to a very slight extent. He always tended to think that there would only be a small number of individuals who varied in a positive direction and gained some advantage in the struggle. There would similarly be only a small number of positively disadvantageous variants. Most of the population remained "normal."

The "Essay" that Darwin wrote in 1844 provides a snapshot of his thinking in the years following the conception of the theory. In many respects it looks forward to *Origin of Species*, adopting the same technique of leading the reader into the argument by explaining first how artificial selection works and then showing that there is an equivalent process at work in nature adapting species to their environment. Darwin gives the following imaginary example to illustrate the action of natural selection:

> let the organization of a canine animal become slightly plastic, which animal preyed chiefly on rabbits, but sometimes on hares; let these same changes cause the number of rabbits very slowly to decrease and the number of hares to increase; the effect of this would be that the fox or dog would be driven to try to catch more hares, and his numbers would tend to decrease; his organization, however, being slightly plastic, those individuals with the lightest forms, longest limbs and best eyesight (though perhaps with less cunning or scent) would be slightly favoured, let the difference be ever so small, and would tend to live longer and to survive during that time of the year when food was shortest; they would also rear more young, which young would tend to inherit these slight peculiarities. The less fleet ones would be rigidly destroyed. I can see no more reason to doubt but that these causes in a thousand generations would produce a marked effect, and adapt the form of the fox to catching hares instead of rabbits, than that greyhounds can be improved by selection and careful breeding. (Darwin and Wallace, 1958, p. 120)

In some respects this is quite a sophisticated description. Darwin notes that the positively unfit individuals will be destroyed, but he speaks in terms of positive advantages increasing the chances of survival and (a very important addition) reproduction. These points reappear almost word for word in the *Origin of Species* (Darwin, 1859, p. 80 and pp. 90–91; 1872 edn., p. 63 and p. 71).

However, elsewhere in the essay and in the early editions of the *Origin* Darwin tends to imply that there will only be a very small number of favorable variants in any generation (Darwin, 1859, p. 80). This raises the very serious issue of the level of fitness needed to make a difference. Given that, as Darwin often insists, the struggle for existence follows from the vast overproduction of offspring in every generation, it follows that many "normal" individuals will die—but if the fittest are only a tiny minority, surely some of the normal individuals will survive by chance too. Darwin was well aware that on the rare occasions when food becomes plentiful, there is a rapid but temporary increase in the population, implying that now almost every individual is able to survive and breed. By visualizing variation as the production of only small numbers of differing individuals within a population composed mostly of "normal" types, Darwin cuts himself off from the opportunity to develop the insight that the degree of fitness influences the chances of survival and reproduction. This is still the case even when (in response to the 1867 critique by Fleeming Jenkin) he insists that small individual differences are enough to make a difference. He had no concept of a continuous range of variation that would affect the chances of each individual surviving and reproducing.

One area where Darwin does clearly recognize that variation affects the chances of reproduction is in the very special case of the relationship between flowers and the insects that fertilize them. Here it is obvious that the shape and color of the flower, for instance, determines not whether the plant lives or dies, but whether it will reproduce successfully—and this is quite enough to let selection operate. Exactly the same point emerges in the theory of sexual selection, developed at length in the later parts of the *Descent of Man* (1871). Here the contrast between selection based on survival and selection based on superior reproductive powers becomes evident in cases such as the peacock's tail. This may be a disadvantage in the basic struggle for existence, but it confers a reproductive advantage which outweighs the risk to the male birds. But even in these cases, if Darwin was thinking in terms of favorable variants being extremely rare, he was unlikely to develop the insight that one might correlate the range of variation with the chances of reproduction.

Returning to the question of what determines who will survive, there was another important development in Darwin's thinking which arose partly from his acceptance of the relentless nature of the struggle for existence even in a stable environment. As Janet Browne (1980) points out, his "principle of divergence" arose from his realization that there might be different ways of becoming "fitter." By changing their behavior to exploit a new way of adapting to the environment, individuals could gain an advantage because they were not competing directly with others seeking a different way of getting access to resources. This would encourage the population to split into distinct varieties, each adapting to a different lifestyle, and the pressure of the struggle for existence would force the varieties to specialize more and more, effectively driving them apart until they became distinct species.

There was an important corollary of this new model of struggle, because Darwin was well aware that the balance of nature in any one area was constantly likely to be disturbed by the immigration of newcomers. He now realized that in these circumstances there would be a struggle between the "invaders" and any of the previous inhabitants which exploited the same resources. The less favored species would gradually diminish in numbers and would eventually go extinct. Here was a tool that could be used to explain many of the facts of geographical distribution, including the general impression shared by Darwin and most of his contemporaries that species evolved in the "superior workshops of the north" would invariably displace southern species if they got a chance to invade their territory (Darwin, 1859, p. 380; 1872 ed., p. 340). Note, however, that he had now shifted the level of struggle back to where Malthus himself had imagined it—between competing groups rather than individuals. Natural selection could explain the rise and fall of animal types in the course of the earth's history in terms of superior forms expanding their territory at the expense of less-developed earlier types, until they in their turn were swept away by newer and higher forms. The implication that "fitness" might mean something more than local adaptation was emerging even in Darwin's own writings.

At the same time, the notion of "group selection" could also be used to modify our conception of what might count as the fittest characters for the individual within the group. This was particularly important for Darwin when he came to explain human moral instincts in *Descent of Man*. At first sight it might seem that cooperative or what we would call altruistic instincts would be eliminated by natural selection because they put the individual at a disadvantage. But Darwin pointed out that if we think in terms of animals living in social groups, altruistic instincts might actually be of benefit because a group whose individual members cooperated together might displace rival, less well-organized groups in the struggle to occupy territory. The benefits of living in a successful group outweigh the disadvantages of occasionally sacrificing your own interests to those of other members. For Darwin such cooperative instincts were the foundation stone of human morality, now overlaid with much rationalization to make them seem something that lifts us above the brutes.

The fact that struggle can be imagined to take place at various levels must also be borne in mind when considering Wallace's descriptions of natural selection. I have long argued that Wallace's original conception of natural selection in his 1858 paper was based on a model of competing varieties, in the sense of sub-species, rather than competing individuals—he openly writes in terms of the increase in the numbers of the favored varieties and the decrease of the unfit until they become extinct. Other writers have argued that Wallace's version of selection also presupposes an absolute standard of fitness required by the environment. On such a view, it would only be when the environment changed that a new standard of fitness would emerge, favoring some varieties at the expense of others. This would be similar to Darwin's original concept, which also postulated selection coming into play only when the environment changed. This view is still evident even in Wallace's later writings, for instance in his *Darwinism* (1889, p. 103). The one area in which Wallace went beyond Darwin was in his recognition of the extent of variation normally existing within any wild population. In the letters quoted at the start of this chapter he openly criticized Darwin for implying that there would only be a small number of favored variants in any generation. Wallace used actual measurements of variation within large samples of animals to show that there was a continuous range of variation for most characters, with most individuals clustered around the mean and smaller numbers at either extreme. He even uses a primitive diagram of a bell-shaped normal curve of distribution. Had Wallace been able to think in terms of this variability affecting the chances of each individual surviving and, more important, reproducing, he would have anticipated the major developments that would soon take place in biologists' thinking about natural selection.

On a wider scale, Wallace played an important role in extending the broader notion of struggle which Darwin had introduced to explain the relationships between species. In particular he engaged in a major project to explain the facts of geographical distribution in evolutionary terms, culminating in his book *The Geographical Distribution of Animals* (1876). Like Darwin, Wallace argued that the fauna of most regions could be seen as a composite of species, many of which had evolved elsewhere and had migrated into new territories, sometimes co-existing with, but often displacing the original inhabitants by driving them to extinction. Also like Darwin, he postulated a general movement of superior types evolved in northern regions into the southern continents. Following this model, late-nineteenth-century biogeography emerged as an active field of research whose explanatory paradigm resounded with the metaphors of imperialism. More highly evolved animals and plants "invaded" or "colonized" new territories, when they "conquered" or "dominated" the original inhabitants (Bowler, 1996, chaps. 8 and 9).

Here the notion of the survival of the fittest was envisioned playing itself out on a global scale, matching the scale of the great empires established by the European powers and (less formally) by America.

MODERN DARWINISM

We shall return to the topic of imperialism when we consider the social implications of Darwinism more explicitly below. But we must first complete the story of how the somewhat loosely defined concept of natural selection introduced by Darwin and Wallace was transformed into the modern synthesis of genetics and natural selection. Darwin's insight that an individual organism's chances of surviving and reproducing are determined by how well adapted it is to its environment and lifestyle was supplemented by Wallace's later recognition that every population exhibits a continuous range of variation for any character. This was achieved initially by the biometrical school of research led by Karl Pearson and W.F.R. Weldon. In the early twentieth century their pre-Mendelian ideas about how an organism's characteristics were transferred to the next generation were reinterpreted in terms of genetics by biologists such as R.A. Fisher and J.B.S. Haldane. In their model it was the genes which determined an organism's characteristics, and the fitness of those characters which determined how effectively the corresponding genes would be copied into the next generation. The survival of the fittest would be translated into something like "the superior reproductive powers of the better-adapted."

It is worth noting in passing that there were many biologists who did not participate in these developments. Indeed, there were many who accepted the general idea of evolution but completely repudiated the Darwinian explanation of it. The fact that Spencer introduced the term "survival of the fittest" has tended to obscure the fact that he saw natural selection as only a secondary mechanism of evolution, less important than the Lamarckian process of the inheritance of acquired characteristics. He continued to defend Lamarckism in the later decades of the century, yet did so within his general worldview which saw the struggle for existence as the main driving force of evolution and human progress. In the Lamarckian theory, animals adapt themselves to changes in their environment and pass their improvements on to their offspring. In the much-quoted example of the giraffe, it's not that the short-necked giraffes are eliminated in every generation—all the animals stretch their necks to reach the leaves of the trees on which they feed and get longer necks as a consequence. In effect, all the animals strive to become fitter, and Spencer seems to have thought that most of them would succeed and would pass their improvements on to their offspring. For him the struggle for existence was a driving force for Lamarkian evolution as well as for natural selection, because it was the pressure of competition which encouraged the animals to make the effort needed to adapt themselves to new conditions.

Other Lamarckians such as the novelist Samuel Butler realized that Lamarckism offered a vision of evolution that would allow one to dispense altogether with the need for the struggle for existence. A whole generation of thinkers, and many biologists too, sought to develop a form of evolution which did not require one to believe that nature was the scene of constant struggle and death. The most extreme versions of this approach denied any role for adaptation in evolution. For the advocates of orthogenesis and of saltationism (evolution by sudden jumps), there was no pressure from the environment and species could evolve characters that were useless or even harmful. If the pressure from whatever organic processes generated new variations was strong enough, it would flood the population with a new character which would become established even if it were positively maladaptive. Paleontologists such as Alpheus Hyatt looked at what appeared to be bizarre structures evolved by some species in the fossil record and concluded that these were useless features being produced by some internal, biological pressure to evolve. Hyatt worked on cephalopods, but another popular example was the huge antlers of the so-called Irish elk, which were supposed to have become so large that the animals could barely hold their heads up. These maladaptive characters may have caused the eventual extinction of the species, but the fact that they could have developed to such an extent was a clear indication that the adaptive pressures imposed by the environment were relatively slight. Advocates of evolution by saltation, including William Bateson (1894) and Thomas Hunt Morgan (1903) also maintained that the whole Darwinian program of explaining evolution in terms of adaptive pressure was an illusion. For them there was no such thing as the survival of the fittest—almost all new characters could survive and reproduce, whether beneficial, neutral, or even mildly harmful.

Bateson and Morgan went on to become pioneers of modern genetics, and it could be argued that their reliance on laboratory breeding experiments (where, of course, there is no struggle for existence) blinded them to the role of adaptation in the wild. But their involvement in the explosion of interest in Mendel's laws of heredity which occurred after 1900 highlights the problems that had to be overcome by the biologists who created the synthesis of genetics and Darwinism. Darwin's formulation of his theory was limited both by his restricted model of variation and by his inability properly to articulate what it was that the organism gained from being "fitter," i.e., better adapted. An interesting commentary on these limitations can be seen in the writings of the evolutionary psychologist Conwy Lloyd Morgan (1890, pp. 91–96). He argued that Darwin had been confusing two very different processes, one in which the most "satisfactory" individuals were selected for breeding, the other in which the most "unsatisfactory" were eliminated.

At one level, Darwin had been aware of the difference because he often thought in terms of the comparison between two modes of artificial selection by humans. In deliberate selection to achieve a preconceived goal, as when trying to create a new breed of pigeons with a particular character, the breeder picks out only those individuals in which the required character is well developed. All the rest are not permitted to breed. Lloyd Morgan argued that the equivalent process in the wild was a genuine natural selection. It could be seen at work in, for example, the development of a particular color of flower because insects favored that color and thus promoted the reproduction of those plants in which it was best exhibited. Here fitness was defined by a particular form of a character which conferred not better chances of survival, but better chances of reproducing. Sexual selection was an even more obvious version of the process, depending for instance on the preference of the hen bird for the most brightly colored males. But Morgan insisted that this was a significantly different process from that which occurred in what he called "'natural elimination,'" when those individuals with a positively harmful character are weeded out in the struggle for existence. In the production of protective camouflage or resemblances, any individual which is more conspicuous than the norm will be picked out by predators. Of course the standard defined by the norm will itself change as evolution proceeds. Morgan correctly identified the key process as differential elimination and reproduction, but he does not seem to have realized that here as well an in natural selection (as he defined it) there might be a continuous variation in the chances of survival. It wasn't just that the least fit were eliminated, with all the rest surviving. In fact throughout the range of variation, the closer an individual is to the most-adaptive end of the range, the better his or her chances of surviving and reproducing. This was the key to the next step in the clarification of how natural selection could be modeled.

The first steps in this process were taken by members of the so-called biometrical school of Darwinism led by the biologist W.F.R. Weldon and the statistician Karl Pearson. Weldon developed the insight hinted at in Wallace's crude models of the continuous variation that could be observed in samples of wild populations. In the course of the 1890s he measured the variation in various characters in a number of species including snails and the crabs in Plymouth harbor. Pearson supplied the mathematical techniques that could be used to analyze the data. They showed that in many cases there was indeed a continuous range of variation that could be represented by the bell-shaped normal distribution—most individuals are clustered around the mean value for the character, with smaller numbers stretching out toward the extremes on either side. The distribution of height in the human population would be an equivalent phenomenon. Weldon and Pearson were both, somewhat unusually for the time, committed Darwinians, and they realized that the variations in the character would have implications for the survival and the reproduction of the individuals. Those best adapted to the environment (the fittest) would tend to survive more easily in the struggle for existence. In a stable situation where the species is well adapted to a fixed environment, those at the mean value of the range of variation will be the fittest, and selection will be constantly whittling away at those on either side of the range, especially those at the very extremes. But if the environment changes to favor one end or the other of the range, the individuals at that end will survive and reproduce more, and that character will be more strongly represented in the next generation. Conversely, those at the maladaptive end of the range will be killed off, or at least will be less likely to reproduce. The result will be that the curve of variation will become skewed toward the fitter end and over many generations the mean value of the whole range will shift as the whole population adapts.

Although Pearson was working with a pre-Mendelian theory of heredity, this did not prevent him being able to show how the differential rates of reproduction would drive the population as a whole toward better adaptation. It didn't matter what the actual mechanism of heredity was: As long as characters were transmitted to some extent from parent to offspring, the selective effect would work. Crucially, the analysis transforms the whole logic of natural selection by placing the focus on the rate of reproduction, not on survival itself. Of course survival is involved, because you can't reproduce if you are dead, but the mechanism of selection will work even if we assume that the mildly unfit are able to survive in some circumstances. So long as their disadvantage interferes with their chances of reproducing successfully, their character will be diminished within the population of future generations. Weldon was actually able to show that the curve of variation for the size of the crabs in Plymouth harbor was skewed, indicating a selective effect that was driving the population as a whole toward increased size. He was reluctant at first to speculate on what was causing the effect, but one obvious possibility was that the harbor was being dredged, thus muddying the water, and bigger crabs were better able to cope with the new environment. The biometricians' work clearly defined fitness in terms of adaptation to the local environment—crabs elsewhere would be unaffected. We shall see that when it came to fitness in the human population, however, Pearson had to adopt different criteria. The problem was that what were conventionally described as "unfit" humans—the poor in the slums of the great cities—were reproducing more than the superior types who made up the professional classes. Whatever his confusions over the definition of fitness, though, Pearson's support for the eugenics movement (discussed later) is a clear sign that he recognized the distinction between survival and reproduction, and understood that it was the latter which drove changes in the population.

The first generation of geneticists assumed that Mendel's model of heredity based on distinct character differences rendered the kind of continuous variability studied by Weldon and Pearson irrelevant for evolution. Pioneers such as Bateson and Morgan thought all significant characters were created as units by mutation and then bred true—and would be reproduced whether or not they conferred any advantage or disadvantage to the organism. The synthesis between the biometricians' more sophisticated formulation of the selection theory and genetics was made possible when a new generation of biologists recognized that most genes code for minor effects, and that many different genes can influence a particular physical characteristic. New mutations feed into the "gene pool" of the population, generating the range of variations that the Darwin and his later followers had always seen as the raw material of natural selection. Applying the biometricians' model of differential rates of reproduction for individuals at each point in the range of variation, we arrive at what Ronald Aylmer Fisher called in the title of his 1930 book *The Genetical Theory of Natural Selection*. Fisher's theory ignored the possibility that genes might interact with one another and worked with a simple model in which each gene coded for a particular variant form of a character. So the range of variation was made up of a spectrum of states coded for by a series of genes. Following the model pioneered by Pearson, it was then a straightforward procedure to assign a degree of "fitness" to each gene corresponding to the adaptive advantage or disadvantage it conferred. The "fitness" determined the organism's chances of reproducing, which in turn determined the rate at which the gene would be copied into the next generation. Natural selection worked by gradually changing the frequency of individual genes within the population according to whether or not they conferred an advantage in reproduction. The idea that it was the superior reproduction of the fittest that counted, not their mere survival, remained central to the new model of natural selection.

There were disagreements over how great the effect of a single gene might be. Fisher assumed a population composed of genes conferring only slight differences in reproductive power, which implied a very slow rate of selection. But J.B.S. Haldane (1932) noted that there were cases where the difference might be much greater, allowing selection to proceed at a rapid rate. The classic example was that of industrial melanism in the peppered moth *Amphidasys* (now *Biston*) *betularia*, where a dark or melanic form had rapidly replaced the normal gray moths in areas where the trees were blackened by industrial pollution. Haldane argued that the melanic form must have had a rate of reproduction 50 percent higher than the normal genes for the replacement to have occurred so rapidly. The presumption was that the dark color served as camouflage protecting the moths against predators—this was what conferred their superior "fitness." It is significant that in later controversies over the validity of the selection theory, H.B.D. Kettlewell's pioneering studies of the peppered moth have been challenged by critics anxious to show that this classic example of how a beneficial gene can rapidly dominate a local population does not hold up to scrutiny. The argument centers partly on the evidence provided to suggest that protection from predators really does benefit the melanic forms. This in turn points to a problem which had bedeviled modern Darwinism: the difficulty of showing that there is an actual difference in "fitness," i.e., some adaptive benefit, which explains why certain genes increase their frequency in the population.

The problem also arose in the context of the theory of sexual selection which, after a long period of neglect, became a major focus of attention within modern Darwinism. Now that it was clearly recognized that survival was only one factor influencing the outcome of the processes determining an organism's fitness, modern biologists could appreciate the insight that lay at the heart of Darwin's concept of sexual selection. A character that was apparently disadvantageous as far as survival was concerned might nevertheless be developed because that disadvantage was outweighed by the benefits it conferred in attracting mates. Hence gaudy displays such as the peacock's tail. But sexual selection only encouraged the temptation to define fitness in terms of the genes' chances of being transmitted to future generations. It was a lot easier to measure gene frequencies than to sort out the complex adaptive and behavioral factors which determined fitness in the original Darwinian sense. It is often surprisingly difficult to identify exactly what benefit a particular character confers—the fact that even the case of industrial melanism (where the camouflage effect seems almost self-evident) has been challenged illustrates this point. In practice, many population geneticists simply assume that a gene that is increasing its frequency must code for an advantageous character, even if they are unable to be sure what it was. Fitness becomes defined in terms of reproductive success.

The most obvious expression of the problems arising from this oversimplification is the well-known argument that the theory of natural selection is meaningless because it is based on a tautology. If we cannot define "fitness" except in terms of actual survival, then "the survival of the fittest" means no more than "the survival of those who survive."

More properly, one might say "the superior reproductive powers of the fit" means no more than "the superior reproductive powers of those who reproduce more." Darwinists retort that fitness is in fact defined by some adaptive benefit in a particular environment, which allows the individuals to survive better and reproduce more. In principle it should be possible to work out what the adaptive benefit is, but since in practice it is often difficult to do this, Darwinism is seen by many critics as vulnerable to the charge that the whole notion of "the survival of the fittest" is meaningless. Fortunately there are more recent studies, including some based on the finches of the Galàpagos islands which Darwin studied, which show how changing conditions do in fact have an immediate impact on the characters which dominate the population.

SOCIAL DARWINISM

At several points in the above story we have noted the social implications of Darwin's theory, and our analysis of the concept of "the survival of the fittest" cannot ignore the extent to which it has been applied to human affairs. Spencer's evolutionary philosophy was as much social as it was scientific, and it would have been surprising if his endorsement of Darwin's selection theory had not encouraged others to apply it to human history. It has been widely assumed that the late nineteenth century was dominated by a "Social Darwinism" which saw the struggle for existence as the driving force of progress in the social as well as the biological sphere. But some historians have challenged the assumption that the late nineteenth century was dominated by those who adopted "the survival of the fittest" as their motto for human interactions. Some forms of so-called Social Darwinism turn out to have more complex roots in biological thinking. Even where the struggle metaphor was applied, it was exploited to justify several different levels of competitive behavior, individual, national, and racial. Even more seriously, the attempted application of the selection metaphor to human affairs raises all the issues about the definition of "fitness" that bedeviled the biological debates.

The first point to note is that we cannot simply equate appeals to the "struggle for existence" as the driving force of progress with the claim that the process works through the "survival of the fittest." Although Spencer coined the latter term, he always assumed that for the higher animals, including humans, natural selection was less important than the Lamarckian mechanism of the inheritance of acquired characteristics. His emphasis on the role of struggle was intended to show that pressure from the limitations imposed by the environment did more than merely kill off the unfit—it also encouraged all members of the population to strive to improve themselves. The benefits arising from their efforts could then be transmitted to their offspring. When Spencer applied his worldview to human progress, exactly the same effects were supposed to be at work. A competitive society was most efficient in promoting progress because it encouraged everyone to try to better themselves—or face the consequences of failure. Only the very worst, the least fit, would actually be weeded out. The wave of enthusiasm for Spencer's philosophy during the late nineteenth century was based on the assumption that competition was both natural and beneficial, but the basis of social evolutionism was more Lamarckian than Darwinian. All too often, though, Spencer's self-help ideology has been misinterpreted by historians as a simpleminded application of the metaphor of the survival of the fittest inspired by Darwinism.

Whether the mechanism of evolution is Lamarckian or Darwinian, the whole process of transferring metaphors between biology and social theory illustrates the problem of what we mean by "fitness." For both Darwin and Spencer, the original definition within biology had to depend on adaptation to the local environment or to the species' lifestyle. But both of them tended to assume that there was also a more general sense of fitness which increased as evolution generated progress toward higher levels of organization. It was easy enough to assume that intelligence (presumably determined by brain size) tended to increase because, in the long run, more intelligent animals were more likely to succeed in the struggle for existence. Human intelligence represented the natural outcome of such a process, and it seemed obvious that further progress in this area might occur if society encourages the same interactions between individuals as were normal in the state of nature. But, as both Darwin and Spencer were aware, we are adapted to living in a social environment as well as a natural one, and the characters developed to facilitate living in a group may be different to those based on survival as an individual. Fitness in human terms becomes defined by those characters we think are best for the individual who is well-adapted to a progressive society. But this has now become a matter of choice, leaving room for disagreements over what counts as fitness fueled by different ideologies. There is also the very real question of whether those with what we think to be superior characters actually have a better chance of transmitting those characters into future generations.

The "robber barons" of American capitalism, including industrial magnates such as John D. Rockefeller, were certainly enthusiastic supporters of Spencer, because his philosophy seemed to endorse the unrestrained free-enterprise competition in which they had come out on top. But their endorsement revealed the artificiality of the analogy between biological and human social and economic competition. We might chose to define fitness purely in terms of economic success, allowing us to argue that struggle promotes progress by weeding out the least efficient firms and individuals. But there is no biological equivalent of the takeover and the monopoly—indeed the whole point of the latter is to eliminate further competition. And at the level of personal characteristics, how can we equate success in society with the success of those animals which survive in the struggle for existence? The inheritance of wealth adds a whole new dimension which cannot be equated with the transmission of biological characters and threatens to undermine the effects of competition in future generations. This is why one high-profile American capitalist, Andrew Carnegie, used his accumulated wealth to benefit ordinary people (e.g., by founding public libraries) instead of passing it on to his own children. They would have to prove their fitness by starting from scratch, rather than by living as parasites on the family fortune.

Serious problems thus arose from the attempt to define what counts as "fitness" in the human situation. In principle, Spencer's social philosophy was a form of the old Protestant work ethic: What made one a successful member of society was application of the traditional virtues of thrift, industry, and initiative. But surely an unrestrained struggle is equally likely to promote purely selfish attitudes, including many that we do not normally think of as virtuous or socially valuable. In a purely selfish society, anything that helps one to succeed is defined as good, equating to fitness in the Darwinian model. Darwin himself responded with dismay to a newspaper critique which claimed that his theory justified Napoleon and every cheating tradesman (letter to Charles Lyell, 4 June 1860, in Darwin, 1984, vol. 8, p. 189). Spencer had argued that such ruthless and underhanded behavior is penalized in the social environment and is thus kept to a minimum, but the comments Darwin reported show that some people remained unconvinced. His own appeal to group selection to explain the emergence of social and altruistic instincts would at best set up a balance between those instincts and more selfish behavior. The late-nineteenth-century thinkers who looked for alternatives to Darwinism were motivated by the fear that in a purely competitive society, the worst kinds of behavior would be promoted far more effectively than the social virtues.

An equally serious problem emerged as the second generation of Darwinists began to realize that it was not the survival of the fittest which counted, but their superior reproductive powers. It was easy enough to imagine that among primitive humans and modern "savages" those who were successful within the tribe would tend to have more children. But this was by no means the case in a modern, industrial society, and by the end of the nineteenth century it was becoming clear that those middle-class persons defined as the "fittest" in terms of the ideal social virtues were not having more children than the less-favored individuals who were banished to the slums of the great cities. On the contrary, it was feared, the least fit members of society were breeding like rabbits, while the middle-class professionals were limiting the size of their families because of the cost of raising the children properly. Far from promoting the biological and mental progress of the human race, the free-enterprise system was allowing the worst kind of characteristics to flourish, threatening to undermine everything that Western society had achieved. This argument was the basis of the eugenics movement, which replaced individualistic social evolutionism as the most active social application of Darwinism in the decades around 1900. The whole point of eugenics was to replace the purely natural system of reproduction with a system of artificial selection applied to humanity. The fittest members of society—assumed to be the middle classes—would be encouraged to have more children, while the least fit—especially the feeble-minded—would be prevented from breeding, if necessary by artificial sterilization. One active proponent of eugenics was Karl Pearson, whose support for the theory of natural selection as the mechanism of biological evolution was linked to a clear appreciation that the process was not working within modern society. Pearson threw his weight behind the campaign to apply artificial selection to limit the damage that unrestricted breeding was doing to the character of the British nation. Nature was no longer to be allowed to determine who was fit and who should breed.

It is important to note that Pearson's enthusiasm for protecting the biological character of the British people by state intervention was coupled with a general endorsement of a highly managed rather than a free-enterprise society. The reason why he was so concerned to have a centralized society and economy was because he believed that in the modern world the real struggle for existence was taking place between nations and races, not between individuals. Pearson was an imperialist who believed in the British empire as a symbol of the nation's superiority over its rivals. This was the period when the European nations were engaged in a race to colonize as much of the world as possible. The resulting antagonisms generated an increasing militarism which culminated in the outbreak of World War I. The Germans too were convinced of the superiority of their own nation and its culture, and the idea that the rival nations were locked into a struggle for existence to determine which was the fittest became commonplace. Fitness in this case would be determined by success in the struggle to dominate Europe and to acquire external colonies. The German general Friedrich von Bernhardi openly appealed to the Darwinian theory to justify his call for Germany to assert her role as the superior European culture, and his position was endorsed by the leading evolutionary biologist Ernst Haeckel. The claim that Haeckel's version of Darwinism played a role in the creation of German nationalism, and in the subsequent rise of Nazism, is one of the most controversial issues in the history of Social Darwinism.

There was another way of applying Darwinian rhetoric in the context of late nineteenth-century imperialism, and this was in the area of the relationship between the human races. The expansion of European power across the globe fueled a growing sense that the white race was superior to the inhabitants of the regions now being conquered and colonized. The rhetoric of Social Darwinism could be used to justify the process by arguing that it was natural for the "higher" forms of humanity to displace the "lower" ones in a global struggle for existence. Progress had always depended on the newer and more highly evolved species displacing the older, less advanced forms. In this way the wider sense of competition embodied in the Darwinian approach to biogeography could be applied to the various branches of the tree of human evolution. Although there were many different ideas about how the human races had evolved, most of them took it for granted that the white race had advanced further from the ancestral ape than the others. In effect, the races were not just geographical varieties of humanity; they corresponded to different stages in the process of upward evolution. Applying the Darwinian model of biogeography to this situation allowed one to see the displacement and even the extermination of the older, less developed races as natural, inevitable, and—if progress was to be maintained—desirable. The fact that it was the inhabitants of the northern regions which were expanding into territories further south reinforced the analogy, since the evolutionists had always assumed that the more challenging climate of the north had stimulated the most rapid development of higher characteristics.

Karl Pearson linked this form of social Darwinism to his eugenic concerns for the biological future of the white race. In the chapter on Darwinism in his *The Grammar of Science* he openly advocated the displacement of races which were incapable of properly exploiting the resources of the territory they occupied by the superior Europeans.

> It is a false view of human solidarity, a weak humanitarianism, not a true humanism, which regrets that a capable and stalwart race of white men should replace a dark-skinned tribe which can neither utilise its land for the full benefit of mankind, not contribute its quota to the common stock of human knowledge. The struggle of civilized man against uncivilized man and against nature produces a certain partial "solidarity of humanity" which involves a prohibition against an individual community wasting the resources of mankind. (Pearson, 1900, p. 369)

There was no doubt as to what counted as "fitness" in Pearson's eyes: It was defined by the values of white, industrial Western civilization, and those races which did not measure up would suffer the consequences.

Others saw the process of racial competition as a continuation of the progressive steps by which humanity had risen above the apes. In the early twentieth century it was widely assumed that human evolution consisted not of a single line of development but of numerous separate lines evolving in parallel. But some lines had certainly advanced further than others, and whenever the members of a higher type were able to invade territory occupied by a lower, the latter had been driven to extinction (on theories of human origins see Bowler, 1986). In his *Ancient Hunters* of 1911, the palaeoanthropologist W.J. Sollas argued that such a process of successive eliminations was one of the driving forces of human evolution, and suggested that the old adage of "might is right" reflected the basic character of the evolutionary process.

> What part is to be assigned to justice in the government of human affairs? So far as the facts are clear they teach in no equivocal terms that there is no right which is not founded on might. Justice belongs to the strong, and has been meted out to each race according to its strength; each has received as much justice as it deserves. What perhaps is most impressive in each of the cases we have discussed is this, that the dispossession by a new-comer of a race already in occupation of the soil has marked an upward step in the intellectual progress of mankind. It is not priority of occupation but the power to utilize, which establishes a claim to the land. Hence it is a duty which every race owes to itself, and to the human family as well, to cultivate by every possible means its own strength; directly it falls behind in the regard it pays to this duty, whether in art or science, in breeding or in organization for self-defence, it incurs a penalty which Natural Selection, that stern but beneficent tyrant of the organic world, will assuredly exact, and that speedily, to the full. (Sollas, 1911, p. 383)

As with Pearson, the definition of what counted as fitness was clear, and gave the European powers the right to displace those races which did not measure up to its own standards.

The classic example of this process of natural selection at the racial level was the extinction of the supposedly brutal Neandertals by the modern humans who had invaded Europe during the Palaeolithic period. The anatomist Arthur Keith used this to explain and, in effect, justify the European expansion into the territories occupied by the "lower" races of today. If one wanted to understand what was happening to the Australian aborigines, Keith argued in his *Antiquity of Man* of 1915, one need only look at what had happened to the Neandertals—they had been wiped out by a "more virile" type of humanity (Keith, 1915, p. 136). The "lower" races of the present were depicted as relics of earlier stages in human evolution driven to the margins of the inhabited world, where they had survived, protected by their relative isolation until the arrival of the triumphant Europeans. Keith went on to develop a whole theory of human evolution which focused on racial and tribal competition as the driving force (Keith, 1948). Despite experiencing both world wars, he still saw war as a natural and inevitable part of the human situation and argued that without some form of conflict the species would stagnate.

This application of the principle of the survival of the fittest to the interaction between the races was probably the most consistent and the most widely accepted form of Social Darwinism. There were always some ideological critics of Spencer's individualist model linking the struggle for existence to the free-enterprise system, but very few Europeans and Americans in the decades around 1900 were able to free themselves from the assumption of white supremacy. Here the idea that one could define an absolute standard of "fitness" which allowed the whites to be ranked above the other races was taken for granted. So too was the assumption that this superiority allowed and justified the elimination or enslavement of the less advanced forms of humanity. The Nazis' blind acceptance of the superiority of the Aryan race was the most aggressive application of this way of thinking, but it drew upon a tradition of white supremacy that had been widely accepted for many decades, and that had widely routinely justified the expansion of European power via the analogy with biological Darwinism. There was only one thing that troubled the exponents of this ideology, and that stemmed once again from the awkward fact that it was the rate of reproduction which drove the process of natural selection, not the achievement of some purely human standard of fitness. Just as the supporters of eugenics worried about the excessive breeding capacity of the "unfit" classes languishing in the slums, so some white supremacists worried about the "Yellow Peril"— the possibility that the rapid expansion in the numbers of the Oriental races might put them in a position to swamp the whites despite the latter's superior technology and civilization. Here was recognition that by nature's standards the values developed by human cultures might not represent the best measure of fitness, at least in the long run.

In the end, the modern synthesis of genetics and Darwinism played an important role in undermining the assumptions that had driven the ideology of white domination. The whole idea of distinct racial types crumbled as genetics showed that there was no biological foundation for the degrees of difference postulated by the older models of human origins. Nor was it plausible to depict the inhabitants of some regions as relics of earlier stage in human evolution preserved like living fossils into the present. There is indeed some genetic variation between the populations of different regions, and some of the differences represent adaptations to the conditions of the areas on which ancestral populations evolved. But the suggestion that there are genetic foundations for higher levels of intelligence in certain racial groups is widely condemned as a relic of the old prejudices. The concept of the survival—or more properly the superior reproductive power—of the fittest remains as the cornerstone of modern Darwinism, but we are now well aware of the difficulties that plague our efforts to define the concept of fitness, both in biology and in human affairs. Strictly speaking, "fitness" means the level of adaptation to the environment or to the lifestyle of the species (which includes the social environment for those organisms which live in groups). Identifying exactly how a certain character provides such an advantage can often be surprisingly difficult in the wild, which is why "fitness" has often been defined instead in terms of the actual reproductive success which, strictly speaking, is a consequence of the adaptive advantage. Such an oversimplification runs the risk of undermining the very basis of the Darwinian theory. It does, however, illustrate the most important transition which took place in the development of the theory: the recognition that it is not the "*survival* of the fittest" that counts but their superior reproductive power.

To the opponents of modern Darwinism, however, the concept of the survival of the fittest encapsulates the theory's reliance on a worldview in which everything is explained in terms of ruthlessness, suffering, and death. The fact that the theory was so often used in attempts to justify policies which violate the traditional foundations of Christian morality is a sign of its origins in a purely materialistic worldview. Modern Darwinians have to work hard to free their theory from an image which associates it with the glorification of attitudes which most of us now find repellent. On the other hand, those who see nature as a manifestation of divine wisdom and benevolence would do well to remember that the inevitability of a struggle for existence arises from the sheer reproductive capacity of living things, which, as Malthus (himself a clergyman) pointed out is a fundamental characteristic of how they were created. Darwin merely borrowed this insight and applied it in a new way—although as a consequence his name has become a symbol of the threat it poses to the biblical account of life's origin and purpose.

FURTHER READING

Bannister, Robert C. *Social Darwinism: Science and Myth in Anglo-American Social Thought* (Philadelphia: Temple University Press, 1979).

Bowler, Peter J. *Evolution: The History of an Idea*. 3rd ed. (Berkeley: University of California Press, 2003).

Browne, Janet. *Charles Darwin: Voyaging* (London: Jonathan Cape, 1995).

Browne, Janet. *Charles Darwin: The Power of Place* (London: Jonathan Cape, 2002).

Desmond, Adrian, and Moore, James R. *Darwin* (London: Michael Joseph, 1991).

Gayon, Jean. *Darwinism's Struggle for Survival: Heredity and the Hypothesis of Natural Selection* (Cambridge: Cambridge University Press, 1998).

The Neandertals

Marianne Sommer

We are here concerned with the Neandertals as an icon of human evolution. In fact, the Neandertals might be *the* icon of human evolution. From the time their remains were discovered, they have been bones of contention. As the notions of evolution and of a human antiquity by far transcending the few thousand years arrived at from the chronology of the Bible had only just begun to dawn on the scientific horizon, Neandertal bones were implicated in the debates around these fundamental issues. It was only three years after the first Neandertal discovery that these notions were consolidated, both in Charles Darwin's *On the Origin of Species* (1859) and the consensus on human prehistory. As the first fossil bones discovered that eventually came to be accepted as such, the Neandertals raised troubling questions about humankind's place in nature, and about our relation to (the rest of) the animal kingdom. They were instrumental in the redrawing of boundaries, such as those between animal and human, nature and culture, and archaic/primitive and modern/advanced. Why are these boundaries so important? And who was and is assigned a place on which side of the line?

Comparable to the great apes, which from the first encounters with Westerners in the course of the age of discovery troubled observers by their similarities to humans, the Neandertals brought to the fore questions about human uniqueness. Seventeenth- to nineteenth-century comparative anatomy was inscribing difference into the bodies of apes, and so were anthropologists often approaching the Neandertals in the nineteenth-century and beyond. Like the great apes and non-*sapiens* hominids (in general), the Neandertals are therefore denizens of a boundary zone that, though shifting, is relevant to our understanding of self. They are uncomfortably similar while being disturbingly different, demanding new answers to the question what it means to be human. Among the morphological and mental criteria for distancing these beings from us, perfect bipedalism, language, self-awareness, and elaborate tool and art production go a long way back. As the philosopher Raymond Corbey observes,

> the history of the anthropological disciplines to a considerable degree has been an alternation of humanizing and bestializing moves with respect to both apes and humans, a persistent quest for unambiguousness and human purity, and an ongoing rebuff of whatever has threatened to contaminate that purity. (Corbey, 2005, p. 1)

In this rebuffing, the medieval conception of the so-called *scala naturae* has played a significant role. The Great Chain of Being is a hierarchical and static ordering of animate nature from the simplest to the most complex living form. In this ranking, humans were just below the angels and sharply differentiated from the animals below by their unique creation in God's image. Even with the advent of a historical understanding of the cosmos, this hierarchical arrangement was not entirely abandoned, although it became temporalized. It is still evident in the evolutionary thinking of the nineteenth century, for example in the notion of a missing link between ape ancestors and human descendents, but it also reverberated in the perception of certain peoples as simultaneously contemporary humans and relics of a more primitive past. "Savage races" were ranked between the apes and modern "civilized races" in the static worldview of the eighteenth century and between "fossil human races" and contemporary "advanced races" in the progressionist thinking of nineteenth-century evolutionists.

It is this demarcation line between "them"—the non-human, sub-human, pre-human, lower human—and "us" that was central for the history of the Neandertals. As implied in the quote above, they were shuffled back and forth between the human and non-human side of the line. What seems to render the Neandertals particular among the boundary beings is their enormous appropriation in popular culture. In the early decades of the twentieth century, they attained a popularity that by far transcended their study, in the laboratory or the field; the typical places where one might find the paleontologist. They became public figures that had their space in newspapers, illustrated magazines, popular science books, museums, and prehistoric science fiction novels. It is this cult of the caveman—and the Neandertals are the cavemen par excellence—that renders them iconic. How did they gain such an iconic status? As we will see, in their role as the slouching and dumb or violently brutish cavemen, an image that stuck to the Neandertals up to this day, they functioned as the foil against which humanness emerged more sharply.

It is therefore not only the animal-human and similar boundaries that are crossed by the iconic "Neandertals;" they also move between science and other areas of culture. In following the Neandertals through history, we thus need an understanding of the relationships between scientists, artists, journalists, and other media representatives, as complex and flexible. The differentiation between "science" and "popularization" is more complicated then we usually think. Between the specialized scientific journal and the mass media there are a plethora of genres such as grant proposals and textbooks that turn the supposed sharp line into a continuum of styles. That the production of knowledge between science, the arts, and the media or their publics is interactive and multidirectional is possibly particularly obvious in paleoanthropology, which is a field of great public interest, and where the practitioners are likely to be involved in the writing of popular books, in the visual reconstruction of hominids, and in the shaping of entire exhibits on human origins. In my Neandertal story, I am therefore interested in a diversity of sites at which scientific knowledge is gained and transformed, including supposedly non-fictional presentations of science in newspapers, magazines, books, television documentaries, and museums, as well as in more strictly fictional uses of scientific knowledge in literature, film, and television.

Throughout the science-fiction spectrum, the Neandertals tended to be defined in comparison to modern humans, be that extant "races" perceived to be more or less evolved or the Neandertals' (near) contemporaries, the Cro-Magnons. In the process, modern humans, and especially the "civilized white races" at their apex, appeared anatomically, culturally, and intellectually superior. The reconstruction of the relationship between the Neandertals and "us" continues to be at the center of investigation, and this quintessential caveman can serve as a (often comic) reminder of the ape within, of our origin in the brute world that might at any time claim its due. As such the Neandertals are political figures. Conceived to be closer to "a natural state of living outside culture," as directly subjected to the powers of nature such as natural selection, the cavemen might seem to tell us something about the way we were and in truth still are, more or less hidden beneath layers of civilization. Through this kind of non sequitur, claims about the "natural"—and by inference right—human society, and about the "nature" of the human "races" or sexes can be empowered by appeals to the moral authority of nature.

Not surprisingly in view of the fact that the Neandertals have been of interest in relation to questions about our own identity, peeps back into their time have often taken the shape of encounters. Relying on experiences with contact zones, these "social spaces where disparate cultures meet, clash, and grapple with each other, often in highly asymmetrical relations of domination and subordination—like colonialism, slavery, or their aftermaths as they are lived out across the globe today" (Pratt, 1992, p. 4), science and fiction have engaged in the mental experiment of what might have happened when the Neandertals and anatomically modern humans met. In fiction, these encounters may be developed along the lines of contemporary persons traveling back in time or space to the Neandertals, or of the interactions between Cro-Magnons and Neandertals of the same time period. In the first instance, the more civilized intruders witness an indigenous people who are perceived as lost in time, as relics and reminders of the modern man's past. As apt for contact zones, the stories resemble ethnological discourses and feed on the experiences of western people as colonizers. The same can be observed for scientific imaginings of the long past encounter between the Cro-Magnon invaders into Europe and the indigenous Neandertals up to well into the twentieth century:

> It is now clear that towards the end of the last ice age an evolutionary tragedy was enacted in Europe very similar in nature to that which is now taking place under our eyes in the Continent of Australia. A vigorous people, the early Caucasians [i.e. the Cro-Magnons], *colonised* Europe and in the process exterminated its native Neanderthalian population. (Keith, 1936, pp. 14–15, emphasis mine)

In their role in the reenactment of encounters and relations between "higher" and "lower races" in past and present, the Neandertals have experienced a turbulent history as scientific and cultural objects, however. Far from unchangeably functioning as "the primitive other" that justifies the ways of "the advanced self," they have at times taken the shape of Dorian Gray's portrait, confronting Western humankind with its true disfigured face. Again alternatively, they have been seen as "one of us" in a positive or negative sense. This might best be illustrated by a contemporary Neandertal image from the front page of *Die Zeit*, one of the more distinguished German newspapers.

The occasion is the 150th anniversary of the discovery of a Neandertal cranium and postcranial remains in the Feldhofer Grotto in the German Neander Valley (the actual meaning of *Neandertal*). In *Die Zeit* of 12 January 2006, the history of the Neandertals in science and culture is presented as a particularly German affair, and the Neandertal reconstruction looks deceitfully real and deceitfully German. At first sight, the man might be taken for a figure of importance in German history. In fact, the headline reads "the most famous German," and in the accompanying article where another reconstruction, this time as a present-day "ordinary German" in red-black plaid shirt and anorak, is reproduced, the Feldhofer Grotto find is referred to as an "icon of German archeology." Indeed, for paleoanthropologists and archeologists 2006 was not the year of Mozart or Freud, but of the Neandertals, whose history was celebrated in exhibitions and with an international conference under the auspices of the United Nations Educational, Scientific and Cultural Organization (UNESCO).

I will now move back in time to this and even earlier discoveries of Neandertal bones and follow the story of their scientific interpretation to a very influential Neandertal reconstruction in the early twentieth century, which was largely responsible for the caveman image of a hairy, apish, club-bearing creature that still haunts popular culture. As it turns out, already at this point the Neandertals were associated with Germany, but in a rather less flattering way than the national icon on the *Die Zeit* cover suggests. We will then want to know how the same hominids could come around full circle and change from ape men to essentially humans with sturdy bodies. In fact, in between, the image of the Neandertals has gone through another cycle of rehabilitation and defamation, this time aided by stunning new technologies such as comparative DNA analysis. Though still unstable, the net development of the scientific Neandertal image from the mid-twentieth century onward has been one of improvement. After all, it seems moot to call a hominid an evolutionary failure who flourished for about 200,000 years up to about 35,000 years ago over an area reaching from Portugal to Uzbekistan, and from Northern Europe to the Near East. Clearly, the Neandertals successfully coped with the hardships of glacial Europe by means of intricate adaptations.

In the history of the Neandertals, the scientific debates essentially revolved around a set of questions that are tackled with knowledge from geology, paleontology, archeology, biogeography, and molecular anthropology and systematics among other disciplines: Are they within the range of modern human variation (*Homo sapiens*), or different enough to count as a separate species (*Homo neanderthalensis/primigenius/*etc.)? Are they ancestral to modern humans or a dead-ending evolutionary side branch? Have they encountered modern humans and if so, what did these encounters look like? Did they interbreed, or, at the alternative end of theorizing, were the Neandertals extinguished by the superior newcomers without contributing anything to our line? Obviously, in the answering of these questions, one's view of the Neandertals' mind and behavior, besides their anatomy, is decisive. Today, the reasons circulating in paleoanthropology and prehistoric archeology for the Neandertals' disappearance are (combinations of) the following: their absorption by the higher number of modern human invaders; their inability to secure sufficient resources in competition with the culturally superior modern humans; their lower birth rate and higher mortality; their overspecialization; their inability to cope with the decline of the climate; their violent extermination through the Cro-Magnons, or the new diseases they brought.

FOSSIL MAN?

What we today consider as osseous and cultural remains of Neandertals had been discovered long before the concepts of a great human antiquity—by far transcending the approximately 6,000 years based on the chronology of the Bible—or that of evolution, had their breakthrough. In fact, it was around these objects, among others, that controversy about human antiquity unfolded. Neandertals were therefore bones of contention. The greatest comparative anatomist of his time, Georges Cuvier (1769–1832) of the Musée d'Histoire Naturelle in Paris, had repeatedly declared that no human bones had been found in association with the remains of animals of the former world, such as the mammoth, the woolly rhinoceros, the cave hyena, and cave bear. In a posthumously published work and in the heads of his followers, this authoritative statement lived on beyond Cuvier's death. Among the latter were such influential figures as the British geologist William Buckland (1784–1856). In a widely read work, Buckland, by then reader of geology at Oxford University and canon in the Anglican Church, added that it was very unlikely that fossil human bones would ever be discovered.

> No conclusion is more fully established, than the important fact of the total absence of any vestiges of the human species throughout the entire series of geological formations. Had the case been otherwise, there would indeed have been great difficulty in reconciling the early and extended periods which have been assigned to the extinct races of animals with our received chronology. On the other hand, the fact of no human remains having as yet been found in conjunction with those of extinct animals, may be alleged in confirmation of the hypothesis that these animals lived and died before the creation of man. (Buckland, 1836, vol. 1, p. 103)

It was clear that the history of the earth extended over immense periods of time. But it was generally held that humankind was a recent creation, having arrived in Europe after its strange animals had become extinct, and after the last significant geological revolution had given the land its present shape. Buckland's verdict was an attempt at reconciling the stunning insights from the young historical geology with a non-literal reading of Genesis. However, also where human history was concerned, this reconciliation grew increasingly controversial. One threat of relevance to the story of the Neandertals was posed to it by the work of the physician and natural historian Philippe-Charles Schmerling (1790–1836) in Liège. Schmerling began to explore the caves on the banks of the River Meuse and its tributaries in 1829, and unearthed human bones of at least six individuals, worked bone, and flint flakes.

Schmerling was confident that the cave sediments had not been disturbed, and that the human remains had been brought into their present

situations at the same time and by the same means as the bones of the
now-extinct animals with which they were associated. In his *Recherches
sur les ossemens fossiles* (1833–1834), he therefore concluded that these
were the traces of a "human race" that had been contemporary to the
strange European fauna of the likes of the mammoth and that had
belonged with them to the state of the world prior to the last geological
revolution (i.e., antediluvian):

> As I dare guarantee that none of these pieces [human-made artifacts] has
> been introduced at a later time [than the fossil animal remains], I attach
> great importance to their presence in the caverns; since even if we had not
> found human bones, in conditions entirely favoring the consideration that
> they belong to the antediluvian époque, these proves would have been
> provided to us by the worked bones and flints. (Schmerling, 1833-34, Vol. 2,
> p. 179, translation mine)

The decomposition, the often broken and rounded state, the color and
relatively slight weight of both the animal and human bones further
strengthened Schmerling's conviction of their great antiquity. While Schmerling's adult skull (Engis 1) dates from the young Paleolithic period,
the second skull is in fact that of a Neandertal child (Engis 2), but it took
about 100 years for it to be recognized as such.

Obviously, in the 1830s the time was not ripe to welcome "Fossil
Man," and Schmerling's discoveries were not taken as insights into the
prehistory of humankind. Even the famous human skullcap and other
bones that were unearthed in association with those of extinct animal
species in the Feldhofer Grotto in the Neander Valley (near Düsseldorf,
Germany) in 1856 initially met with the same fate. Nonetheless, these
Neandertal remains that were handed over to the German Johann C.
Fuhlrott (1803–1877), teacher and natural historian, eventually became
the first accepted fossil human bones. Fuhlrott as well as Hermann
Schaaffhausen (1816–1893), professor of anatomy at Bonn University,
who helped describe the find, saw in the bones a primitive human type
with certain ape-like traits.

Schaaffhausen applied a crude dating test, the tongue test, which had
already served Buckland for a preliminary estimation of the age of bones.
The stronger a specimen would adhere to the tongue of the experimenter,
the older it was likely to be. The outcome of this test, as well as the den-
drites visible under a lens that covered the human bones, were indicative
of their fossilized state and therefore their considerable antiquity. This
impression was supported by the fact that even though Schaaffhausen
observed that the greater part of the cartilage was still present, it had
been transformed into gelatin (denatured collagen). While Schaaffhausen
tentatively placed the Neandertal find in an evolutionary framework,
others were more skeptical about its meaning. One of his colleagues at

Bonn University attributed the seemingly strange limb anatomy to rickets, and suggested that the discovery was the pathological remains of a Cossack cavalryman who had been wounded in the pursuit of Napoleon's army through Prussia.

Despite the skepticism of influential German anatomists, some of the greatest geologists and natural historians of the time came to follow Schaaffhausen in his estimate. After all, by the time the description of the Feldhofer find reached Britain, the situation was different from the one Schmerling had found himself in. In 1859, Darwin's *On the Origin of Species* produced a stir in Victorian society, with waves soon to follow in other countries, and the same year, the scientific community reached a consensus on the acceptance of human coexistence with the Pleistocene fauna such as mammoth and woolly rhinoceros. Most significantly, the British geologist Charles Lyell (1797–1875) visited the Feldhofer Grotto with Fuhlrott in 1860 and brought a cast of the skull to Thomas Henry Huxley (1825–1895), who, inspired by Darwin, was one of the most prominent early advocates of evolutionism.

Huxley concurred with Schaaffhausen on the contemporaneity of the Fuhlrott remains with the now-extinct fauna; he regarded them as of prehistoric origin. Although Lyell in contrast to Huxley was skeptical about the application of evolution to humans, he conceded that if the Neandertal skull were really of high antiquity, it would support the transformation theories of Jean-Baptiste Lamarck (1744–1829) and Darwin. At the other end of the scientific spectrum, the Berlin pathologist-anthropologist Rudolf Virchow (1821–1902), possibly to frustrate the cause of evolutionists, adhered to an interpretation of the Feldhofer Neandertal as a pathological aberration of living humans. He rejected the idea that they were dealing with a newly found human type of a former world, and his opinion carried great authority.

Although Huxley and Schaaffhausen to the contrary regarded the Feldhofer specimen as belonging to a primitive human type, they considered it not too primitive to be positioned within the hierarchy of existing "races." Thus, while Huxley reached the conclusion that the Neandertal from the Feldhofer Grotto was "the most pithecoid of known human skulls" (Huxley, 1894, p. 205, see also pp. 168–207), he classified the Feldhofer Neandertal as a mere fossil variant of modern humans because the skullcap fell within the range of normal cranial capacity. Huxley's analysis of the Neandertal cranium led him to believe that it was closer to the Australian aborigine, whom he thought to occupy the lowest rung in the hierarchy of extant human types, than to the apes. In his view, the Neandertals might have represented the direct evolutionary precedents of modern human "races."

With the rapidly increasing European record of ancient human skeletal and cultural remains, it became clear that one was dealing with more than one prehistoric human type, and attempts at classification were made. The *Crania ethnica: Les crânes des races humaines* (1882) by the French zoologist and anthropologist Armand de Quatrefages (1810–1892) and the anthropologist and ethnographer Jules Ernest Théodore Hamy (1842–1908) expanded on the description of "prehistoric human races" and cultures that had been initiated by the *Reliquiae Aquitanicae; Being Contributions to the Archaeology and Palaeontology of Périgord and the Adjoining Provinces of Southern France* (1875). Among the fossil human evidence from Europe, de Quatrefages and Hamy distinguished three "races," that of Canstadt (including Neandertal remains), that of Cro-Magnon, and that of Furfooz.

The Furfooz cave had been excavated among other Belgian sites by Édouard Dupont (1841–1911), who discovered human crania. The "Canstadt race," which represented their oldest and most primitive type, was named after a skullcap of uncertain origin—possibly discovered as early as 1700 near Cannstatt (close to Stuttgart in Germany)—that in retrospect showed close similarities to the Feldhofer Neandertal. With their description of the "Canstadt race" they thus took another step towards the recognition of Neandertals as a distinct fossil form of humans, even though they also included modern human remains in the group. Perpetuating the tradition of establishing hierarchies of "races" in which recent as well as Paleolithic types figured, the *Crania ethnica* was subdivided into two parts, the first representing a classification of the fossil human "races," and the second a classification of the extant ones.

Further Neandertal finds had been and were made at La Naulette (near Dinant) and Spy (region of Namur) in Belgium in 1866 and 1886, at Krapina (Croatia 1899), and eventually at La Chapelle-aux-Saints, Le Moustier, La Quina (1908), and at La Ferrassie (1909) in France. Earlier discoveries, such as Schmerling's child cranium from Engis and the cranium from Forbes Quarry (Gibraltar, 1848), were reevaluated, and it became clear that the Neandertals were indeed a distinct and fossil human type. That the controversy around the Neandertals as a distinct fossil human type lasted some time becomes evident in remarks by the Belgian anatomist Julien Fraipont (1857–1910) in "Current Notes on Anthropology," published in *Science*, that gave an account of the meeting of the German Anthropological Association where anthropologists, among them Vichow, called the "Neandertal and Canstadt races" *Fantasiegebilde* (creatures of the imagination).

Already the first discoveries of Neandertal bones had been accompanied by press coverage, and from their birth as "Fossil Man," the Neandertals were public figures. In 1873, an anonymous visual reconstruction of Neandertals (in relation to the Feldhofer remains) was circulated through the pages of *Harper's Weekly*.

Although in the text Neanderthal Man was described as "ferocious looking, gorilla-like," he was also said to be a "human being" (Anonymous,

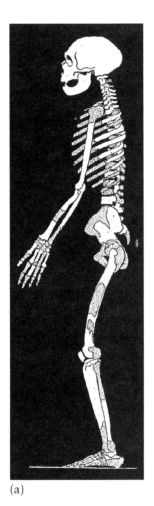

(a)

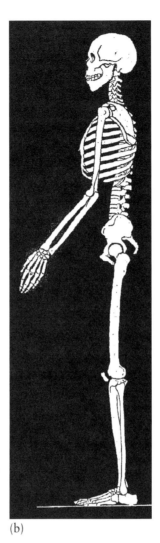

(b)

(a) "Reconstruction of the skeleton of the Man from La Chapelle-aux-Saints" and (b) "Skeleton of an Australian." From Boule, Marcellin (1913), "L'Homme fossile de la Chapelle-aux-Saints," *Annales de Paléontologie* 8:209–279, pp. 232, 233.

Harper's Weekly, 19 July 1873, p. 617). This savage though human description was supported by the iconography that emphasized the strong brow ridges, receding front, flat nose, protruding jaw, and stout and muscular body, while portraying "the wild man" as a hunter of the mammoth who had a hairless body, was fully upright, and capable of human facial expressions. Similarly, the cultural surroundings of the two figures, whose depiction is laden with Victorian gender stereotypes, render the Neandertals human. They have dogs as domestic companions, know fire, wear a primitive kind of clothing, and have rather elaborate compound tools. Thus, even though the Neandertal of *Harper's Weekly* is clearly more primitive and savage than earlier representations of prehistoric humans (see, for example, Louis Fugier's *L'Homme primitif*, 1870), at the time the article was published, the Neandertals had not yet achieved their peak as beastly cavemen.

The anonymous author of the brief report in *Harper's Weekly* discussed the scientific views of the Neandertal find as representative of a "fossil human race" held by Huxley, Schaaffhausen, and other evolutionists, while he did not give much credence to the idea that the peculiar anatomy of the skull was abnormal or pathological. The illustration, too, expressed the notion that though primitive, the Neandertals were within the imaginable range of human beings. The same impression is conveyed by the Neandertals in H.G. Wells's "Stories of the Stone Age" (May–September 1897), who, though showing the anatomical differences to modern humans also found in the *Harper's Weekly* image, have stone axes and spears and even practice a kind of shamanism.

As hinted at above, in tandem with the birth of the Neandertals as a fossil human type, another type that was to accompany them throughout their history of interpretation was born, the Cro-Magnons. Furthermore, the description of the Neandertals as a specific type of Paleolithic humans did not mean the end of controversy. Rather, the image of the Neandertals was soon to undergo a major turn. In fact, this turn to the worse was foreshadowed in the interpretation of the Krapina finds. The Croatian paleontologist Dragutin Gorjanovi -Kramberger (Karl Kramberger) of the University and National Museum in Zagreb observed that the several hundred bone fragments, all of which he attributed to the Neandertal type, had been broken for marrow and occasionally burned. He outspokenly argued a theory that had been associated with the Neandertals for some time—that the prehistoric men from Krapina had been cannibals.

FROM RACE TO SPECIES, FROM ANCESTOR TO FOIL

By the early twentieth century, the belief that the Neandertals had evolved into (some of) the "modern human races" was widely accepted. This role of the Neandertals had been strengthened by the German anatomist Gustav Schwalbe (1844–1916), who classified the Neandertals as a separate species (*Homo primigenius*), rather than as a "fossil human race" as done by Huxley, thereby enlisting them unambiguously for an evolutionary scenario of hominid development. He suggested a linear line of descent that now went from *Pithecanthropus erectus* via the Neandertals to modern humans.

Alas, this unilinear phylogeny was soon challenged, most powerfully by Marcellin Boule's (1861–1942) reconstruction of the Neandertals. The work of the powerful paleontologist at the prestigious Musée d'Histoire Naturelle in Paris on the La Chapelle-aux-Saints specimen was crucial for the rejection of the Neandertals as human ancestors by the anthropological community. The relatively well-preserved Neandertal skeleton was discovered as part of a burial in a cave in the village of La Chapelle-aux-Saints (Department of Corrèze, in southwest France) in 1908. Boule had the fortune of receiving the specimen among other things because it was discovered by clergymen. The obvious alternative to Boule at the museum would have been the École d'Anthropologie, which was however associated with radical politics, materialism, and anticlericalism.

A founder of the École had been the prehistorian Gabriel de Mortillet (1821–1898), who had also been an active politician, holding a seat in the Chamber of Deputies for the extreme left. Striving for progress in a humanist sense, de Mortillet had reasoned that the political left would eventually prevail by necessity. He had predicted an inevitable succession from the reign of the nobility, to the reign of the bourgeoisie, and finally to the reign of the socialists. These politics had been strongly interwoven with his view of human evolution. On the basis of a universal law of morphological and cultural progress, paleoanthropology and Paleolithic archeology were political weapons for radical socialist aims, with human history as integral part and logical consequence of human prehistory.

In de Mortillet's scenario of hominid evolution, the law of progress had guided the human ascent from a hypothetical missing link between the highest anthropoid ape and "the lowest savage," to which he gave the genus name *Anthropopithecus*, via Neanderthal Man to the anatomically modern Cro-Magnons. This morphological progress had been paralleled by advancement through the material cultures as found in the archeological record, and had been accompanied by changes in geology, fauna, and flora. Among de Mortillet's greatest achievements was the establishment of a classification system that in part is still used. He subdivided the Paleolithic period into cultures named after French cave sites: The oldest, the Chellean (also Acheulean), consisted of tools made by chipping off small flakes to put a sharp edge on the original core of the flint; the Mousterian, the culture associated with the Neandertals, referred to tools made from the flakes rather than the core; the Solutrean showed beautifully shaped blades of chipped stone; and the last Paleolithic culture, the Magdalenian, was characterized by a sophisticated artistry in bone.

But the new Neandertal find from La Chapelle-aux-Saints was not destined to substantiate such a linearly progressive scenario of human development. On 14 December 1908, Boule's interpretation of the La Chapelle-aux-Saints skull was presented to the Académie des Sciences. In the paper and succeeding monographs on the find that would prove formative of the scientific and popular views of the Neandertals for many decades to come, Boule emphasized the simian traits of the skeleton. He also continued an earlier suspicion that Neanderthal Man had bent knees and a stoop. Since according to Boule's interpretation the Neandertals had differed significantly from modern humans, morphologically as well as culturally, he agreed with Schwalbe that the Neandertals were a separate species and not simply a pathological form of modern humans or a "fossil human race."

However, Boule contradicted Schwalbe, de Mortillet, and others by rejecting the Neandertals as ancestors of modern humans. He dated the La Chapelle-aux-Saints Neandertal to the Middle Pleistocene, which according to his interpretation put it in close temporal proximity to the modern human type, both of which it thus appeared had inhabited Europe at the same time. The expulsion of the Neandertals from the line leading to modern humans also put into question the ancestral status of *Pithecanthropus*, which seemed to show Neandertal specializations even more markedly. Hominid evolution must therefore have had more than one line of descent, and the ancestors of modern humans were pushed further back into the as yet unknown evolutionary history.

To confer weight to his conclusions, Boule used imagery and rhetoric that distanced the Neandertals from living humans on the one hand, and from Paleolithic *Homo sapiens* on the other hand. Appealing to national pride, the Neandertals were verbally denigrated through juxtaposition to the noble contemporary Cro-Magnons:

> It has to be remarked that this human group of the Middle Pleistocene [the Neandertals], so primitive with regard to physical characters, must also, judging from the standards of Paleolithic archeology, be very primitive from an intellectual point of view. When, during the Upper Pleistocene, we find ourselves, *in our country*, in the presence of industrial manifestations of a higher order and of true art, the human skulls (race of Cro-Magnon) have acquired the principle characteristics of true *Homo sapiens*, which means beautiful foreheads, large brains and faces with little prognathism. (Boule, 1908, p. 525, translation and emphasis mine)

In the 1913 paper, Boule furthermore visually juxtaposed his Neandertal reconstruction to a modern Australian aborigine skeleton with the expectation, it seems, that the viewer would immediately notice the obvious differences. Even "the primitives at the peripheries of the earth," the logic went, were considerably more advanced than this brute, which was closer to the apes than any "human race."

When thus contrasted to the morphology and culture of the "Cro-Magnon race," whose presence in Europe Boule claimed to have overlapped with the Neandertals', the backwardness of the latter seemed even more striking. After all, as the caves in southern France amply illustrated, the Cro-Magnons had possessed modern human anatomy and had enriched their elaborate tool culture with veritable pieces of art, such as engravings on stone and bone. Thus, in Boule's presentation of the Neandertals, both comparisons—with the "most primitive living human race" and with the contemporary prehistoric "race of the Cro-Magnons"—served the purpose of expelling them from the human line of descent. The Neandertals were far too primitive in morphology and culture to serve as our direct ancestors.

In the attempt to explain Boule's brutish reconstruction morphologically, it has been pointed out that the bones of the Old Man of La Chapelle-aux-Saints had been afflicted with osteoarthritis. However, Boule was aware of the deforming illness, even if he did not take it sufficiently into account in his reconstruction. Today we know that, while the pathology of the Old Man may well have forced him into something of a stoop, the classic Neandertals in a healthy state were fully human in posture. Where the non-morphological reasons are concerned, Boule could have aimed at applying the view of mammalian evolution as dendritic held by his direct predecessor at the museum to hominid paleontology, and as claiming priority in this field vis-à-vis the École. Alternatively, Boule's work has been interpreted as standing mainly in the Cuvierian tradition, and therefore as having been influenced by antievolutionism and catastrophism. Boule's delegation of the Neandertals to a dead-ending side branch left *Homo sapiens* without immediate evolutionary precursors, and their sudden replacement by anatomically modern humans would thus represent such an instance of catastrophic change.

That Boule intended his work on the Neandertal specimen to strengthen his position as *the* authority in matters of human paleontology in France can be inferred from his interactions with the press. Although Boule distanced the Neandertals from everything French, the prestige the discovery implied for French science and the French nation was used to sell his work to the press and the public. Nationalism was easily kindled at the time, and it was in this context that the Neandertals became associated with Germany while the Cro-Magnons were further instrumentalized for the particularly French. In an article in *L'Anthropologie* with the title "La guerre" (1914), Boule dehumanized the Germans in a manner reminiscent of his dehumanization of the Neandertals of prehistoric times. In stark contrast, Boule argued, France had been the land of progress since Paleolithic times. The sculptures and murals of the Cro-Magnons that decorated the French caves gave clear testimony to an advanced aesthetic sense. These morphologically modern humans had been the pioneers of true culture. And just as the artistic Cro-Magnons had once triumphed over the brutish sub-*sapiens* Neandertals, so would the French of today triumph over the degenerate Germans in World War I.

Clearly, Boule's reconstruction of the La Chapelle-aux-Saints Neandertal had a great impact, and his work was widely publicized in newspapers. The French press of the Third Republic turned the caveman into a veritable beast: "It has not only the *appearance*, but a detailed examination establishes that if this 'man' sometimes stood upright, he must nonetheless more often have lived 'on all fours'" (Alphonse Berget, *Le Figaro*, "À l'Académie des Sciences," 15 December 1908, translation mine). In view of such beastliness, the *Journal des débats* reassured its distinguished readership that there was nothing extraordinarily shocking in the Neandertal's primitive aspect. Even at this moment, there coexisted "radically lower forms of the human species" in the remote corners of the earth: "There have always been inferior humans on earth, even very alive at this moment. There is nothing astonishing in the fact that one finds at the beginning of the ages beings that betray a coarse constitution, a primitive state destined little by little to disappear" ("Revue des sciences," *Le Journal des débats*, 24 December 1908, translation mine). If anything, Neanderthal Man just like "the extant savage human races" made the march of progress of such civilizations as the French even more apparent.

Le Journal even invented a jingle on the La Chapelle-aux-Saints specimen's primitiveness: "He was not very pretty, this old hunter of the quaternary forests. With his protruding jaws, receding chin, his pronounced eyebrows that formed a bar joined above the nose, his receding forehead and flat skull, the appearance of his face took after the ape" (Remy Perrier, "L'Homme de la Chapelle-aux-Saints," 6 June 1913, translation mine). The verse no doubt gained in impact through the brutish pictorial reconstruction of the Neandertal that accompanied the article. It was the work of the Czech painter, engraver, and illustrator Franz Kupka (1871–1957).

There could be no clearer indication of the change the Neandertal image had undergone between science and the press than this image, which had circulated in the French magazine *L'Illustration* and in the British *Illustrated London News* (Honoré, 20 February 1909, pp. 128–129; Anonymous, 27 February 1909, pp. 312–313). The difference to the *Harper's Weekly* illustration is striking. It is immediately obvious that the Neandertal has changed from being perceived as within the range of "the same" to being identified as "the other." This hairy creature, in a barren environment that seems to reflect its dull mind, is marked by an expressionless face, an uninventive club, bent knees, and a forward stoop of the upper body. His arms are long, his legs are short,

"The Man of Twenty Thousand Years Ago," by Franz Kupka, from "The Most Important Anthropological Discovery for Fifty Years," the *Illustrated London News*, 27 February 1909, pp. 300–301, 312–313, on pp. 312–313. (Courtesy Illustrated London News Picture Archive.)

and his chest is of incredible dimensions. In a large reproduction of the image, one may recognize the shape of an ape in his shadow. Evidently, he has not progressed far from such a stage.

Kupka's apish image of the La Chapelle-aux-Saints Neandertal, in the production of which Boule himself had his share, has since been reproduced so often (and this chapter is good proof of it) that it has been inscribed into our cultural heritage. In general, the outcome of the co-construction of the Neandertals between science, the arts, and the press in the early twentieth century was such that it is small wonder that the Neandertals were soon no longer welcome in the human pedigree, and disappeared in most diagrams of hominid phylogeny from the line leading to modern humans. The new model of phylogeny, according to which the Cro-Magnons were the result of a long independent prehistory of anatomically relatively modern humans, came to be dubbed the Praesapiens Theory. It was associated with evolutionary scenarios that envisioned the Cro-Magnons as invading Europe and violently disposing of the inferior Neandertal type.

THE BIRTH OF AN ICON

With Kupka's still widely reproduced image of the abominable caveman, we enter the phase of the story of the Neandertals where they have a very active and changeable public life. This is not solely due to Boule's science, however, but has to do with wider cultural developments. By the early twentieth century, newspapers had become more specialized and internationally distributed, as news spread faster due to the telegraph, the postal system, and the railway. Mass production had become possible during the nineteenth century through the introduction of steam printing, the rotary press, and linotype typesetting machines, all of which radically increased the scale of production and decreased production costs. Finally, regular use of photographs began with the perfection of the halftone process for facsimile reproduction. Around the turn of the century, newspapers incorporated photography for reporting topical events, and the profession of newspaper illustrator gradually became obsolete. The success of these industrialization technologies depended on the relative freedom of the press and combined with the fruits of free and compulsory elementary education, which expanded the number of potential readers, initiated the belle epoque of the newspaper.

But the newspapers were not the only kind of print media that peaked. As the popularization of science boomed, the number of poplar books on human evolution written by scientists increased considerably. At the other end of the popular science spectrum, science fiction entered its golden age in the Anglophone world. Serialized and mass-printed on the cheap paper of the pulp magazines, writers of scientific romances such as the American Edgar Rice Burroughs (1875–1949), creator of Tarzan, reached an unprecedented number of readers. In between the two genres, science journalism became institutionalized with the turn to the mass-democratic and mass-media publics of the interwar years. The diversification of mass-distributed media were accelerating life, and with the possibilities generated by radio, cinema, and later on home television, the times were ripe for the Neandertal caveman to become a multi-media icon.

Among the most successful prehistoric science fiction romances serialized in the pulps that dealt with imaginative encounters between modern humans and Neandertaloids were Jack London's *Before Adam* (October 1906–February 1907), J.H. Rosny-Aîné's *La guerre du feu* (1982 [1911]), Sir Arthur Conan Doyle's *The Lost World* (April–October 1912), and Burroughs's trilogy of *The Land That Time Forgot*, *The People That Time Forgot*, and *Out of Time's Abyss* (August 1918; October 1918; December 1918). These narratives are comparable insofar as they employ the imaginative space of fiction to think about the new insights from paleontology and archeology. The fictive encounter between twentieth-century moderns and evolutionary relics may take place in as yet unknown geographical areas. Simultaneously or alternatively, the stories re-create the encounters between anatomically modern humans and Neandertals of the Paleolithic era that so troubled anthropologists. More often than not, these were envisioned as of a violent nature.

Although London's Folk People and Doyle's hominid relics in South America are arguably Neandertaloid, Rosny-Aîné and Burroughs most directly drew on the paleoanthropology of the time. In Burroughs's trilogy, the twentieth-century Americans, Brits, and Germans, stranded on the crater of an ancient volcano, literally travel though time on Caspak, where the geological epochs are arranged in linear succession from south to north. For reasons of dramaturgy, Burroughs makes hominids the contemporaries of dinosaurs, but the overall series is made up of flora and fauna of an ever higher order from jungle to arid mountain regions. Thus, the expeditions on the crater first encounter a mixture of *Pithecanthropus* and Piltdown, then a "race" similar to "the so-called Neanderthal Man of La Chapelle-aux-Saints," succeeded by one resembling the Grimaldi skeletons, a Cro-Magnon type, and a Neolithic Nordic type.

Burroughs's Neandertals are of the Kupka blend: "There was the same short, stocky trunk upon which rested an enormous head habitually bent forward into the same curvature as the back, . . . the knees bent forward and never straightened. This creature . . . carried heavy clubs" (Burroughs, August 1918). It is only in the Cro-Magnons that we meet a people more like the white invaders: "Here again was a new type of man—a higher type than the primitive tribe I had just quitted. They were a taller people, too, with better-shaped skulls and more intelligent faces. There were less of the ape characteristics about their features, and less of the negroid, too" (Burroughs, August 1918). Burroughs's anthropological gaze on these prehistoric hominids betrays the same kind of hierarchy of fossil and living "races" that we have met with in science, but contrary to Boule's view of human evolution, in the Burroughsian Caspak fiction hominids climb the ladder of types from ape to perfectly modern human in the impatient, accelerated fashion of the times of the conveyor belt.

The apogee of the Neandertal–Cro-Magnon contrast, again accompanied by a comment on "racial" relations of the present, was realized in another format that reached 100,000 people a year—the Hall of the Age of Man exhibit at the American Museum of Natural History in New York. Its president, the paleontologist Henry Fairfield Osborn (1857–1935), began work on it in 1915, but due to World War I it was not opened to the public until 1924. Osborn's vision of the museum was educational. He intended it as a re-creational substitute for direct nature experience for the big town citizens. The Stone Age "races" of man were particularly apt for the education of the public through entertainment, since they represented a life in harmony with nature. The Cro-Magnons functioned as the "super race" in the exhibit, cultivated just enough to express their beautiful minds in art but maintaining an integrated lifestyle that subjected them to the forces of nature.

For the new museum hall, Osborn directed Charles Knight's (1874–1953) production of the beautiful and infamous murals of life scenes of the Neandertals, the Cro-Magnons, and the Neolithic Nordic type. Seemingly counter-intuitively, Osborn intended the images, which might easily be read as a picture story about a linear progress from the Neandertals via the Cro-Magnons to the Nordics as in Burroughs's fiction, as a warning against degeneration. Never again, was Osborn's message, had there been such a high human type as the Cro-Magnons had represented at their apex. Even the Neolithic Nordics, while they were clearly the great hunters of the north, were perceived as second to the Cro-Magnons of the Reindeer Age in art. Besides, the Cro-Magnons had larger brains than that of the average modern European, and brain size was still used as a correlate of intelligence.

Alas, and here comes the warning, the Cro-Magnons' Golden Age (the parallel to the Hellenic culture was explicitly drawn) came to an end with their giving up their nomadic lifestyle. The Mesolithic Azilian culture was interpreted as demonstrating the downfall of the Cro-Magnons, an anticlimax from the preceding high culture of the Reindeer Age. As becomes clear from *Men of the Old Stone Age* (1915), Osborn pointed his finger at the supposedly common phenomena that "races" of high industrial and artistic development are prone to stagnate and decline.

For Osborn, this was a comment on what he perceived as the urban, feminized, mongrelized, and artificial life led by more and more Americans.

In *Men of the Old Stone Age*, another Neandertal drawing created by Knight under Osborn's direction had been published. It continued the iconography of the primitive, implementing what are by now the visual stereotypes associated with the Neandertals, such as a bad hairdo and general hairiness, a dumb facial expression, a stoop, large feet, a thick neck, a club, and animal skin garment, some of which were already found in Kupka's reconstruction. Interestingly, none of these icons was based on anatomical or archeological evidence. It might well be that Burroughs had also been aware of this visual reanimation of the Le Moustier site from Osborn's successful book, besides Kupka's of the La Chapelle-aux-Saints site. As we have seen for Boule before, Osborn in book and exhibit, visually as well as rhetorically, juxtaposed the Neandertals to the Cro-Magnons to render the first more primitive and the second more advanced in the contrast. For Osborn it seemed clear that the superior Cro-Magnons, who had invaded Europe from the east, had caused the Neandertals' most likely violent extinction.

In Osborn's view of humankind's rising to Parnassus, the Neandertals did not play the role of a stage in the ascent. Rather, within his idiosyncratic expression of the Praesapiens Theory, anthropoid apes, sub-*sapiens* hominids, and even contemporary "non-Aryan races" were distanced from the "Nordic/Anglo-Saxon line." Extinct human types became means to the end of advocating against miscegenation and other supposed causes of degeneration. However, even within the more strictly academic world, there was no unanimity about the Neandertals and their place in the hominid family tree. A minority of scientists, most importantly the Czech-American Aleš Hrdlička (1869–1943), believed in their ancestral status. Furthermore, new discoveries of Neandertal remains also outside Europe initiated a change in the outlook of some paleoanthropologists in the 1930s. Nonetheless, the scientific image of the Neandertals as very different from and deficient in comparison to the modern human newcomers, as an evolutionary failure driven to extinction, remained dominant until after World War II.

THE FLOWER CHILD AND ANOTHER BACKWARD SWING OF THE PENDULUM

From the 1930s onward, breathtaking finds were made in Mediterranean coastal regions (then Levant). In particular, an Anglo-American team found skeletal remains of some ten people and Mousterian stone tools in the Skhūl and Tabūn caves at Mount Carmel (Israel, then Palestine).

As these appeared to show considerable variation but were generally rather modern in anatomy, several possible interpretations emerged: They represented a transitional form between Neandertals and Cro-Magnons; they were hybrids between the two; or, conversely, they were the parent stock of the classic European Neandertals and the Cro-Magnons. For this and for additional reasons, many anthropologists within the next decades came to regard the Neandertals once again as within the range of "us" and readmitted them to the taxon *Homo sapiens* as the subspecies *neanderthalensis*, and to the human family tree as our direct ancestors, even if within otherwise differing models of human evolution.

The re-inclusion of the Neandertals not only in the species *Homo sapiens* but also in our direct line of descent gained momentum in the aftermath of World War II. During the 1950s and 1960s, the impact of the Darwinian synthesis was felt in paleoanthropology, and such mechanisms of organic change as use-inheritance and orthogenesis still strongly relied on by such figures as Osborn increasingly gave way to explanations by natural selection, adaptation, and genetic drift. The new synthesis also meant a shift from a typological to a populational approach. In contrast to a typological thinking that encouraged the creation of a new species if not genus for new fossils, the biological concept of species emphasized variation within species.

This new theoretical approach not only led to explanations of Neandertal morphology in terms of adaptation, but an understanding of species as polymorphic populations also allowed to think of evolution as multi-regionally unilinear within the single-species concept. The concept comprised that at any one time in the course of hominid evolution, there had been only one species. On the basis of potential or actual gene flow between geographical populations throughout evolution, the systematist Ernst Mayr (1904–2005), who also included the australopithecines within the range of variation of *Homo*, thus regarded all as yet known hominids as representing a single line of descent (*Homo transvalensis-erectus-sapiens*).

This was at a time when, in the aftermath of the inconceivable atrocities committed during World War II, the views on human diversity were indeed political, and the UNESCO statement on race had been put together to prevent the instrumentalization of anthropology for fascist and racist politics. As before hierarchies of extant "races" based on the perception that there were considerable and deep-rooted differences between them in biology as well as culture had encouraged a view of the Neandertals as even more different from "whites" than the "lowest living race," the suspicion towards racialist thinking as expressed in the Family-of-Man paradigm now supported the Neandertals' re-inclusion as the same. Nevertheless, the most popular among the Neandertal interpretations that incorporated ideas from the evolutionary synthesis was the belief that in the Western Asian populations one had proof of a local transition from Neandertals to Cro-Magnons, the end product of which had entered Europe, signaling the demise of the classic Neandertals. In this scenario, the Neandertals were both our ancestors and, in the cold-adapted Southern and Western European form, our victims.

Despite continuing disagreements, a new understanding of the Neandertals as similar to us, or indeed as some kind of noble savages, had emerged that reached its climax with the discovery of a child grave at the Iraqi cave site of Shanidar. The scattered pollen was interpreted as the remains of flowers used in the burial ceremony. Just as spectacularly, the bones of a seriously disabled man who must have lived with his handicaps for some time were also discovered. As appropriate for the time, these Neandertals were labeled by Ralph Solecki *The First Flower People* (1971). The Neandertals now seemed to have been a Paleolithic people who cared for their old and injured, and who had ideas about the afterlife that led them to engage in burial rituals. The trend toward a more positive image of the timeless other was also signaled by novels such as Isaac Asimov's *The Ugly Little Boy* (1958) and William Golding's *The Inheritors* (1955), where the perspective is no longer that of the modern humans, but from the experiences of the innocently childish Neandertals. While for the Wells of the 1920s ("The Grisly Folk," 1921), as for many around that time (see quote from Keith above), the replacement of a supposedly morphologically and intellectually inferior type (here the evil beast Neandertals) by a more advanced one (here the heroic invader Cro-Magnons) was seen as necessary for the progress of mankind, in Golding's fantastic encounter story the modern human newcomers bring violence and corruption of a kind unknown to the Neandertals. After all, in the aftermath of World War II, the prices that had been paid for technological and scientific advancements were horribly obvious.

In science, the so-called multiregional continuity model, with local *Homo erectus* populations giving rise to the present human geographical varieties, which includes at least some Neandertal populations in the line leading to modern humans, has remained among the competing hypotheses for human evolution. However, already in the 1970s, not least due to the complication of the situation through the discovery of new fossils, the linear line of descent and the claim—which had been in favor since the synthesis—that the hominid fossil record indicates little taxonomic diversity, were challenged. The approach of comparative multivariate statistical analysis was applied to large samples of fossils, pointing out meaningful morphological differences between Neandertals and recent humans. Furthermore, a new evolutionary theory, that of punctuated equilibria, was brought to bear on the human data.

The hypothesis postulates that in the absence of environmental pressure a population can remain static over long periods of time, while it may respond to environmental change by a comparatively fast process of speciation. Speciation does therefore not have to be gradual, but may be episodic, which could also account for the gaps in the fossil record. Rather than assuming smooth evolution within one or a few hominid species, it was now suspected that human evolution had been marked by speciation and extinction, and that the clusters and gaps in the fossil record indicated real taxonomic proximity and distance. This suspicion seemed confirmed by the results from cladistics, the method that became popular for the establishment of evolutionary relationships and patterns in the hominid fossil record. The perspectives from morphometric analyses, cladistics, and the punctuated equilibrium hypothesis suggested great taxonomic diversity and numerous dead-ending side branches in hominid phylogeny, one of which was *Homo neanderthalensis*.

In addition, new dating techniques became available that in contrast to radiocarbon dating could be applied to material as old as the early Neandertal associated remains. By the late 1980s, thermoluminescence dating and electron spin resonance were used on burned flints and teeth to get a clearer picture of Neandertal chronology. But in particular where the interpretation of the large sample of what by then seemed to be *Homo sapiens* as well as *neanderthalensis* remains from various sites in the Eastern Mediterranean was concerned, the new dating methods revealed a complex picture of alternating and overlapping occupations. Combined with fossil discoveries in Ethiopia that indicated a very early date for anatomically modern humans there, the idea took shape that modern humans had first arisen from archaic forms in Africa, and that they subsequently migrated outward, replacing all archaic humans and Neandertals across the globe.

This out-of-Africa scenario received unexpected support when stunning news came out of the molecular biology laboratories. Since the discovery of the molecular structure of DNA and the mechanism of its self-replication by James Watson (1928–), Francis Crick (1916–2004), and Rosalind Franklin (1920–1958) in 1953, revolutionary techniques such as comparisons of proteins and DNA (nucleic and mitochondrial) of living peoples have been applied to the question of human phylogeny. Famously, the results of mitochondrial DNA sequencing of approximately 150 people from African, Asian, Australian, Caucasian, and New Guinean populations supported the view that the modern human geographical varieties are the descendents of a population that left Africa relatively recently. What was tellingly dubbed Eve Theory or Recent African Evolution model postulated that all human mtDNA could be referred back to a female who had lived in Africa about 200,000 years ago. Most important in this context, according to the African Eve scenario, once modern *Homo sapiens* had evolved from archaic *Homo sapiens* in Africa some 100,000 to 140,000 years ago, they spread across the globe and completely replaced the archaic *Homo sapiens* and Neandertals they encountered.

In association with these grand picture revolutions, the Neandertals of the 1980s and 1990s experienced a downgrading in anthropological imagination. It was questioned whether they had been able to think ahead, their stone culture seemed marked by a lack of invention, and they appeared generally inadequate to have coped with the competition from the truly human Cro-Magnons. Under the pen of some, the one-time mammoth slayers were even reduced to scavengers.

However, the comparatively large paleontological and archeological Neandertal record and the wonderful new methods and technologies that have provided alternative ways of approaching questions of chronology, phylogeny and biogeography have not settled the Neandertal question. The results from comparative DNA analyses of recent populations as well as of comparisons of DNA extracted from Neandertal remains with that of archaic and living humans are most often taken to suggest that "the modern newcomers" from Africa did not interbreed (significantly) with "the local ancients." But the technologies face such serious problems as uncertain mutation frequencies, inconsistency of results from different gene types, the possibility of recombination, the rapid deterioration of DNA, the survival of only tiny fragments, and the likelihood of contamination by bacterial, fungal, and modern human DNA. Furthermore, "very old and non-African" elements have been made out in modern human DNA that might indicate some local development, and the results from paleogenomics are elsewhere read as indicating genetic swamping due to the superior number of moderns entering Europe. In other words, even if there had been intermixture between the Neandertals and Cro-Magnons, this might not have left any recognizable genetic traces.

NO END IN SIGHT

The story of the Neandertals as told above is streamlined. At no point in their history has there been consensus in science, or only one way of fantasizing in art. Although the caveman moron is not (and arguably never was) part of the more specialized scientific discourse, opinions still diverge on whether they, like us, had (complex) symbolic communication, buried their dead ritually, and were altogether rather human, or whether they lacked the inherently human capacity for freeing their minds from the here and now, for the use of language and strategic planning. Now as then, the Neandertals are viewed in relation to the Cro-Magnons and to humanness per se. Can they tell us what makes us human through their lacking? Will the sequencing of the entire Neandertal genome finally provide an answer to the question of whether they are "us" or "them," whether we harbor the beast (or the innocent) within?

Despite the growing archeological and paleontological record and new technologies such as paleogenetics and computer-based virtual 3D-reconstruction (tomography and laser surface scanning), we carry the baggage of the cultural meanings of the Neandertals, which influence approaches to and readings of the data. A double standard has in fact been identified for our expectations of Neandertal/archaic versus modern human finds. It has for example been pointed out that in the case of the question of graves as testimony of intentional and ritual burial, different standards are applied to Neandertal graves than to those containing Cro-Magnon remains. In general, the archeological and biological record of the Neandertals and archaic humans tends to be explained by inferring minimal capacities, while the record of the so-called anatomically modern human group is more easily seen as evidence for complex mental and physical functions.

Anthropologists are generally aware of these issues, and the standard popular books about the Neandertals inform their readers of the associated uncertainty: "It is simply a story—as accurate a tale as we can make it, but one that is shaped by the recognizable imprint of our own prejudices and biases" (Trinkaus and Shipman, 1993, authors' note); "These ancient folk are so bound up with questions of identity and heritage, breeding and potential that scientific observation is even less likely than normal to be able to divorce itself from present contemporary opinions. Forget claims of scientific objectivity" (Stringer and Gamble, 1993, p. 8); and "There's no doubt that what we think today is inevitably shaped by what we believed yesterday, and this is something we should keep firmly in mind as we go on to look at the Neanderthals from our current perspective" (Tattersall, 1999, p. 119). While this self-reflexivity is testimony of a mature science that is aware of its being part of a particular history and culture, there can be no mistake about the fact that these experts, like their predecessors in the nineteenth century, agree that the Neandertals are incontrovertible evidence of human evolution and can only be explained in this way. Their disagreements and sometimes outspoken arguments have to be viewed as a sign of a healthy scientific community and of the fact that the field engages our emotions.

That the issues of paleoanthropology and human evolution more generally engage all our emotions has become clear in the course of the historical outline. The Neandertals have never been under the exclusive control of science. Rather, from the late nineteenth century onward, cavemen have been used for more or less good-humored comments on socio-cultural phenomena, and in current everyday life they are more present then ever for example in the figures of *The Flintstones*, one of the most successful animated television series of all times that started in the

1960s, and their multi-media avatars. Also Gary Larson's *The Far Side* comic strips that from the 1980s onward were syndicated worldwide, published in many collections, and turned into animated films have become part of our cultural heritage. The 1980s furthermore brought forth the film *The Caveman* (1981)—my favorite—that features the Beatle Ringo Starr as Neandertal anti-hero. All of these comic adaptations of the Neandertals work with our familiarity of the caveman iconography, with the last mentioned adding a sarcastic comment on our constant policing of the boundary between "them" and "us," animal and human, ancient and modern, Neandertal and Cro-Magnon, when it turns out that a simple stretch of the back suffices to turn the slouching caveman into the upright modern.

At present, the iconic status of the Neandertals is paid homage through their appropriations in children's learning tool kits, computer games, as plastic figures, on stamps, T-shirts, stickers, and in the visual arts. Neandertals also continue to be the objects of inquiry and speculation in every genre from the scientific journal article to the science fiction novel, and as we have seen at each point on the spectrum both science and fiction are involved. This is exemplified in "documentaries" such as the BBC series *Walking with Cavemen* (2003), which succeeded the award-winning *Walking with Dinosaurs* and *Walking with Beasts* series. It combines a certain scientific take on, among others, the Neandertal question ("The Survivors" episode) with a fictional reanimation of the dead bones and stones by means of the newest cinematic technologies. The animals are computer generated or animatronic, while the hominids are played by actors wearing makeup and prosthetics. The reality-fantasy mix is in the form as well as in the content, and the viewer is uncertain about the genre: Is it a natural history documentary in good Attenboroughian tradition or prehistoric science fiction involving time travel? But such series like *Walking with Cavemen* and its Channel 4 predecessor *Neanderthal* want to be taken seriously, and scientific authority may not only be conveyed through the consultation of many scientists, the docustyle, and technology and makeup, but possibly also through accompanying popular books that explain "the informed guesses" made (Palmer, 2000). Thanks to these strategies of collateral authentication, fictive elements such as Neandertal social structure and behavior, including gender relations, and their fatal competition with the Cro-Magnons appear as scientifically more robust than they actually are.

Also, the self-proclaimed fictional genres continue to intrigue due to their place between science and fiction. They, too, see in the Neandertals a tool for thinking about us, and the subject is oftentimes approached with a particular agenda. Jean Auel's *Earth's Children* series of five paleofiction novels, which has even been termed feminist, is a case in point. The message communicated through her imaginative portrait of

the Neandertals and Cro-Magnons and their interactions is that, if anything stood and stands in the way of human progress, it is male chauvin-
ism and male power based on rigid traditions. Though this vision is as
far from being politically neutral as any other, Auel's neverland of the
Neandertals is not so easily dismissed as mere fiction. In *The Clan of the
Cave Bear* (1980), for example, she makes clear use of Solecki's discoveries at Shanidar. For her novels that sold close to 35 million copies in
twenty languages, Auel visited important sites, experimented with Stone
Age technologies, and attended congresses on human evolution.

The literary bridge between science and fiction is also crossed from the
other end, for example by the evolutionary paleontologist Björn Kurtén,
who wrote *Dance of the Tiger* (1978). His incoming Cro-Magnons interact in various ways with the European Neandertals, whose fictional
musicality has retrospectively been substantiated by the discovery of a
flute at a Neandertal site in Slovenia. Finally, like films and literature,
static visualizations of Neandertals are co-productions of science and
art. Current 3D reconstructions such as reproduced in *Die Zeit* in 2006
often seem deceivingly real because they are bafflingly human. They sig-
nify that despite continued disagreement on Neandertal behavior and
mind, the scientific picture of the Neandertals is a far cry from Kupka's
ape-man. Clearly, the image and article in *Die Zeit* highlight that the
Neandertals have lost none of either their popular or their scientific
appeal. At least in Germany, the 150th anniversary of the name-giving
find from the Neander Valley was enthusiastically echoed in the media's
celebration of the hero of the Stone Age, the pop star from the Paleolithic. In 2006, Austria might have had Mozart and Freud to celebrate,
but nothing like the most famous German.

The Many Faces of the Neandertals

You will hardly find a recent book on human evolution and the history of
anthropology that does not include some remarks or even a whole chapter on the meaning and influence of pictorial reconstructions of prehistoric times. From a few skeletal remains the scientists try to re-create the
look of a dead individual and of an extinct species. Especially when it
comes to the Neandertals, the images have become increasingly important. Since the discovery of the first Neandertal skeleton, the reconstructions of this species have taken on many shapes and faces. Sometimes
Neanderthal Man looks like an apish, hairy brute, sometimes "he" appears smart and modern, and sometimes—but not very often—"he" is a
woman. No other creature shuffled from ancestor to cousin and back so
many times. Accordingly, the looks and characteristics of the Neandertals fluctuated between embodying the essence of what it meant to

belong to us and what it meant to be the other, depending on the place they were given in our family tree.

By establishing a visual language for images of humankind's prehistory, anthropologists shaped our conception of what it is that distinguishes modern humans from those "less evolved." In the nineteenth century, distinct visual markers for primitiveness and progressiveness were developed, which help us to decode a reconstruction of prehistoric life: Naked, apish people devoid of culture or language, surrounded by wild animals and in exotic landscapes were typical for primitive prehistoric life. In contrast, a small family of modern looking humans, domesticated animals, sewn clothing, stone axes, fine art (i.e., cave paintings or adornments), and tents to live in provide an image of a rather sophisticated and progressive prehistoric life. Where do the Neandertals fit in?

In most reconstructions of Neandertal life, they are neither naked, nor do they wear pants and a shirt; they wear loincloths, sometimes a plaid of animal skin. They do have cultural artifacts, but "only" a club or a simple stone tool, no complex tools such as axes—although anybody who ever tried to make a stone tool knows that it is quite an accomplishment. Neandertals do not have fine art such as adornments or cave paintings, but they know burial objects. That shows that the visual markers often locate the Neandertals somewhere between the extremes of the culture-devoid animal and a truly human culture. We may turn to an examination of the visual setting for further clues. In which kinds of behaviors do Neandertals engage in the pictures? Clearly, it makes a huge difference whether scientists and artists imagined the Neandertals in a burial scene, caring for their dead in the belief in an afterlife, or if they showed them in a cannibalistic scene with some kind of morbid barbecue. While the material achievements most common in images of the Neandertals can be ambiguous, their way of living may tell a clearer story.

A reconstruction of prehistoric life or prehistoric humans will always be hypothetical. We have to fill the gaps in our knowledge with imagination. Because we are the only humans today, our reconstructions of other human species necessarily draw on and influence what it means to be human.

Konstanze Weltersbach

FURTHER READING

Palmer, Douglas. *Neanderthal* (London: Channel 4 Books, 2000).

Schwartz, Jeffrey H. *What the Bones Tell Us* (Tuscon: University of Arizona Press, 1993).

Sommer, Marianne. "Mirror, Mirror on the Wall. Neanderthal as Image and 'Distortion' in Early 20th-Century French Science and Press," *Social Studies of Science* 36(2) (2006): 207–240.

Sommer, Marianne. *Bones and Ochre*: *The Curious Afterlife of the Red Lady of Paviland* (Cambridge, MA: Harvard University Press, 2007).

Stringer, Chris, and Gamble, Clive. *In Search of the Neanderthals. Solving the Puzzle of Human Origins* (London: Thames and Hudson, 1993).

Tattersall, Ian. *The Last Neanderthal. The Rise, Success, and Mysterious Extinction of Our Closest Human Relatives*, 2nd ed. (Boulder: Westview, 1999).

Trinkaus, Erik, and Shipman, Pat. *The Neanderthals. Changing the Image of Mankind* (New York: Alfred A. Knopf, 1993).

Java Man

Pat Shipman

Missing links are unknown—missing—by definition. Between 1887 and 1900, a Dutch anatomist, Eugène Dubois (1858–1940), searched for and found one of the most famous missing links, a transitional ape-man fossil that he called *Pithecanthropus erectus*. (In modern classifications, Dubois's specimens are known as *Homo erectus*.) These were the very first specimens to provide tangible proof of the controversial hypothesis that humans had evolved from ape-like ancestors. The very idea that humans were subject to evolution, like other creatures, was startling to some. Others already believed humans had evolved, but were surprised because *Pithecanthropus*'s combination of physical features showed that bipedalism and upright posture preceded the enlargement of the brain to modern size. Many scientists had expected that humans' large brain came first because they saw the brain as the feature that made us special, that gave us intelligence, language, the ability to make tools, and moral sensibilities. But *Pithecanthropus* said otherwise.

Dubois's analysis and description of the fossils as a transitional form between apes and humans was received skeptically, both because of the species' unexpected features and, perhaps, because Asia as the birthplace of the human lineage had not yet come into vogue. Ironically, even though Dubois had accomplished the unprecedented feat of finding the missing link between apes and humans, he was widely disbelieved. Dubois responded to this criticism with publications, lectures, and displays of the fossil that made it and human evolution central issues in science for the first time.

Dubois and his *Pithecanthropus erectus* are iconic because they permanently changed the way people thought about evolution. The discoverer and his discovery brought humans irrevocably into the evolutionary framework that had been constructed and established by earlier scholars.

EVOLUTION AND THE MISSING LINK

From the time that evolutionary ideas were first expressed, the evolution of *humans* was the most controversial point, as it is still today. Long prior to Dubois's life and work, scientists, philosophers, and laypeople alike debated whether or not humans were part of the animal world. Were humans subject to the same natural laws as other species, or were they something above or separated from the natural world? Darwin's book *On the Origin of Species* (1859) provided a specific focus for debate, though Darwin was not the only scholar wrestling with the ideas of selection and transmutation. Still, his book was a masterful statement of evolutionary principles as they were, and could be, known by nineteenth-century naturalists with no knowledge of the mechanisms of inheritance.

Darwin was famously slow to develop his ideas and produce the book, which he worked on for twenty years after he returned to England from his five-year voyage on the *H.M.S. Beagle*. Part of the difficulty lay in conceptualizing *how* evolution might occur. Another major difficulty slowing Darwin's productivity was his personal abhorrence of controversy. He foresaw that the theory of evolution—*as applied to humans*— would cause an uproar in religious circles and his beloved wife Emma held orthodox religious views.

For example, his reluctance was clearly demonstrated in 1857, when Wallace asked Darwin in a letter if he intended to discuss human origins in the book he was working on, then tentatively titled *Natural Selection*. Darwin replied, "I think I shall avoid the whole subject, as being so surrounded with prejudices, though I fully admit that it is the highest and most interesting problem for the naturalist" (Burkhardt and Smith, 1985, pp. 515–516).

When Alfred Russel Wallace (1823–1913) independently developed ideas on natural selection and sent a draft article to Darwin for his comments in 1858, Darwin was appalled. He felt he had taken too long in writing up his own ideas and that Wallace would publish first and take credit for the ideas. Darwin was prodded into taking action. Friends and colleagues of Darwin arranged that Wallace's article and some notes of Darwin's were presented together at a Linnean Society meeting in July of 1858. Since D comes before W, this maneuver established Darwin's priority to the ideas and yet gave Wallace—who was not nearly as well-connected as Darwin—entrée into the English scientific world. After the Linnean Society meeting, Darwin produced the manuscript of *Origin of Species* in just over a year.

Darwin prudently confined his discussion of human evolution to a single pregnant sentence: "Light will be thrown on the origin of man and his history." Yet he feared this caution was not drastic enough to save him from criticism. During the process of negotiation with John Murray to publish *Origin*, an anxious Darwin wrote to his friend Charles Lyell and queried timidly, "Would you advise me to tell Murray that my book is not more *un*-orthodox than the subject makes inevitable? . . . I do not discuss the origin of man. I do not bring in any discussion about Genesis, &c &c" (F. Darwin, 1887, vol. 2, pp. 151–154).

Darwin hoped to minimize controversy but in this aim, he was unsuccessful. The first edition of *Origin of Species* sold out immediately and the book was widely reviewed and hotly debated, precisely because of evolutionary theory's implications for human origins. Adam Sedgwick, Darwin's former tutor at Cambridge, felt that Darwin's theory would overthrow Christian belief and endorse the ruthless and selfish struggle for survival. In a later printing, Darwin added the daring word "much" to his timid sentence, "Light will be shed on the origin of man and his history" (Darwin, 1859, p. 449). He did not discuss human evolution further in that book.

The crux of the famous exchange on the subject of evolution between Huxley and the bishop of Oxford, Samuel Wilberforce, clearly concerned *human* evolution. At a meeting of the British Association for the Advancement of Science on 30 June 1860, Wilberforce mounted an open attack, ridiculing Darwin's book and Huxley's favorable review of it. Perhaps in an attempt to lighten his rather long and pompous speech, Wilberforce asked Huxley slyly if it was on his grandfather's or grandmother's side that he was descended from an ape. The dominant Victorian view was that women—ladies—were the delicate and genteel upholders of morality and should be protected from anything harsh or vulgar. Implying that Huxley's grandmother had intercourse with an ape was a jest in decidedly poor taste.

Huxley seized the occasion and replied,

a man has no reason to be ashamed of having an ape for his grandfather. If there were an ancestor whom I should feel shame in recalling, it would be a *man* [i.e., Wilberforce], a man of restless and versatile intellect, who, not content with an equivocal success in his own sphere of activity, plunges into scientific questions with which he has no real acquaintance, only to obscure them by an aimless rhetoric, and distract the attention of his hearers from the real point at issue by eloquent digressions, and skilled appeals to religious prejudices. (L. Huxley, 1900, p. 199)

Evolutionary theory gained some converts that day thanks to Huxley's quick wit and Wilberforce's ponderous jest.

Considering the furor that *Origin* provoked, it is important to realize that Darwin did not use evolutionary theory as a theoretical framework for evaluating human fossils. At the time of his writing, the only fossils of human ancestors (hominins) known to the English-speaking world had modern anatomy. Thus, evolutionary theory as applied to humans was fossil-free at the outset. Darwin's work depended upon observations and comparisons of living animals and on his knowledge of selective breeding in domestic animals, not upon fossil hominin specimens.

Even though the first Neandertal fossils had been discovered in Germany in 1856, they were not known in England when Darwin was writing *Origin of Species*. In 1861, the description of the Neandertal fossils by German anatomists was translated into English, but the problem of human evolution was still avoided. One of the early, detailed treatments of these fossils in English was in Huxley's *Evidence as to Man's Place in Nature* (1863). Huxley presented three main types of evidence attesting to the phylogenetic connection between apes and "man." He reviewed the discovery of nonhuman primates and summarized the general features of chimpanzees, gorillas, orangutans, mandrills (now considered monkeys), and gibbons. He explained the numerous, detailed anatomical resemblances that unite humans and apes into a group that shared a common descent from an ancestral stock that was, at that time, unknown. But when Huxley came to consider the known fossil record of humans, he did not suggest they were transitional forms linking apes to humans.

Many scientists of the day, Huxley included, were preoccupied with attempting to catalogue and arrange the various races of mankind into a hierarchy based on their skulls. He pointed out that apes, although possessing the same anatomical parts of as humans, were nonetheless quite different; even "primitive races" were unmistakably human. One key issue was in brain size:

> It must not be overlooked . . . that there is a very striking difference in absolute mass and weight between the lowest human brain and that of the highest apes. . . . This is a very noteworthy circumstance, and doubtless will one day help to furnish an explanation of the great gulf which intervenes between the lowest man and the highest ape in intellectual power. (Huxley, 1863, p. 140)

In the Neandertal cranium, Huxley recognized many pithecoid or ape-like characters, such as its massive brow ridges; long, flat cranial profile; and sloping occiput, or back of the skull.

> Under whatever aspect we view this cranium . . . we meet with ape-like characters. . . . But Professor Schaaffhausen states . . . that the cranium holds 1033.24 cubic centimeters of water. . . . So large a mass of brain at this, would alone suggest that the pithecoid tendencies, indicated by this

skull, did not extend deep into the organization; and this conclusion is borne out by the dimension of the other bones of the skeleton given by Professor Schaaffhausen, which show that the absolute height and relative proportions of the limbs, were quite those of a European of middle stature. (Ibid., p. 205–206)

To Huxley, the Neandertals were not a missing link. "In no sense, then, can the Neanderthal bones be regarded as the remains of a human being intermediate between Men and Apes. At most, they demonstrate the existence of a Man whose skull may be said to revert somewhat towards the pithecoid type" (Ibid., p. 206).

In short, though Huxley found Neandertals more ape-like than other "races" he thought they were still *Homo sapiens,* like all living races of mankind. Thus, Huxley did not address the primary issue under contention: that humans had evolved from an ape-like ancestor. In Neandertals, Huxley did not believe he had an appropriate fossil on which to hang such a claim.

Similarly, a contemporary British scientist, C. Carter Blake, titled one of his articles on the Neandertal remains "On the cranium of the most ancient races of man." In the nineteenth century, racial origins—not human origins—were the dominant focus of study when Neandertal fossils were the subject.

In Germany, the reaction to these fossils was generally similar. Biologist Ernst Haeckel was an avid promoter of evolutionary theory and wrote a series of books on evolution that influenced scientists and laypeople alike. In *Natürliche Schöpfungsgeschichte* (1868), or *The History of Creation* (1873), Haeckel explicitly proposed that there was an evolutionary link between humans and apes. Though he had never seen one, Haeckel dared to predict what that missing link would be like once it was found. He even dared to give this link a name:

The Men, or Pithecanthropi, very probably existed towards the end of the Tertiary period. They originated out of man-like apes, or Anthropoides, by becoming completely habituated to an upright walk. . . . Although these Men must . . . have been much more akin to real Men than the Man-like Apes could have been, yet they did not possess the real and chief characteristic of man, namely, the articulate human language or words, the corresponding development of a higher consciousness, and the formation of ideas. (Haeckel, 1892, p. 398)

Elsewhere in the book, he hypothesized:

Those processes of development which led to the origin of the most Ape-like Men out of the most Man-like Apes must be looked for in the two adaptational changes which, above all others, contributed to the making of Man, namely *upright walk* and *articulate speech*. The two *physiological* functions necessarily originated together with the two corresponding *morphological* transmutations, with which they stand in closest correlation, namely, the *differentiation of the two pairs of limbs and the differentiation of the larynx*. (Ibid., pp. 405–406; emphasis in original)

Like Huxley, Haeckel didn't think Neandertals were the missing link.

A similar attitude was taken by Belgian scientists Julien Fraipont and Maximilian Lohest, when they published a monograph on additional newly found Neandertal skeletons from the cave site at Spy, Belgium. The nearly complete skull closely resembled that of the original Neandertal specimen from Germany; it was long, low, and adorned with brutish brow ridges and a big face. The leg bones, the authors felt, indicated a bent-kneed shuffling gait rather than true, upright movement as in humans—a point about which they were mistaken. The preoccupation of the time with human races prevailed. Fraipont and Lohest concluded that Neandertals were an example of an ape-like race of humans. Indeed, they titled their monograph *La race humaine de Néanderthal ou de Canstadt en Belgique* (The Human Race of Neanderthal or of Canstadt in Belgium).

The most significant historical issues here are that evolutionary theory was formulated by Darwin in the absence of fossil hominins and that, once those fossils began to be discovered, the first ones to be discovered were considered to be human: members of a primitive race, but human. Prior to Dubois's discovery, no known fossil had ever been postulated to represent the transitional ape-man that Haeckel hypothesized had existed.

DUBOIS AND THE DELIBERATE DISCOVERY

Unlike the scientists mentioned in this article up to this point, Eugène Dubois grew up in an era in which both evolutionary theory and a few fossil hominins were already familiar. Born on 28 January 1858, Dubois was the son of a pharmacist and sometime mayor of the town of Eijsden, in far southern Holland. He was an extraordinarily bright, ambitious, and athletic child. When Dubois was ten years old, Karl Vogt, a renowned German biologist, gave a lecture locally on the subject of evolution. Dubois was not permitted to attend but eagerly absorbed the news that the lecture had caused an uproar in the conservative region.

One of the pointed questions from the audience was: "Dr. Vogt, tell me this: Do apes have churches? Do they have libraries?" Those in the audience with traditional religious views thought the quip settled the point: Apes were apes and humans were unutterably different. But the science students in the audience found the remark hilarious, observing loudly that neither the illiterate peasants of Russian nor the inhabitants of darkest Africa had libraries or churches, yet they were still human. Clearly, human evolution as a concept was not yet accepted everywhere.

The debate fascinated Dubois. He persuaded his father to send him not to the local school but to a technical high school in Roermond that specialized in science. There he was able to study chemistry, zoology, mineralogy, and botany with the most modern laboratories and equipment. At the age of nineteen, Dubois graduated from Roermond and enrolled at the University of Amsterdam to study medicine. Medical studies involved a general and thorough grounding in anatomy, biology, chemistry, and botany in those days. During his education, Dubois abandoned the Catholic faith of his family because, he later said, he felt the Church hindered free thinking and that it was wrong to mix philosophy with natural science. This decision to renounce his faith was a blow to his family, especially his sister who became a nun.

Dubois was an extraordinarily hard-working student who rose at 3 or 4 a.m. to study and his professors recognized his brilliance and his successful career in anatomy seemed assured. In 1881, he graduated and was offered an assistantship under Max Fürbringer, an internationally renowned anatomist and the chair of the department of anatomy at the University of Amsterdam. By 1884, Dubois had qualified as a physician and was promoted to prosector of anatomy, in charge of human gross anatomy and anatomy for art students. He began a research program on the embryological development and comparative anatomy of the larynx, a clear indication that he had remembered Haeckel's words about the importance of speech in the missing link. Late in 1885, Dubois was offered a lectureship at the University of Utrecht, but he turned the position down after being assured by Fürbringer that Dubois would succeed to the chair of anatomy in a few years' time when Fürbringer retired.

In 1886, Dubois published two articles on the comparative anatomy of the larynx which marked the beginning of his research career. Under pressure from Fürbringer, Dubois reluctantly cited his superior's ideas in his articles, but he felt he had been forced to do so. He became extremely sensitive to any suggestion that he was not independent of Fürbringer and feared his mentor was trying to take credit for his ideas. In the same year, Dubois was promoted to lecturer at the University of Amsterdam and married one of his former art students, Anna Lojenga, but these happy events did nothing to assuage his professional insecurity. He felt overshadowed by Fürbringer and anxious to establish his own reputation. The publication of the monograph by Fraipont and Lohest on the new Neandertal fossils from Spy reawakened Dubois's interest in human evolution and he decided to change the focus of his research.

Specifically, Dubois decided to find the missing link, setting an unusual goal for himself and using unusual methods to attain it. With the logic and thoroughness that had served him well as an anatomist, Dubois began working out where he had the best chance of finding such a fossil. By reading and synthesizing the scientific information available to him systematically, Dubois constructed a novel plan based on five primary points.

1. The missing link had to be more ape-like than the Neandertals found in Europe and was therefore most likely to be found in the tropics, where modern apes lived. Asia was a probable location because gibbons were the living ape closest to humans according to Haeckel, and because one of the only fossil apes known at the time was a genus from the Siwalik Hills of British India. This fossil ape, called *Anthropopithecus*, had been identified as a chimpanzee ancestor by the British paleontologist Richard Lydekker.

2. Fossils of other animals were known from the island of Java in the Dutch East Indies (then known as the DEI and now Indonesia). These animals were thought to be contemporaneous with the Siwalik fauna from India and quite similar to their counterparts in India.

3. Thus, the fossils from the DEI were the right age, they were Asian in character, and the Indonesian fauna was likely to include an undiscovered missing link.

4. Limestone caves were excellent places for the preservation of fossils; most European fossils had been collected from limestone caves. Sumatra, the island next to Java, was riddled with limestone caves that had not yet been examined for fossils.

5. The untouched caves of Sumatra had the correct geology, topography, climate, fauna, and antiquity to preserve the fossils of the missing link. Therefore, Dubois concluded, he must go to the DEI to find the missing link. Dubois later published these points, expounded at much greater length.

As Dubois began welding these facts into a rigorous framework, he discussed them with a colleague, Max Weber, expecting support. To Dubois's surprise, Weber told him bluntly, "Don't throw away all you have for a figment of your imagination, like the foolish boy in Grimm's fairy tale." After hearing Dubois's carefully reasoned arguments about why the missing link had to be in the DEI, Weber added, "For all that, you will not find the missing link, and you will give up your future." Dubois protested, "But if I find the missing link. . ." Weber replied, "Ah, then you are in one strike a famous man" (Dubois, n.d.; Bernsen, 10 December 1930).

Dubois applied to the secretary-general of the Colonial Office for a grant to go to the DEI to find the missing link. He was advised "don't let yourself be carried away by that crazy book of Darwin's." But Dubois was entranced with his novel hypothesis and deeply unhappy working under Fürbringer. Despite this discouragement, he decided to leave his promising university career to seek fame and fortune in the DEI. In the end, his friend Weber realized that Dubois would not be dissuaded from the plan. He wrote in a letter:

> After thinking more about your plans it has become even clearer to me . . . that whether or not you go to the East Indies, you have lost your link with anatomy. . . . From your point of view of teaching and its value and from your pessimistic ideas of science and from your outspoken dislike of teaching I think that, for you, happiness can never lie in the theoretical explanation. Posts of this kind are so hard to come by that they ought not be occupied by people who are not completely content in this working environment. . . . You know the work demanded in anatomy and you didn't like it, unless it involved your own research; you like it so little that you want to choose another working environment. I think that you have already decided in this matter in your heart. (Weber to Dubois, undated letter, probably 1887)

Despite receiving little encouragement, Dubois quit his university post and enlisted in the Royal Dutch Indies Army as a military surgeon in order to get to the DEI and support his family while he was there. He was required to sign on for an eight-year tour of duty. His wife Anna and their infant daughter, Eugenie, accompanied him to the DEI.

This was a bold and potentially dangerous move. Resigning a top university position to accept an ill-paid, temporary position in the military was a high-stakes gamble for a married man with a growing family. Dubois had no interest in a military career and his chances of returning to an academic post after publicly resigning from a coveted position were slim. His colleagues were likely to regard him as unstable and difficult. (Indeed, when he returned from the DEI with the missing link fossil in his hand, he still had some difficulty obtaining a professorship.)

Further, going to the East Indies involved a real risk of death or disease. Even as late as the nineteenth century, the European death rate in the DEI was 50 percent during times of epidemics and in normal times was still very high. For example, between 1816 and 1845, one in fifteen Dutch soldiers died in the DEI, which was almost twice the death rate of soldiers in Amsterdam and 1.5 times the rate of death of indigenous Indonesian soldiers during the same period. Malaria, cholera, smallpox, typhoid, dysentery, and diarrhea were the primary killers and afflicted most long-term European colonials; syphilis was rampant particularly among military men. Disease was not the only problem. Between 1873 and 1904, there was an active and bitterly contested war in the DEI between the Dutch colonial rulers and the local people in the northern Sumatran province of Aceh. Many soldiers were wounded or killed in the Aceh war, but still more soldiers died of disease than of battle wounds.

Of course, physicians were still at risk of disease, since there were no known cures for most tropical diseases. For example, quinine was known to work against malaria but the effective dosage and duration of treatment were not known. Physicians were in very short supply in the colony. In 1870, there were 49,000 Europeans in the entire DEI and only 156 medical officers. At that time, only half of the medical officers survived their first five years in the colony and very few survived the twenty years of service that was required for a pension.

It was 1887 when Dubois enlisted and he, his wife Anna, and their young daughter Eugenie sailed to the DEI. At the time of sailing, Anna Dubois was pregnant with their second child. She faced the prospect of delivering a baby in a colony with few white women, fewer medical facilities, and many tropical diseases.

The reality of these risks can be seen in the story of the Dubois family. Before they returned to the Netherlands in 1895, Anna had given birth to three more children. The two boys, Jean and Victor, and Dutch-born Eugenie survived their years in the Indies, but a fourth child died at birth. There is no way to know if that child would have lived had it been born in Amsterdam. Dubois and his family frequently battled various diseases and Dubois nearly died of malaria on several occasions. Further, Dubois's father never approved of his son's wild scheme to quit his job and go off to the colonies to find the missing link. Dubois longed to return home in triumph; he wanted desperately to "make good" in his father's eyes. Sadly, his father died while the young family was in the DEI; Dubois never got to say goodbye. The sacrifices Dubois made for science were real and terrible.

FINDING THE MISSING LINK

How did Dubois find the missing link? Stationed in Sumatra, Dubois used his time off from the hospital to look for fossils. In 1888, Dubois published the article based on his ideas proving that the Dutch East Indies were the place that the missing link would be found. Cleverly, he ended his article with provocative words:

> It is obvious that scholars from other countries will soon realize the promise of the East Indies. They will come and search for important fossils here and will find them, unless the Dutch authorities do something more to support such scientific work. And will the Netherlands, which has done so

much for the natural sciences of the East Indian colonies, remain indiffer-
ent when such important questions are concerned, while the road to their
solution has been signposted? (Dubois, 1888, p.165)

Of course, he was erecting a clear signpost to the possibility that
important fossils could be found in the DEI with his article. He hoped
that the fear of being beaten to the discovery by other nations would
spur the government into supporting his work. He had already found a
good many fossils and this scientific publication showed how important
and feasible this scientific work was. In March 1889, the Governors-
General of the Dutch East Indies passed a resolution granting him a year
of leave from his medical duties and seconding him to theDirector of
Education Religion and Industry, Governor Kroesen, in order to search
for fossils. Given how badly the Dutch army needed physicians, Dubois
must have been very persuasive to be relieved of his medical duties. In
addition to his regular pay, Dubois was allotted an extra 250 guilders a
month (which probably increased his salary by 30 percent), free transport, and the use of two civil engineers, Gerardus Kriele and Anthonie
de Winter, to supervise a crew of up to fifty forced laborers. After a year,
Dubois had found thousands of fossils but no missing links in the Sumatran caves. He received permission on 14 April 1890, to move his operations to Java, the territory of Governor Groeneveldt.

Dubois's plan was for the men to search caves in the wet season, a
standard fossil-hunting technique in Europe. Innovatively, he also
decided they would prospect the exposed riverbanks in the dry seasons.
Once the men had settled in to excavate in a promising area, Dubois
returned home to await regular reports and shipments of fossils from
Kriele and de Winter. Soon Dubois requested and received a three-year
commitment to his research program in Java.

In August 1891, Dubois marched his men along the banks of the
Bengawan Solo River, looking for likely areas to excavate. Responding
to an intuition that Dubois himself admits a religious man might call the
providence of God, he chose a point bar that extended into the river.
Almost immediately, they begin finding so many fossils that Dubois
referred to the site as his charnel house. In September 1891, the men
excavated a molar tooth of an ape-like primate. In his report to Governor Groeneveldt, Dubois gave this find the same name that has been
used for the chimpanzee-like fossils from India:

The most important find was a molar (the third molar of the upper right
side) of a chimpanzee (*Anthropopithecus*). The genus of anthropoid apes,
occurring only in *West-* and *Central equatorial Africa* today, lived in British
India in the Pliocene and, as we can see from this discovery, during the
Pleistocene in Java. (Dubois, 1892a, pp. 13–14)

In October, the team turned up the skullcap of a higher primate from the same level of the site. Dubois wrote a triumphant report to Groeneveldt.

> Near the place on the left bank of the river where the molar was found, a beautiful skull vault has been excavated that, undoubtedly (like the molar), has to be ascribed to the genus *Anthropopithecus* (*Troglodytes*). . . .
>   As far as the species is concerned, the skull can be distinguished from the living chimpanzees: first because it is larger, second because of its higher vault. . . . The height of the frontal part is not lower than the human skull from Neanderthal [*sic*] or the first skull from Spy; but the fossil chimpanzee deviates from these because its parietals are flatter and the occipital is less developed. With the Pliocene *Anthropopithecus siva-lensis*, of which there is only an incomplete mandible known, only the molar can be compared directly. Probably there is a close affinity between them . . . Besides, the most important fact is that the living chimpanzee, in his teeth, approaches humans more closely than does the gorilla or the orang-utan, which is found in the same region of Java, while this Pleistocene chimpanzee approaches the human more closely because of its skull. (Dubois, 1892b, p. 13)

On 30 December 1891, Dubois also wrote to Governor Kroesen in Sumatra about his find, saying, "The creature to which this skullcap had once belonged was truly a new and closer link in the largely buried chain connecting us to the 'lower' animals."

Hoping to find more of his fascinating creature, Dubois and his men resumed excavation at Trinil in May 1892. Dubois immediately contracted a life-threatening case of malaria and, as soon as he was able, left the crew to work under Kriele and de Winter. In August, the men found a fossil femur (thigh bone) of an animal they have not encountered before; the specimen reached Dubois at his home in Toeloeng Agoeng in September.

The exact progression of Dubois's assessment of the fossil finds can be traced with unusual precision because of documents preserved in the Dubois Archives in Naturalis, a museum in Leiden, the Netherlands. Dubois immediately recognized that the femur was from a large, strongly-built higher primate and identified it as belonging to the same individual of *Anthropopithecus* as the molar and skullcap in his third quarterly report to Groeneveldt, written on 23 November 1891. After preliminary analysis, Dubois presented an unexpected but firm conclusion:

> This being was in no way equipped to climb trees in the manner of the chimpanzee, the gorilla, and the orang-utan. On the contrary, it is obvious from the entire construction of the femur that this bone fulfilled the same mechanical role as in the human body. Taking this view of the thigh bone, one can say with absolute certainty that *Anthropopithecus* of Java stood upright and moved like a human. (Dubois, 1892a, pp. 12–13)

The fossil femur proved one of Haeckel's predictions about the missing link to be correct. Later in the same report, Dubois took an even stronger position:

> Because of this find, a surprising and important fact has been brought to light. The Javanese *Anthropopithecus*, which in its skull is more human than any other known anthropoid ape, already had an upright, erect posture, which has always been considered to be the exclusive privilege of humans. Thus this ancient Pleistocene ape from our island is the first known transitional form linking Man more closely wit his next of kin among the mammals. *Anthropopithecus erectus* Eug. Dubois, through each of its known skeletal elements, more closely approaches the human condition than any other anthropoid ape, especially in the femur . . . the first step on the road to becoming human taken by our ancestors was acquiring upright posture. Consequently the factual evidence is now in hand that, as some have already suspected, the East Indies was the cradle of mankind. (Dubois, 1892a, p. 14)

Within a few days, Dubois realized he had miscalculated the skullcap's cranial capacity and wrote to Groeneveldt on 4 December 1892: "The brain of this transitional form was considerably larger than one would gather from the report . . . nearly 1,000 cc." This made the brain of the new creature almost twice as large as any living ape but only three-quarters the size of a modern human brain.

On 19 December, Dubois received a chimpanzee skull, which had been sent by Weber at his urgent request. Dubois wanted it so he could compare the Trinil skullcap with the skull of the modern ape directly. These more detailed studies caused Dubois to reach a new assessment of his fossil form. On 28 December, he wrote a letter to Groeneveldt which began: "I have, Your Excellency, the honor of offering the first installment of the description of some of the fossils I have collected. This installment deals with only one species, *Pithecanthropus erectus*."

As the original handwritten document reveals, Dubois first wrote the letter *A*, intending to continue using the name *Anthropopithecus*, then lifted his pen and

This photograph of the skullcap from Dubois's 1894 monograph shows the superb quality of his illustrations. When the matrix was removed from inside the skullcap, Dubois estimated the total cranial capacity in life as 855 cc: about 75 percent of that of an average modern human. (Credited to Naturalis, © Nationaal Naturhistorisch Museum, Leiden, The Netherlands.)

rewrote over it the *P* of *Pithecanthropus*. Remarkably, the very moment of his decision is still visible.

By adopting Haeckel's generic name for the missing link, Dubois made a bold statement of his opinion that his fossil was indeed the missing link he had sought. By abandoning Haeckel's trivial name, *alalus* (meaning "speechless"), Dubois admitted that he did not have the specimens that might have told him if the species could speak, which is an exceedingly difficult point to confirm in any case. By substituting *erectus* for *alalus*, Dubois both took a conservative approach to naming his specimen yet simultaneously declared that he had found an upright, bipedal, or erect form.

THE TRANSITIONAL FORM

While Dubois struggled to write his monograph on the three fossils that were already in hand, excavation continued at Trinil. As before, thousands of mammalian fossils were recovered but none were recognized as belonging to *Pithecanthropus* at the time.

Before Dubois finished his monograph, professional criticism began. On 6 February 1893, an Indies newspaper, *Bataviaasch Nieuwsblad*, printed a letter to the editor which was almost certainly written by the editor himself. The letter mocked Dubois's last quarterly report and his claim that the Indies were the cradle of mankind. The author charged that

> The esteemed D. Dubois, prejudiced because he has completely swallowed Darwinism, has gone too far; he has constructed a connection between the human femur and the monkey skull and molar where none has ever existed. . . . In the meantime, this publication of Dr. Dubois will create a furor, especially in the "Land of Intellectuals," and it appears to me that the facts must be reviewed by an impartial committee of experts, before the government should endorse such a report. (Anonymous, 1893)

The letter was signed "Homo Erectus," the first use of this name in print.

This public accusation of professional incompetency was closely followed by dual personal tragedies. On 11 April 1893, Dubois's father died suddenly in the Netherlands, without seeing his once-promising son achieve his seemingly impossible aim. On 30 August, Dubois's wife Anna miscarried and the child died. Dubois was the physician in attendance. The sole record of this event is in Dubois's pocket calendar for the year. On the date, he wrote "Anna abortus," using the Latin word *abortus* to mean a natural abortion or miscarriage. The rest of the week's spaces were blank. Later, he crossed the words out with pencil lines pressed so hard that they nearly ripped the page. Still later, he rewrote the harsh words.

Dubois completed his manuscript and carried it to the printer in Batavia in January 1894. His thirty-nine-page monograph was entitled in German, *Pithecanthropus erectus, eine menschenaehnliche Uebergangsform aus Java*. In English, the title reads, *Pithecanthropus erectus, a manlike transitional form from Java*. In case readers overlooked the implications of his using Haeckel's generic name, *Pithecanthropus*, Dubois made his position clear in his title and throughout the text, writing: "*Pithecanthropus erectus* is the transitional form which, according to the theory of evolution, must have existed between Man and the anthropoid apes; he is Man's ancestor" (p. 31).

Not only was Dubois's monograph unusual for its bold scientific stand, it was also unusual because he used a new approach. He did not compare his fossils to the skeletons of the various human races to discover which one they most closely resembled. He put his fossil into an evolutionary dichotomy, with humans on the one hand and living apes on the other. This procedure demonstrated his belief that *Pithecanthropus* lay between apes and humans and not among the human races.

Because he believed that *Pithecanthropus* was ancestral to humans, Dubois made another inflammatory declaration. He created a new zoological family, the Pithecanthropidae, to accommodate his new genus and species—a family that lay between the apes (the Simiidae in his terms) and the humans (the Hominidae in his terms). He had found fossils that, by his analysis, proved the origin of humans from an ape-like ancestor.

The anatomical regions represented by the original *Pithecanthropus* specimens were especially fortuitous. Given Haeckel's prediction about the upright posture of the missing link, having a femur that was so clearly adapted for bipedal walking was important. Another common expectation was that the brain evolved and enlarged early in human evolution, because it was intellectual capacities—including linguistic ones—that set humans apart from apes. Thus, having the skull that contained the brain and the femur that propelled the body in walking was crucial in reaching and demonstrating the conclusion that these fossils represented a transitional form. The tooth, though not nearly as commonly the focus of attention as the skullcap or femur, was an important supporting element that made clear the fossil's identity as higher primate.

Dubois's descriptions, illustrations, and photographs were superb; the comparative techniques and measurements were careful and some innovative. Intending his work to be anatomical, Dubois paid little attention to the geology of the site or to the reasons that he had judged the molar, femur, and skullcap to belong to a single individual. These omissions left room for skepticism and criticism, which soon followed. Dubois sent copies of his monograph to friends and colleagues in Europe, so that his ideas would be known when he returned to Europe in 1895. He did not anticipate extensive criticism once his full description and analysis was laid before the scientific community.

Before Dubois returned to Europe, he did two important things. First, he had a special monument made and placed at Trinil, to mark for all time where his *Pithecanthropus* had been found. Since he was a man with a large ego and great self-confidence, it is especially significant that neither his name nor his initials appears on the monument. The thing to be celebrated and commemorated was not his effort, but *Pithecanthropus erectus* (which he called *P.e.* here) itself.

In autumn 1895, Dubois traveled widely in Europe, speaking at meetings and showing his fossils—trying to persuade his colleagues that he had, indeed, found the missing link between apes and humans. When he went to Paris, France, he spoke in front of the École d'Anthropology (the School of Anthropology), where Léonce-Pierre Manouvrier was a prominent professor. Working closely with Manouvrier, Dubois compared the femur of his *Pithecanthropus* to 400 human femurs, to document the differences and similarities between human thigh bones and that of *Pithecanthropus*.

According to a Dubois family story, Manouvrier and Dubois finished their work late in the evening and went together to a nearby café to eat and discuss their results. They compared measurements and anatomical features, talking intensely over their food. Finally, when the café was closing, they left to walk to Dubois's hotel, still arguing and discussing. After walking several blocks, Dubois suddenly grabbed Manouvrier by the arm and cried, "Où est *Pithecanthropus?*" (Where is *Pithecanthropus?*) He had left the suitcase containing his precious fossils in the café!

Dubois dashed madly through the streets, frantic to reclaim his life's work. He arrived back at the café as the proprietor was locking the door. Yes, the man remembered them, they had been the last two patrons in the café, but a suitcase? No, he knew nothing about a suitcase. Dubois forgot his command of the French language and shouted in Dutch. The fossils, the fossils must be there! He must have them back! Finally Manouvrier, who was more suave and calmer than the nearly hysterical Dubois, managed to make the problem clear and their waiter cheerfully produced the old valise he had found under their table.

Dubois quickly opened it to check that his specimens were there, as they were. His heart pounded with relief and fear. What if he had lost the specimens he had sacrificed so much to find? What would happen to his reputation, his ideas, his theories? When Manouvrier left Dubois at this hotel, Dubois was still so stunned by the near-disaster that he was very distracted. Manouvrier recommended that Dubois sleep with the fossils under his pillow that night.

Dubois also undertook a prolonged trip to India to examine the original specimens of *Anthropopithecus* in the Indian Museum in Calcutta and to search the Siwalik Hills where Lydekker had found them. During this trip, Dubois began to receive critical reviews of his monograph that had appeared in scientific journals. The first to reach him, sent with a box of photographic equipment was a copy of a German journal including a scathing article by Paul Matschie who dismissed Dubois's reasons for linking the three specimens into one species, suggesting that Dubois had wrongly put an orangutan skullcap with a human femur. Many others would chastise Dubois for associating the various specimens into one species.

More criticism was soon forthcoming. Lydekker, the finder of *Anthropopithecus*, admitted that Java was a place where "the remains of a connecting form between man & the higher apes would be extremely likely to occur" but then launched into denigration of Dubois's work.

> Zoologists have naturally been attracted to the title of the work before as it claims in no uncertain terms that such a missing link has actually been discovered. A feeling of disappointment will, however, probably come over the student, when he finds how imperfect are the remains on the evidence of which this startling announcement is made and when he has submitted them to a critical examination he will probably have little difficulty in concluding that they do not belong to a wild animal at all. (Lydekker, 1895, p. 291)

Ironically, Lydekker had had much less of *Anthropopithecus* when he declared it to be a chimpanzee ancestor than Dubois did when he declared *Pithecanthropus* to be a human ancestor. Lydekker concluded that Dubois had found a microcephalic human and had mistaken it for a normal ape-man.

Other scholars criticized Dubois for comparing his skullcap only to apes and not to the various primitive races of humans. Several scholars emphasized the resemblances of parts of *Pithecanthropus* to a gibbon or another ape and suggested Dubois had mistaken a giant ape for an ape-man. Very few accepted his ideas at first without disagreement, only the American C. Othniel Marsh, the German Ernst Haeckel, and the Dutchman Hugo de Vries.

Dubois determined that he must face his accusers as soon as possible in order to show them his specimens and their features. In a letter written in April 1895, Dubois said,

> For *Pithecanthropus* my early arrival in Europe is urgently needed at present. This form has thoroughly awakened a large and general interest,

however, my description is wrongly understood by so many that it seems to be necessary to give a further, oral explanation and especially to display the specimens as soon as possible, by which means, I am fully confident, all reasonable doubt about the [transitional] character of the fossils I described will be removed.

Dubois's confident response to the criticisms changed many opinions. In his first few years back in Europe, Dubois lectured and attended conferences in many European cities, including Leiden, Amsterdam, Paris, Liège, Brussels, London, Edinburgh, Cambridge, Dublin, Berlin, and Jena. Those who saw his fossils and listened to his lectures often changed their opinion of *Pithecanthropus*.

Even those who did not agree with Dubois were forced to deal with his fossils. Between 1895 and 1900, Dubois's colleagues published a total of ninety-five articles on *Pithecanthropus* and Dubois himself published nineteen articles. It seems probable that, if Dubois had ignored the criticisms of his work or had not been able to counter them, the debate and discussion would have died out rapidly. Instead, by his perseverance and his continuing research, Dubois turned *Pithecanthropus* into an icon of evolution. The evolutionary transition from apes to humans became one of the most important scientific topics of the day. Partly because of the enormous interest in *Pithecanthropus,* Dubois was able to keep the excavation at Trinil going until 1900 under the supervision of Kriele and de Winter. The first textbooks on human evolution began to appear, all mentioning *Pithecanthropus,* and journals devoted to human evolution were founded, signaling the birth of human evolution as a distinct field of scientific study.

After his initial whirlwind of visits around Europe, Dubois grew bitter that, despite his publications and learned lectures, the scientific community still did not agree wholly with his interpretation of the fossils. He was particularly offended by the actions of Gustav Schwalbe, a German anatomist. Dubois had kindly sent Schwalbe a cast of the skullcap and had allowed him to study the fossils for some days in 1895. Then in 1899, Schwalbe published a very lengthy article about *Pithecanthropus* without asking permission of the man who had found the specimens and permitted him access to them. Dubois felt his kindness had been sorely misused and that Schwalbe had appropriated his material unethically. Dubois's old fear of having his ideas stolen began to resurface. This feeling was aggravated by a sudden announcement that Madame Selenka, the wife of a German zoologist, was taking an expedition back to Trinil to excavate at Dubois's own site. The implication was that Dubois had missed important specimens, despite the years his men had worked at Trinil and despite the vast volume of dirt and rock they had excavated. Though the Selenka expedition found no more specimens of *Pithecanthropus*, this invasion of Dubois's site contributed to a period of withdrawal.

From about 1910 until 1922, Dubois published nothing about human evolution, though he worked on other research projects and published many articles. He had become deeply reluctant to let other scholars view his fossils. He felt such a personal connection to his *P.e.*, and had sacrificed so much to find his specimens, that he did not want to give others the chance to write anything about them until he himself had done every analysis he could think of. There were rumors Dubois had gone mad or reconverted to Catholicism and destroyed the fossils. As he had no proper facilities in which to store and protect his fossils, he kept the *P.e.* specimens in a bank vault for a while, then in a cabinet, and finally in a safe in his house.

In 1912, Dubois turned physical anthropologist Aleš Hrdlička, of the Smithsonian Institution, away at his front door without seeing him. Hrdlička, who had written in advance and traveled far to see these important fossils, was furious and went to the newspapers about the matter. Others were also turned away. The problem was aggravated when Henry Fairfield Osborn, a paleontologist and the immensely powerful head of the American Museum of Natural History, wrote to Dubois asking if J.H. MacGregor, a former student of Osborn's, might come to Amsterdam to study *P.e.* and make casts of the specimens. Osborn also wanted casts of the fossils to be used in a new human evolution exhibit. Dubois telegraphed Osborn that the fossils were under study and "inaccessible." He had no technicians to make casts to be distributed to museums and did not want other to have access to casts anyway.

In 1922, Osborn mounted a frontal attack, circulating a petition among the scholars in the scientific community protesting Dubois's refusal to allow other scholars access to his fossils. The signed petition was sent to the Royal Academy of Sciences of the Netherlands, as the most august representatives of scientific life in the Netherlands. The Academy took the complaints very seriously; Dubois's behavior had caused them serious international embarrassment. Wisely, they refrained from censuring Dubois or instructing the government to stop Dubois's salary or repossess the fossils, but the threat was clear. The Academy sent a formal letter to Dubois indicating that he was to make the fossils themselves and casts available to fellow scientists within a reasonable period of time. They agreed he might wait until he had published some additional works on the fossils within the year. Finally, in 1923, Dubois was forced to allow access to his fossils once again. In a gracious gesture, he invited Hrdlička and a group of his students to be the first to see them. He became involved once again in human evolution, particularly once Davidson Black found fossils similar to *Pithecanthropus* in China. In fact, Dubois's and Black's fossils are today considered to be the same species, *Homo erectus*.

Dubois sculpted this life-sized reconstruction of Pithecanthropus for the Paris Exhibition of 1900, using his eleven-year-old son as a model. He put a divergent big toe on each foot because he believed his transitional form would have ape-like feet, but it also made it fully erect and upright. This photograph shows the statue in the basement of Naturalis, the natural history museum in Leiden, The Netherlands, where it is kept. (Credited to Naturalis, © Nationaal Naturhistorisch Museum, Leiden, The Netherlands.)

From 1895 until his death in 1938, Dubois was honored widely. He received the Prix Broca in France, an honorary doctorate from the University of Amsterdam, an honorary fellowship from the American Museum of Natural History, and the Order of the Knights of the Netherlands Lion. However, his arguments and feuds with fellow scholars on many topics—primarily concerning *Pithecanthropus*—extended throughout his life. The independent mind, the enormous self-confidence, and the stubbornness that brought him success in finding the missing link also made him an intensely difficult man.

A TALE OF TWO ICONS

Dubois's dominant role in bringing humans into the evolutionary framework is clear in this brief historical review. He had original ideas, acted on them boldly, and met with unusual success, though he risked much and fought fierce criticism to persuade his opponents. For this, he can be considered iconic.

In this special case, it would be neither fair nor historically accurate to separate the finder from the find. Dubois felt a deep connection to his find and referred to it affectionately as *P.e.* throughout his life. He had a brass stamp made up with the initials *P.e.* on it, which he used on his correspondence with sealing wax. For the famous Paris Exhibition of 1900, Dubois sculpted a life-sized statue of *P.e.* based on the physique of his eleven-year-old son, Jean, who posed nude for him. The statue is rich with implicit, semiotic meaning. The quizzical expression on the statue's face suggests uncertainty, some intelligence, and confusion. Tellingly, he also portrayed *P.e.* holding a primitive antler tool in his hand to symbolize his more human qualities. Dubois gave ape-like features to the statue that were expressed through parts of the body he never saw. His statue had elongated, gibbon-like arms and fingers and a divergent big toe. Of course, speculation is necessary in making such a statue.

Using his son as *P.e.* might be dismissed as convenience and not a meaningful choice, but a Dubois family story about the exhibition of the statue in Haarlem, the Netherlands, suggests otherwise. Dubois's son Jean watched an elderly, countrified couple come up to the statue and stop, bewildered, in front of it. "And who," the woman wondered querulously, "is that?" Jean stepped toward them and blurted out, "That is my father!" The couple hurried away, doubtless wondering if the boy was mad. Jean meant that his father was the scientist and sculptor responsible for the statue and perhaps also that the statue was his ancestor. However, in a very real sense, *P.e.* and Dubois were the same person.

FURTHER READING

General information on Dubois's life is summarized in English in Shipman's biography of Dubois (Shipman, P. *The Man Who Found the Missing Link* [New York: Simon & Schuster, 2001]), Shipman and Storm (Shipman, P., and P. Storm. [2002] "Missing Links; Eugène Dubois and the Origins of Palaeoanthropology," *Evolutionary Anthropology* 11:106–116), and Theunissen's book (Theunissen, B. *Eugène Dubois and the Ape-Man from Java* [Dordrecht: Kluwer, 1989]). Many original documents about his life and a collection of thousands of Dubois's letters and manuscripts can be found in the Dubois Archives in the Naturalis, Leiden, the Netherlands. Source information was also taken here from the unpublished diaries written by Father J.J.A. Bernsen, who was Dubois's assistant between 1930 and 1932. These diaries are in the hands of the Dutch Franciscan order and contain a wealth of information about their conversations, Dubois's thoughts, and Dubois's life history.

For a good look at ideas about evolutionary theory in the mid-nineteenth century, nothing is better than Huxley, T.H. *Evidence as to Man's Place is Nature*, reprinted in 1900 as *Man's Place in Nature and Other Collected Essays* (New York: Appleton, 1863). It is perhaps the shortest and sweetest summary of the evidence linking apes and man into a single evolutionary tree that is accessible to modern readers.

Information about evolutionary ideas prior to Dubois's work can be found in many works. Desmond and Moore (Desmond, A., and J. Moore. *Darwin* [New York: Warner Books, 1991]) is lively reading and reveals much about Darwin and his colleagues. Brackman's book (Brackman, A.C. *A Delicate Arrangement; The Strange Case of Charles Darwin and Alfred Russel Wallace* [New York: Times Books, 1980]) on the coincident discoveries by Darwin and Wallace provides a fascinating insight into Victorian manners and Darwin's personality.

A broader overview of evolutionary concepts and discoveries can be found in Eiseley (Eiseley, L. *Darwin's Century* [Garden City: Anchor Books, 1961]) and Rudwick (Rudwick, M. *The Meaning of Fossils: Episodes in the History of Palaeontology* [New York: Neale Watson Academic Publications, 1972]).

The book by Trinkaus and Shipman (Trinkaus, E., and P. Shipman. *The Neandertals* [New York: Alfred A. Knopf, 1992]) focuses on the important

Neandertal discoveries and their impact on thinking about human evolution.

Anonymous. (1893). "Palaeontologische onderzoekingen oop Java," *Batavia-asch Nieuwsblad*, February 6, no. 57.

Blake, C. C. "On the cranium of the most ancient races of man," *Geologist*, June 1862.

Boomgaard, P. "Morbidity and Mortality in Java, 1820–1880: The Evidence of the Colonial Reports," in Norman G. Owen (ed.). *Death and Disease in Southeast Asia; Explorations in Social, Medical and Demographic History* (Singapore, Sudeny: Oxford University Press, 1987), 48–69.

Burkhardt, F., and Smith, S. (eds.). *A Calendar of the Correspondence of Charles Darwin, 1821–1882* (New York: Garland, 1985).

Busk, G. "D. Schaafhausen [*sic*], On the Crania of the most Ancient Races of Man," *Natural History Review* (1 April, 1861), no 2. Translation of Schaaffhausen, H. "Zur Kentniss der ältesten Rassenschädel," *Archiv. Verbindung Mehreren Gelehrten* (1858): 453–488.

Darwin, C. *On the Origin of Species* (London: Murray, 1859).

Darwin, F. (ed.). *Life & Letters of Charles Darwin*, 3 vols. (London: Murray, 1887).

de Moor, J. A. "An Extra Ration of Gin for the Troops; The Army Doctor and Colonial Warfare in the Archipelago, 1830–1880," in G. M. van Heteren, A. de Knecht-van Eekelen, M.J.D. Poulissen (eds.), *Dutch Medicine in the Malay Archipelago 1816–1942: Articles Presented at a Symposium Held in Honor of Professor De Moulin* (Amsterdam-Atlanta, GA: Rodopi, 1989), 133–152.

Desmond, Adrian, and Moore, James. *Darwin: The Life of a Tormented Evolutionist* (New York: Warner Books, 1992).

Dubois, M.E.F.T. "Over de wenschelijkheid van een onderzoek naar de diluviale fauna van den Nederlandsch Indië, in het bijzonder van Sumatra," *Natuurkundig Tijdschrift voor Nederlandsch-Indië* 48 (1888): 148–165.

Dubois, M.E.F.T. "Palaeontologische onderzoekingen op Java," *Verslag van het Mijnweizen. Extra bijvoegel der Javansche courant.* 3rd Quarterly Report of 1891, 1892a.

Dubois, M.E.F.T. "Palaeontologische onderzoekingen op Java," *Verslag van het Mijnweizen. Extra bijvoegel der Javansche courant.* 4th Quarterly Report of 1891, 1892b.

Dubois, M.E.F.T. *Pithecanthropus erectus, eine menschenaehnliche Uebergangs-form aus Java*, (Batavia: Landsrukkerij, 1894).

Fraipont, J., and M. Lohest. "La race humaine de Néanderthal ou de Cannstadt en Belgique—Recherches Ethnographiques sur des ossements humains dé-couverts dans les depôts quaternaries d'une grotte à Spy et détermination de leur âge géologique," *Archive de Biologie* 7 (1887): 587–757.

Gardiner, Peter, and Oey, Mayling. "Morbidity and Mortality in Java 1880–1849: The Evidence of the Colonial Reports," in Norman G. Owen (ed.), *Death and Disease in Southeast Asia; Explorations in Social, Medical and Demographic History* (Singapore, Sudeny: Oxford University Press, 1987), 70–90.

Herfkens, J. W. F. *De Atjeh-Oorlog van 1873 tot 1896 omgewerkt en aangevuld door J.C. Pabst* (Breda: De Koninklijke Militaire Academie, 1905).

Huxley, L. (ed.). *Life & Letters of Thomas Henry Huxley* (New York: D.Appleton, 1900).

Huxley, T. H. "*Evidence as to Man's Place in Nature,*" in T. H. Huxley, *Man's Place in Nature and Other Anthropological Essays* (New York: D. Appleton, 1900).

Kerkhoff, A. H. M. "The Organization of the Military and Civil Medical Service in the 19th century," in G. M. van Heteren, A. de Knecht-van Eekelen, M. J. D. Poulissen (eds.), *Dutch Medicine in the Malay Archipelago 1816–1942: Articles Presented at a Symposium Held in Honor of Professor De Moulin* (Amsterdam-Atlanta, GA: Rodopi, 1989), 9–24.

Knecht-van Eekelen, A. "The Debate about Acclimatization in the Dutch East Indies (1840-1860)," in "Public Health Service in the Dutch East Indies," *Medical History*: Supplement v. 20 (2000): 70–85, 80.

Lydekker, R. "Review of Dubois' *Pithecanthropus erectus*, eine menschenähnliche Uebergangsform aus Java," *Nature* 51 (1895): 291.

Matschie, P. "*Anthropopithecus erectus* E. Dubois," *Naturwissenschaftliche-Wochenschrift* 9 (1894): 122–123.

Schulter, C. M. "Tactics of the Dutch Colonial Army in the Netherlands East Indies," *Revue Internationale d'Histoire Militaire* 7 (1988): 59–67.

Shipman, P. *The Man Who Found the Missing Link* (New York: Simon & Schuster, 2001).

Shipman, P., and Storm, P. "Missing Links; Eugène Dubois and the Origins of Palaeoanthropology," *Evolutionary Anthropology* 11 (2002): 106–116.

Theunissen, B. *Eugène Dubois and the Ape-Man from Java* (Dordrecht, Boston, London: Kluwer, 1989).

Van Marle, A. "De Groep der Europeanen in Nederlands-Indië, iets over ontstaan en groei; III" *Indonesië* 5 (1951–1952): 97–121.

Peking Man

Brian Regal

One of the enduring mysteries in the history of evolution studies is the disappearance of Peking Man. When it was found in China in 1929, Peking Man was immediately seen as an artifact that had enormous implications for understanding human evolution and antiquity. Actually an example of *Homo erectus*, Peking Man's spectacular appearance was followed by its equally spectacular disappearance in the fog of the opening days of World War II. Originally hailed as an icon of evolution for the answers it was bound to give about human ancestry, its disappearance turned it into an icon of another sort. Some point to the disappearance of Peking Man as evidence that the whole thing was a hoax. What most people do not know is that after Peking Man was lost a considerable amount of subsequent material was and continues to be found at the same site that more than make up for those original missing pieces. While the disappearance of Peking Man is an intriguing mystery—one yet to be solved—the overall story is not one of an insidious cover up, but of Chinese scientists (as well as those from around the world) doggedly working under sometimes threatening conditions to explain and discover some of the most important artifacts in the story of our species.

In the late nineteenth and early twentieth centuries more and more attention was being paid not just to the mechanics of human evolution but to the place where the first humans appeared. The discovery of the Neandertalls and the Cro-Magnons caused great excitement and discussion, and increased interest in where our earliest ancestors came from. At the time, most who studied this question would have told you that the first humans had appeared in Asia. This answer was based on almost pure theoretical structures. (The notable holdout to Asia as the cradle of humankind was Charles Darwin who argued that Africa was more likely.)

Peking Man as the Chinese saw him in 1950.

The theory of the Asian origins of humans would not have anything like evidentiary support until Java Man was found in the 1890s. The changeover from the Western/Christian view that humans had first appeared in the Garden of Eden where God had placed Adam and Eve came during the Enlightenment when scholars began the modern search for answers about natural history and politics that did not need to resort to theological superstition or miracles.

PRELUDE IN THE WEST

Peking Man did not appear out of nowhere, nor did the idea to even look in Asia for the earliest humans. Once the literal story of Eden had been called into question, intellectuals were free to begin looking for other gardens. Attention began to move away from the traditional Holy Lands of the Middle East to India, then to central Asia and China. This shift coincided with the growing fascination with Asia, a place little known to Westerners. Besides being free of theological baggage, central Asia became a substitute for the Garden of Eden because it seemed to solve many of the problems associated with human origins without demanding too much in return. In Asia Western scholars had a region onto which they could project a history of the human race that Judeo-Christian dogma and tradition had no hold over.

The search for the origins of humankind, as well as what came to be known as the Asia hypothesis, began in seventeenth-century Europe as a result of a desire for national identity and a growing dissatisfaction with a literal reading of the Genesis story. It had little to do with Asia itself. Intellectuals and politicians in fledgling Western states looked for a heroic past to build up self-esteem and patriotism. In the late seventeenth century some French intellectuals claimed their people to be descendants of the Trojans, even the Titans of Greek mythology. It seemed logical to some that the open skies and clean cool air of high mountains would be the ideal locality for the first humans and so suggested the Himalayas of Tibet. Voltaire (1694–1778) thought that Adam must have taken his culture and society from India, Denis Diderot's (1713–1784) pioneering encyclopedia likewise argued that the oldest science came from the subcontinent, and Immanuel Kant argued for Tibet. In his *Histoire Naturelle* (1749–1767), Count de Buffon argued that man's birthplace must have been in a temperate zone because a

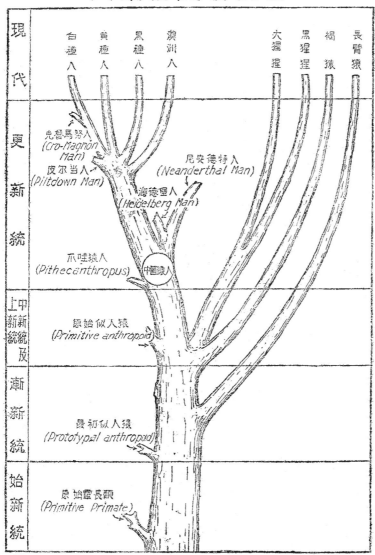

圖 53. 人類發展系統樹(在尼安德特人未發現新證據之前,我們未將這
        一枝列入人類的直系祖先的一枝上)

A Chinese drawn evolutionary tree showing Peking Man's place in
human evolution. This illustration is from 1950: note that Piltdown
Man is still being included in the human family line.

good climate was necessary to produce good men. Buffon theorized that
the Caspian Sea region and adjacent Caucasus Mountains would be a
logical place to look for evidence of their existence, hence the term Caucasian. Johann Friedrich Blumenbach (1752–1840), who had done so

much to advance the study of human biological relationships, also believed in the Asian origins of man. Indeed, central Asia became especially popular with Germans.

German philosophers were unhappy with what they saw as the cold, soulless views of the Enlightenment and so moved to a more romantic idealization of humanity's past. They sought to bring spirit, emotion, and passion back into philosophy, with nature study at its core. Friedrich von Schlegel (1772–1829) created stories of a glorious Germanic past that he believed better suited these lofty dreams. This new history became widely popular especially among the young. This led him to develop the notion of an ancient Indian-Germanic race for which he coined the term *Aryan* in 1819. The Aryans were believed to have originated in the regions north of India and to have swept down to conquer all of Eurasia. After Schlegel, others proposed central Asia as the home of the Aryans. The great philosopher G.W. Hegel (1817–1830) argued that the Aryans were the conquerors and masters of Europe. Some European Christians had grown uncomfortable with the traditional Semitic origin of man as told in the Bible and so were happy to look elsewhere. Indian origins were readily accepted because they seemed to link the Germans to a higher non-Jewish culture. By the middle of the nineteenth century, the assumption that Asia was the cradle of man and the belief in the existence and superiority of the Aryan race were well entrenched in European thinking.

Frenchman Joseph-Arthur de Gobineau (1816–1882), in *The Inequality of the Races* (1853), codified these ideas and argued that the Aryans were the world's superior race and that they originated in central Asia. Gobineau was a polygenist and as such believed each human "race" had a separate origin and so were not directly related to one another. He articulated the idea that history was a racial struggle and that to study it was a form of biology. He feared that what he considered mongrel races were overflowing the world and threatening to swamp the racial aristocracies he believed should be running it. His work became a major inspiration for later race theorists.

Although the Germans considered themselves descendants of the Aryans, they also considered themselves Nordics. The living embodiment of the Nordic ideal, Ernst Haeckel (1834–1919) loved sports, nature, and the "Fatherland." The creator of the first genealogical tree of life, Haeckel coined the term *phylogeny* in 1862 to argue that the embryonic forms of a species showed the stages that an organism went through during evolution. Performing dissections of a wide range of organisms, he said that in the stages of development a fetus went through prior to birth, one could see a playing out of the entire evolutionary process. He said that a human embryo went through a series of more complex forms, from fish to amphibian to reptile, before achieving its final mammalian form. In other words, the short embryonic development of an individual organism (ontogeny) mimics the longer evolutionary development of an entire group (phylogeny). Haeckel also believed that certain races—the Nordics and Aryans, naturally—were able to experience a oneness, a special connection, to nature. The affinity for nature Haeckel claimed for the Aryans, what he called Monism, was just one more reason he believed them superior. Haeckel was an outspoken advocate of Aryan superiority and a mystical German past and believed Germans' origins could be traced back to Asia. He sought to bring to public attention what he considered the dangers of racial decline through miscegenation (race mixing). At the same time his conglomeration of nature worship, mysticism, and science gave "rational" support to the notion of racial superiority. While a proponent of evolution, Haeckel twisted Darwin's idea into what became a justification for racism, a justification that was not in the original idea.

Haeckel and others embraced evolution in general and the Asian origins of man in particular partly because it seemed a way to argue for racial supremacy. He combined science and religion and equated the steady march toward perfection he saw in nature with man's evolution. British naturalist T.H. Huxley was so concerned with Haeckel's mix of science and mysticism that he cautioned the German against going too far with his speculations. Not to be deterred, Haeckel gave free rein to his mystical, occult proclivities and conjectured that the ancestors of the Aryans had come from the lost continent of Lemuria in the Indian Ocean. In 1865 Huxley weighed in on human origins with *On the Methods and Results of Ethnology*. He took a middle road, saying that humans undoubtedly began from one ancestor, but he was not completely thrilled with the idea of the Aryans, calling them a myth and poking fun at those who held the myth and who would use evolution theory to do something it could not do nor was ever intended to do. Charles Lyell generally supported the central Asian origins of man but doubted that the Aryans were still around or that their descendants could really be separated from the rest of the population. If they had existed at all, he surmised, they had intermingled so much with the indigenous peoples of Eurasia that they ceased to exist as a separate group.

Asia as the birthplace of humans also figured prominently in the thinking of the late-nineteenth-century occult revival because devotees of the movement saw it as the starting point for their construction of a non-Christian creation story. This flowering of interest in the mystical aspects of life was popular among artists, poets, writers, and others in the West who felt bored and abandoned by the modern world. The occultist most interested in the Asian origins of humans was also one of the pivotal characters of the revival, the Russian aristocrat Madame Blavatsky. Known as HPB to her friends, Helena Petrovna Blavatsky (1831–1891) landed in New York in 1874 after a stormy youth in Russia and founded the Theosophical Society. The society drew many trendy members attracted to the faddish and exclusive nature of it. Though most members eventually drifted away, a core remained loyal. Hers was a philosophical system through which adepts could allegedly gain knowledge of transcendent reality through revelations and other occult techniques. Blavatsky was a prolific writer, and her most important work was *The Secret Doctrine* (1888). In it she argued that evolution worked through a series of cycles, or rounds, that were divided into seven root races. She worked out a complex and overwrought evolutionary system with the Aryans as the current superior race. She argued that the previous race of Mongolians was guilty of miscegenation and had precipitated their fall from grace by breeding monsters. These monsters were still visible as vestiges of the "lesser races" that inhabit the earth. HPB claimed that man had not descended from the apes but the other way around. Apes, she argued, appeared because of the inbreeding of the Mongolian-Atlantians, which created the missing link in central Asia. Her form of evolution was, like many evolution theories of the nineteenth century, progressive, goal oriented, nonrandom, and always leading to a higher physical form or level of spiritual consciousness. By her own accounts, however, Blavatsky did not develop this system but was instructed in its mysteries by the hidden masters. The masters themselves had masters who were the priest-kings of Atlantis who lived in the hidden city of Shamballah in the Gobi desert of central Asia.

Aside from Eugène Dubois none of the Western human origin theorists discussed so far in this chapter actually went to Asia to determine if their ideas were valid. Although Dubois was the first to search Asia for fossil humans his effort was modest. He quietly excavated in one region for several years. In the 1920s a completely different type of expedition came to Asia to search for ancient humans. Where Dubois's operation was small, sedate, and limited in scope, this new venture was something altogether different.

The Central Asiatic Expedition steamrolled through China and Mongolia for a decade and brought the industrial, organizational, and economic resources of the United States to bear upon the problem of human origins. The expedition was sponsored by the American Museum of Natural History in New York and was the brainchild of expedition leader Roy Chapman Andrews (1884–1960). Born in Wisconsin, Andrews was so determined to work for the museum that he began his career there in 1906 as a menial laborer. He quickly worked his way up by showing a flair for exploration and specimen acquisition. By the late 1910s he had spent many years roving Asia on specimen-collecting expeditions. He had also become enamored of the theory of human origins put forward by his boss, Henry Fairfield Osborn, who had developed a theory of human evolution that Andrews wanted to prove by mounting a major expedition. In 1921 Andrews went to "Outer Mongolia" and the vast Gobi desert to search for the origins of man taking dozens of scientists with him.

Henry Fairfield Osborn was from a wealthy Yankee railroad family and called financier J.P. Morgan uncle. He studied fossils, particularly the horses and their extinct relatives, the titanotheres. These "thunder mammals" were large lumbering creatures with ornate horns and head ornaments that had inhabited the western United States. When looking at fossil horse teeth and the horns of the titanotheres, Osborn came to believe that they showed a slow, steady, progressive march of evolution, with new characteristics appearing when and where the creatures needed them to survive. The evidence also suggested to Osborn that evolution was not the random, chance affair some said it was but something with purpose and divine order. He saw what were for him clear and unambiguous lines of descent from earlier simpler forms to later more complex ones, all moving in a progressive, optimistic, path toward perfection. Osborn was one of many scientists of the period who accepted evolution as a fact but were uncomfortable because it did not involve an ultimate creator or propose.

The underlying framework of Osborn's theory was a belief that when a type of organism first appeared, it was generalized in structure. It was not adapted to any specific environment but was flexible and adaptable to many types of conditions. As they radiated out from their point of origin new versions appeared that were more specifically adapted to the environments they encountered. Osborn believed a more generalized species was superior because it could adapt, whereas specialized species were inferior because they had lost the ability to adapt. Building on what he thought he saw in the fossil record, Osborn believed that highly specialized mammals represented the outer reaches of mammal radiation around the world. The more specialized a type of mammal, the farther it was geographically from the center of original evolution. Mammals in South America and Africa, he thought, were at the extreme edge of their adaptation. So, if one were to follow a line backward from hyper-environmental specialization to more generalized forms, you would find the point of original mammal and thus human evolution. For Osborn that point was Asia.

Osborn was also a believer in Nordic superiority. He justified this by saying that of all the human species and races, the Nordics were the most generalized. Nordics could, he believed, overcome environmental obstacles more readily than any other type of human. He gathered a large collection of human fossils and casts so he could base his conclusions on direct observation of physical evidence. Because of his belief in the notion of racial superiority and purity, however, he refused to see a direct line of descent from Neandertals and Cro-Magnons to moderns. He separated all fossil, as well as modern, humans into different lines of descent in a model called orthogenesis. He wanted to be able to say modern humans, Nordics in particular, had evolved separately from all other groups. He was also uncomfortable with primate ancestry. To avoid all these difficult ancestors, Osborn created a complex circumlocution beginning with a hypothetical creature he called Dawn Man. These Dawn Men evolved directly into modern humans. This way he could argue that all humans, whether living or extinct, appeared first in central Asia but from different base stocks, making them separate species.

In 1900 Roy Chapman Andrews read Osborn's theory and was intrigued. He wanted to do a large-scale interdisciplinary, systematic survey of the region. He put together a well-equipped expedition of specialists—geologists, paleontologists, archaeologists, and others—who would do a comprehensive study of Mongolia's flora, fauna, and geology. Although they wanted to find evidence of human origins, they also wanted to record a wide range of data about the history of the region so that the fossils could be seen in a detailed context. When Dubois worked the Solo River, he did little contextualizing so that when fossils were found, it was difficult to make them speak with the authority they could have. This lack of a wider view led many to dismiss Dubois's findings. Andrews was determined that the Central Asiatic Expedition was not about to fall into such a trap. They arrived in China in 1921 and made their base in Peking. The Central Asiatic Expedition included, along with the American scientists, many local Mongolian and Chinese workers. They would head out into the Gobi desert during the summers and winter in the city. Intent on doing as much as possible in the time allotted, Andrews employed a small fleet of cars and trucks to cover the vast and unforgiving desert, instead of using the traditional camel caravans. They battled bad weather, armed bandits, revolutionaries, intransigent and corrupt politicians, and other difficulties for most of the decade before political breakdown and Chinese suspicion forced them out. Andrews and Osborn bankrolled this expensive undertaking by generous donations from financier J.P. Morgan and corporations like U.S. Steel. It was also heavily publicized, with newspapers around the world covering it. The expedition discovered hundreds of tons of fossils, particularly dinosaurs (including the first known dinosaur eggs). What they did not find were human fossils. For all their work, Andrews and his team never proved Osborn's theory. Though Osborn's theory fell apart, hominid fossils were found in Asia, just a little bit north of where he was looking.

PEKING MAN

Despite what Westerners may have thought, there was an interest in Asia in fossils. Chinese tradition held as far back as the Han dynasty that "dragon bones," as fossils were known, could be ground up and used for a wide range of medicinal and culinary purposes. The dragon has a long history in Chinese culture as a powerful omen, a deity, and a national symbol of political power. A similar thing happened in ancient Greece where fossils were found in abundance but Greek philosophers had no concept of evolution or extinction. It is argued that many of the famous myths of the Greeks, still so popular around the world today, began as ways of interpreting the fossils found around the Mediterranean. Legends of the Titans and other mythical creatures began with the discoveries of fossils. As such it should not be seen as unusual for the people of China to have viewed the strange bones they discovered as being the bones of fantastic creatures—which they in fact were—or that the myth of dragons began as a way to explain the discoveries of dinosaurs and other extinct creatures.

Peking Man and Pop Culture

Though they are now known to be an example of *Homo erectus* the hominid fossils from China have always been referred to by their popular name, Peking Man. From the Chinese city known today as Beijing (Northern Capitol), Peking is the Europeanized version of the city's name. It was first used by French missionaries in the 1600s and is based upon their interpretation of the original Mandarin. The city's name has changed over time as the city's role as a capitol has changed. It has been called Yanging, Khanbaliq, and more often Peiping (Northern Peace). It was renamed Beijing in 1949 when the city resumed its role as the Chinese national capitol.

The Peking Man mystery has turned up in pop culture. In the 1970s the American television police drama *Hawaii Five-O* aired an episode with a storyline about a footlocker full of fossils turning up in Hawaii in the hands of one of the Marines, now an old man, who smuggled it out of China. The 1998 novel *Lost in Translation* by Nicole Mones has the heroine, Alice Mannegan, join a team of archaeologists searching for the lost fossils. Mones's novel, however, is less about the fossils than it is about racism and the more intimate personal relations between East and West.

One of the first Westerners to begin collecting the dragon bones was the German physician and amateur natural historian K.A. Haberer in 1899. Haberer amassed a large collection by going to different drugstores and apothecaries around China and buying them. Druggists bought them from farmers and peasants who dug them up. The druggists in turn sold them to customers along with instructions on how to make them into teas and various remedies. Haberer left the country during the Boxer Rebellion and gave his collection to a colleague, Professor Max Schlosser of Munich, who published them as *Fossil Mammals of China* (1903). Schlosser noticed that one of the pieces in the collection was a decidedly primate-like tooth. Because of the manner in which the tooth was collected, Schlosser had no idea exactly where it came from, so he could only speculate.

During the early twentieth century it was common for Europeans with scientific and technical expertise to be employed by the Chinese government. One of them was a Swedish geologist Gunnar Andersson (1874–1960), who went to China in 1914 at the outbreak of World War I to help the Chinese find iron ore deposits. Andersson was also a fossil collector, so he naturally became interested in a local site called Jigushan (Chicken Bone Hill) outside the village of Zhoukoudian itself some forty miles outside the capital city of Peking. Local people had been finding strange stoney bones which looked to them like modern chicken bones. He went there in March 1918 and saw that the strata from which the fossils were being taken were Pliocene and Pleistocene in age.

Following the Great War, Andersson went to work for China's newly formed National Geological Survey. Andersson was a central figure in the training of the first generation of modern Chinese geologists and did extensive archaeological digs around Henan province. Though it was ostensibly a Chinese operation and had a Chinese director, the survey was dominated by Swedish interests and was largely controlled out of the University of Uppsala. Carl Wiman (1867–1944), the first professor of paleontology at Uppsala, was unhappy with Andersson's collecting techniques and so sent him an assistant in the form of the young Austrian paleontologist Otto Zdansky (1894–1988).

Zdansky quickly set to work at the Chicken Bone Hill site. In late summer 1921 Zdansky found a human-like molar tooth. Because of the harder materials they are composed of, teeth are usually the parts of an organism, mammals especially, that stand the best chance of surviving as fossils. Because of their complexity, teeth can tell much about the creature they come from: whether they are carnivores or herbivores, what their general overall size is, what conditions they lived in, and other valuable clues to their existence. The tooth Zdansky found turned out to be the first piece of a creature later known as Peking Man. Strangely, Zdansky did not immediately publicize this important find. It was not until the prince of Sweden, a main benefactor of the Geological Survey, paid an official state visit in 1926 that Zdansky made it public. At the official scientific meeting of the festivities, Zdansky unveiled the tooth along with a second he had since found. An American, A.W. Grabau, working in Peking at the time is said to have first used the term *Peking Man*. Now public, the discovery was a sensation and was hailed as a major find.

Behind the unveiling of the Peking Man teeth, and the wide-ranging fascination and interest they generated, was an intriguing series of events connecting Zhoukoudian and the Central Asiatic Expedition. Part of the rush by Roy Chapman Andrews to get his expedition going was that Gunnar Andersson had told him that he was planning a large expedition of his own. Andrews then quickly mobilized Henry Fairfield Osborn and the American Museum of Natural History to try to get there first. To justify his desire for an outside expedition, Andrews wrote an article for *Asia* magazine in which he stated that there was no institution in China with the resources to mount such an undertaking. Andrews's chief scientist was Walter Granger, an accomplished paleontologist, field man, and friend of Gunnar Andersson.

When the Central Asiatic Expedition entered China in June 1921, Andersson invited Walter Granger to visit Zhoukoudian. Granger, Andersson, and Zdansky lost no time in getting to work at the site. They soon turned up promising fossil mammals. As Granger was technically working for the American Museum and not the Chinese Geological Survey, he left after a few days of giving the relatively inexperienced Zdansky pointers on field technique. Andersson hinted to Granger that Zdansky still had important work to do at the site. A few days after Granger left, Zdansky found the tooth at the part of the Zhoukoudian site called Locality I. He kept the discovery secret for some time, even from Andersson.

Walter Granger, during his short stay may have seen something at Zhoukoudian that piqued his interest and may account for him telling both Roy Chapman Andrews and Osborn that China might prove a better place to search for the first humans. From some quick reconnaissance he did there, Granger thought the Sichuan region looked promising. Granger's hunch was right. In 1984 at a Sichuan site called Longgupo Cave, Chinese scientists discovered dental fragments suggestive that archaic *Homo erectus* had been there. Indeed, Granger wanted the Central Asiatic Expedition to turn its sights on China instead of Mongolia, but it was not to be. Not long after Granger left Zhoukoudian the Chinese Geological Survey and the American Museum of Natural History worked out an agreement that the Americans would stay in Mongolia and away from Zhoukoudian.

In the audience when Otto Zdansky first presented the Zhoukoudian teeth was a young Canadian anatomist, Davidson Black (1884–1934), of the nearby Peking Union Medical College. He too was interested in human origins. Born in Toronto, Black was an enthusiastic outdoorsman.

He received his medical degree in 1906 and then studied comparative anatomy in England. This period seems to have generated in Black his interest in human origins and the Asia hypothesis. Just as Eugène Dubois joined the Dutch army so that he could get to Sumatra, in 1919 Black took a position at the Peking Union Medical College, largely to gain access to Asia and its fossils. In 1921 he briefly joined Roy Chapman Andrews and the Central Asiatic Expedition before returning to his official duties at the medical school.

Chinese paleontologist Jia Lanpo (1908–2001), who was intimately involved in the Peking Man story, remembers that Zdansky asked Black to write an article describing the Peking material. Originally appearing in the *Bulletin of the Geological Survey of China*, it was reprinted in the Western journals *Nature* and *Science*. Black based his article on photos of the fossils, along with Zdansky's written description (the actual teeth themselves had already been shipped to Sweden for safekeeping at the University of Uppsala, where they remain today). Black argued that the teeth proved the Asia hypothesis and that they were undoubtedly from the genus *Homo*. Zdansky's take on the fossils was more circumspect. He thought that a handful of teeth, as important as they were, did not constitute enough material to make such a grand statement. Zdansky's treatment of the Peking Man material is a bit peculiar, almost as if he was not quite sure what to make of it. Once they were publicized, he wrote several technical monographs on them in 1928 and then left Asia behind for a teaching post in Egypt, never being part of the study of human origins again.

The departure of Otto Zdansky from the scene ended the Swedish dominance of the Peking Man story. Davidson Black then received financial backing from the Rockefeller Foundation to work the Zhoukoudian site along with the Chinese Geological Survey in 1927. To replace Zdansky, the University of Uppsala sent the paleontologist Birger Bohlin (1898–1990) who set to work at Zhoukoudian and found a Peking Man tooth of his own. Black was so convinced of his position that he took the critical step of giving a technical name to the fossils: *Sinanthropus pekinensis*.

Few paleontologists initially accepted Black's assertion that the teeth represented a new genus of the *Homo* line. Fortunately, Birger Bohlin and the Chinese scientists he worked alongside kept plugging away and turned up more material. In 1928 they found a jaw with several teeth in it. The major breakthrough came in 1929 when Pei Wenzhong found a large piece of a skull including the brow ridges (known as Skull III). He was working in the very back of a long, dark, wet, cramped cave when he saw it protruding from the rock. When Pei showed the skull to Black he realized that it was akin to that found by Eugène Dubois in Java years before, a point that strengthened the Asia hypothesis. Other scientists soon acknowledged Black's assertion that *Sinanthropus* was a human ancestor.

Like Gunnar Andersson before him, Davidson Black wanted to mount a major expedition to study the geology and paleontology of China. His idea was undermined when Swedish explorer Sven Hedin (1865–1952) proposed the idea to the Rockefeller Foundation first. Hedin spent years in Asia and had other agendas: He was shipping Chinese artifacts out of the country and may have been helping the German airline Lufthansa map out a commercial route to Asia. Hedin was politically active but erratic. He was initially sympathetic to the Nazis—meeting with Hitler several times—and strongly anti-Soviet; he then became enamored of Chairman Mao, after a flirtation with the Chinese Nationalists under Chiang Kai-shek.

Black did secure funding to start a new research center, the Cenozoic Research Laboratory, attached to the Chinese Geological Survey and housed in the anatomy department of the Peking Union Medical College. Unlike the original Swedish operation, which sent its finds back to the University of Uppsala, the Cenozoic Research Laboratory kept all materials in China. Black also made extensive use of Chinese scientists and technicians. Tragically, however, Black died of heart failure in 1934. He was replaced by the French Jesuit paleontologist Pierre Teilhard de Chardin. Two years later de Chardin was replaced by the German-American Franz Weidenreich (1872–1948).

Pierre Teilhard de Chardin went to China in 1923 after being invited by another Jesuit, Father Emile Licent, who discovered paleoliths in Gansu Province and the Ordos plateau of Mongolia. De Chardin was sponsored by the National Museum of Paris. Together, de Chardin and Licent explored the region, winding up back at the Great Wall in what is now the Ningxia Hui autonomous region. There they found more paleoliths similar to the Aurignacian artifacts of Europe. Continuing to work in China and a nearby section of Mongolia, the "French fathers," as the Chinese called them, found much evidence of early human habitation. They found a suspicious tooth that could have been Pleistocene that was promptly dubbed the Ordos human tooth by Davidson Black and Ordos Man by the Chinese scientists (more parts of "Ordos Man" were unearthed in the late 1970s, verifying its existence). De Chardin, like many involved in the search for fossil humans in Asia at this time, also briefly accompanied Roy Chapman Andrews into the Gobi desert.

It is easy to think of the discovery of Peking Man as the result of European and North American scientists working in the wastes of Asia to find what they were looking for. If Chinese scientists are mentioned, they appear almost as secondary characters moving about in the background. They did, in fact, play the major role. For example, when Birger Bohlin joined the excavation team he was the only Westerner. The chief geologist and administrator at the site was Li Jie, and the chief excavator was Liu Delin who had been with the Central Asiatic Expedition as Walter Granger's assistant. The first skullcap (Skull III) was discovered at Locality I by paleontologist Pei Wenzhong. The Geological Survey itself was under the direction of Weng Wenhao. Most, if not all, the actual digging and fossil preparation was done by Chinese workers, while Davidson Black wrote the papers, and received the lion's share of acclaim for them.

When Pei Wenzhong discovered the crucial first skull, he was assisted by a young paleontologist named Jia Lanpo who joined the team in 1931. Jia was in his early twenties and a graduate student at the time. Jia's job was to search the site for man-made artifacts—called paleoliths—in the form of quartz points, cutters, and scrapers. The site had an abundance of these quartz chips. Jia eventually took charge of the site in 1935. By then, however, he was more or less alone. All the old hands had moved on to complete their educations or to new positions. Funding was also growing scarce because of the unstable political climate. The Zhoukoudian site did not remain open for long. In 1937 fighting between communist and nationalist Chinese forces and then the Japanese invasion made the location untenable. By then fourteen Peking Man skulls in various stages of completion, 147 teeth, seven thigh bones, two upper arms, a collar bone, and parts of a wrist had been unearthed. They were dated to between 250,000 and 500,000 years old. Franz Weidenreich had detailed drawings, photos, and plaster casts made of all the fossil pieces. In 1941 Weidenreich took the photos and casts with him to the United States to the American Museum of Natural History in New York.

THE VANISHING

It is easy to tell the story of the disappearance of the Peking Man fossils as an exciting mystery, and a "whodunit." The events naturally lend themselves to such things: a war-torn country; a collection of priceless, irreplaceable artifacts; and desperate scientists trying to hide them under dangerous conditions. In 1937 invading Japanese troops occupied the Zhoukoudian area and took control of the dig site. They also occupied the main campus of the Peking Union Medical College as well as other sites throughout the capital. The Japanese seemed to show an interest in the fossils which made the Chinese scientists rather nervous. The research team decided they had to do something to secure the fossils and safeguard them from the enemy. Jia Lanpo immediately began smuggling drawings of the fossils out of the Peking Union building.

Meanwhile, Geological Survey director Weng Wenhao and Franz Weidenreich worked out a plan for the fossils themselves. Weidenreich rejected the idea of putting the fossils in his personal luggage to take back to the United States. He was afraid the fossils would be broken, lost, or confiscated. All involved believed that under the dire circumstances the rule about not taking artifacts and antiquities out of China could be lifted and that the United States was the best place for the relics to go. It was then suggested that the U.S. Embassy take the fossils in the diplomatic pouch that was not subject to customs inspections. Embassy staff, under Ambassador Nelson Johnson, was wary of the law about antiquities, so they turned the fossils over to the U.S. Marine Corps detachment in Peking that was about to evacuate to the Philippines. This is the plan that was executed. However, somewhere along the line the fossils vanished.

The loss of the Peking Man fossils is one of the great mysteries of science. There is little consensus as to what happened to them. There have been books and articles written on the subject, and no two points of view are the same. There was disagreement from the start. Claire Taschdjian, a German working as Weidenreich's assistant, claims that she wrapped and boxed the fossils for transport making her the last known person to see them. In *The Story of Peking Man* (1990), Jia Lanpo dismisses Taschdjian's story. He claims that it was a pair of Chinese laboratory technicians, Hu Chengzhi and Ji Yanqing, who did the packing. Jia states that Taschdjian's book, *Peking Man Is Missing* (1977), "is fictional and the details are therefore exaggerated" (Lanpo, 1990, p. 160).

Whoever packed them, the boxes (and the number of boxes is in dispute) were sent to the Marine barracks in Peking, Camp Holcomb. There they were put under the control of a Marine officer who was to take them out as secret materials. This was all in the days just before the Japanese attack on Pearl Harbor and the onset of World War II proper. At one point, when the Marines prepared to defend their tiny garrison from the Japanese onslaught, one man used the boxed fossils as a prop for his Lewis light machine gun. The Marines were meant to board the *USS President Harrison* for evacuation, but the *Harrison* was run aground in a fight with a Japanese destroyer just after 7 December 1941. The Marines themselves were eventually taken prisoner and their luggage train ransacked.

There are many theories about what happened. Did American staff at the Peking Union Medical College take them? Did the Marines hide them and not tell anyone? Did Japanese officers take them secretly back to Japan? Did Japanese enlisted men, unaware of their importance, discard them as useless rocks when they looted the American baggage train? Did they fall overboard while being loaded on the *President Harrison*?

An Associated Press item in 1945 said that when the Japanese occupied the Peking Union Medical College they found only plaster casts. The *Peiping Chronicle* said in 1946 that the fossils were in Japan at

Tokyo University. A series of books appeared in the 1970s that fanned the flames of conspiracy and mystery. Harry Shapiro, a longtime curator and anthropologist at the American Museum of Natural History, released *Peking Man* (1974), which was followed by *The Search for Peking Man* (1975) by self-styled adventurer Christopher Janus and co-author William Brashler.

Shapiro knew Davidson Black personally and wondered why the Japanese seemed to show such an interest in the Zhoukoudian site. He argues that "obviously, the fossils had a high priority [for the Japanese] and the looting must have been planned well in advance" (Shapiro, 1974, p. 19). Shapiro recounts the great sense of loss felt by Franz Weidenreich when news of the tragedy arrived in New York. Later, the casts that Weidenreich had brought back to the American Museum were mistaken for the originals, and rumors started that the museum had them. Shapiro remembers the reports being contradictory and confusing. In 1971 he came across Dr. William Foley, a heart specialist from New York City who claimed to be the Marine officer in charge of evacuating the fossil-filled footlockers from Peking. Foley was to take the fossils during the fall of Peking to the Philippines, then on to the United States. When the war started he was captured.

The Janus and Brashler book is on the sensational side. Janus met with Shapiro and tried to involve him in his scheme. Although Shapiro was intrigued by some of the things he saw, he was skeptical of Janus's motivations and evidence. In a breathless account Janus interweaves his own personal search for the fossils with the historical one: including the meeting with a shadowy female atop the Empire State Building. Janus was an amateur relic hunter who took comments Shapiro made about some of the evidence he was shown, out of context, as endorsement of his theory. Janus was later alleged to have defrauded funds from a Peking Man foundation set up to find the fossils. Jia Lanpo argued that all these stories were sheer fantasy and wishful thinking.

In the early twenty-first century a new flurry of interest in finding the lost Peking Man fossils emerged. In 2001 Jia Lanpo's son Jia Yuzhang was approached and told that the fossils were still in China as his father believed. The younger Jia sent this information to the Ancient Human Research Institute who in turn contacted the original source and was told the story of a man who had been in the Chinese underground during World War II (what the Chinese call the War of Resistance against Japanese Aggression). The man, named Li Zhoudong, was now elderly and in prison on fraud charges. He claimed to others that during the war he knew the Japanese had confiscated the fossils before they could be shipped to America. Li and his compatriots stole the fossils from the Japanese and secreted them away in a secret spot in Peking, where they remain to this day. Institute officials tried to interview Li but were unable to. This was just one of a number of stories that came up yet ended in dead ends.

The two most popular stories in China also include a tale told by a dying old man and one involving an American submarine. In 1966 an aging Japanese soldier who claimed to have been part of the infamous Japanese biological research team known as Unit 731 surfaced with a tale. Not well known outside of Asia until decades after the war, Unit 731 was a Japanese army unit which performed gruesome germ warfare experiments on Chinese as well as other captives. They injected prisoners with various diseases and left them untreated—the test subjects dying in extreme agony—so they could chart the progress and feasibility of various biological weapons. After the seizure of the fossils by Japanese intelligence troops, this man was given the job of looking after them while at the Peking Union Medial College. At the end of the war he was given instructions to get the fossils out of the college. In the chaos he took the boxed-up fossils and buried them under a pine tree in Ritan Park a short distance from the campus. He then left a mark he could recognize on the tree and fled. China's Institute of Vertebrate Paleontology and Paleoanthropology (IVPP) searched the park and found the tree, but digging brought up nothing.

The other story is a variation which picks up where the American troops guarding the fossils story leaves off. Some researchers believe that after confiscating the fossils from the Marines, Japanese troops held onto them but eventually put the fossils on the passenger liner *Awa Maru*. This was a medium-sized vessel masquerading as a Japanese medical ship but which in realty was often used to carry contraband and allied POWs out of China back to Japan. Those prisoners unfortunate enough to have spent time aboard the *Awa Maru* called it a "hell ship" because of the ghastly conditions they were forced to endure while in the ship's squalid holds. On 1 April 1945 the *Awa Maru* was intercepted and torpedoed by a prowling American submarine, the *USS Queenfish*, and sent to the bottom of the Taiwan Straits along with most of its 2,000-odd Japanese passengers. The *Queenfish*'s commander said at his trial that he thought the *Awa Maru* was a military transport and thus a legitimate target. The incident became infamous as the "*Titanic* of Japan" (though it might be more accurately called the *Lusitania* of Japan). After a number of attempts, the wreck was found in 1986 by a salvage–treasure hunt team. In all the talk about the "valuables" on board, mostly in the form of bullion and Chinese art and artifacts looted from the country, there was no mention of the Peking Man remains being there.

That the fossils are somehow, somewhere in Japan is held partly because of suspicion in China of Japan itself. The Peking Man fossils are considered important parts of China's cultural as well as ancient history and as such are held in high esteem. Some Chinese scholars had been calling for a renewed effort to find the lost fossils since the 1990s. Then a group of journalists made a public appeal in 2000. They saw the recovery of the fossils as part of the wider push to reclaim Chinese antiquities pilfered from the country by Westerners for centuries and the Japanese during the war. There is a long complex history of conflict between Japanese and Chinese culture. Many Chinese are still deeply resentful of the way their country was invaded and brutally ravaged by the Japanese. During the war of the 1930s and 1940s the Japanese considered Chinese people beneath them ethnically and culturally and behaved in racist and even genocidal ways during the occupation. There has also been a long-standing argument about Japanese cultural origins. Conventional wisdom says that Japanese culture is predated by that of China and that not only did Japanese culture come from China but the people as well. There are those in Japan, however, who are uncomfortable with the idea of China as their antecedents and prefer to view their culture as older and superior. Peking Man then is not just a set of bones in China, they are the revered remains of ancestors and were worthy of respect for that alone. Many Chinese paleoanthropologists believe that Peking Man and his kin show that modern Chinese people are the direct descendents of *Homo erectus* in Asia. As such, all Asian *Homo sapiens*, as well as culture, spring from China and its *Homo erectus* ancestors. This attitude has led many Chinese scientists to embrace the multiregional theory of human evolution (see below). This concept supports the notion that modern Asian people find their biological origins in Asia not in Africa.

It is not unusual that Chinese scientists would subscribe to a form of Neo-Asia hypothesis. Fossils found in Longgupo Cave in the mid-1990s call into question the long-standing Out-of-Africa theory which is currently the most accepted. These fossils—a tooth and part of a jaw—are dated to 1.8 to 2 million years old. While these dates would make them too old to be *Homo erectus* like Peking Man they might represent an Asian form of the more primitive *Homo habilis* or even an entirely new species (*Homo habilis* is thought to have existed only in Africa). A number of Chinese scientists dispute the African origins theory, at least for Asian people. They point to a number of hominid fossils (in addition to the ones just mentioned) found in China that have what they consider the rudiments of human tooth structure. Fossils like Yuanmou Man (1.7 million years old), New Cave Man (100,000 years old), Upper Cave Man (18,000 years old), and Julai Nur Man (10,000 years old) all have progressively more modern human skull characteristics and suggest that a continuity of genetic structure prove the early appearance of hominids in Asia in parallel to those in Africa. This idea of "regional evolution" is known as "regional continuity" and "multiregionalism" in the West.

Supporting the idea of the Asian origins of Asian people is the discovery of the Shu Ape (First Light) in Jiangsu Provence in the early 1990s. A mouse-sized primate of the middle Eocene Age (4 to 4.5 million years ago), this creature could mean that the earliest primates also appeared in Asia. Professor Qi Tao, who discovered the Shu Ape along with colleague Wang Jingwen, argues that the Shu Ape proves that both the first apes as well as the first modern Asiatic humans appeared in Asia.

Further finds in Myanmar (Burma) in 1999 also support the Neo-Asia hypothesis. *Bahinia pondaungensis*—found by a joint Myanmar and French team—is a form of anthropoid and thus is older than the more recognizable primates. The most primitive members of the primates are the anthropoids which are found in the Fayum desert of Egypt. An early Chinese anthropoid, called *Eosimias* (Dawn Ape), is so primitive that some question whether it should be classified as a primate at all. The *Bahinia* fossils are similar to *Eosimias* but are more numerous allowing for a greater level of study. If both *Bahinia* and *Eosimias* are genuine primate ancestors this would raise many questions about origins as they are only found in Asia. They would push the origins of the primates back to 55–60 million years ago.

If all the hominids of Asia are what some think they are then it shows a level of diversity there which rivals that of Africa of the same period: a period in which *Australopithecus*, *Homo habilis*, and *Homo ergaster* were thought to be contemporaries. It could mean that our African ancestors were more capable of traveling from Africa into Asia more easily and much earlier than ever thought before. It could also mean that these Chinese hominids arose in Asia then entered *into* Africa and replaced the Australopithecines like "Lucy" and went on to become modern Africans in the same way those who stayed in Asia became modern Asians and those who traveled west became modern Europeans.

In the West, David Begun of the University of Toronto too sees Asia as a strong candidate for the cradle of humans and primates. He argues that the only place fossil great apes (those represented by the large chimpanzees, orangutans, and gorillas) were found was Eurasia, not Africa. He says our most distant primate ancestors therefore must have formed in Asia. Around 9 million years ago, populations of primates existed in Asia, some of which were evolving generally in the hominid direction. Some of them, because of climatic change, entered Africa and kicked off the trend that led to primates and hominids. He points to the mosaic characters of the early African hominids *Orrorin* and Toumai as just the sort of creature that would result from such a scenario.

An example of the intellectual conflict over Asian origins is the recent case of Shinichi Fujimura. Since the 1970s Fujimura has produced a string of spectacular archaeological finds which steadily pushed back the origins of Japanese culture. In 1981 he found artifacts in 40,000-year-old strata—the oldest ever found in Japan. As deputy director of the Tohoku Paleolithic Institute (a private foundation) he was known as the "divine digger" for his uncanny ability to find artifacts which revolutionized the study of Japanese prehistory. His most recent discoveries were artifacts dating back 600,000 years—from *before* the time of Peking Man. This would have brought Japanese history back farther than China's laying to rest any idea that Chinese culture was older than Japan's and that Japanese culture finds its origins in China. The divine digger was caught, however, burying faked artifacts. Now publicly disgraced Fujimura made an emotional public apology claiming he hoaxed his discoveries because he was under great pressure to produce older and older finds to push Japanese history back as far as possible to compete with China. His actions placed a considerable number of archaeological sites and discoveries under a cloud of suspicion as to their authenticity. Like the Piltdown Man case of 1912 Fujimura's hoaxing went largely unchallenged in part because they supported a national desire to promote Japanese cultural antiquity. There are some in Japan who resent the common idea that Japan's culture finds its origins in China.

Interest is still so high in China to find the lost Peking Man fossils that in 2005 a new effort was launched. The local Fangshan District (in which Zhoukoudian lays) office of the Committee of the Communist Party in China put together a group of local officials and scientists to find the Peking Man remains. Paleontologist Zhou Guoxing, who in addition to his normal duties has been hunting down the lost fossils since the 1970s, became an advisor to the committee. He said that part of why he agreed to become part of the undertaking was that people will give up anything but their ancestors. Peking Man was too important to Chinese national identity to simply forget about. He also had asked Japanese officials if he could set up a foundation in Japan to coordinate efforts to search there. His requests were consistently denied. The Japanese officials told him that the bones were not in Japan and that he should look for them in America. Zhou was skeptical of the Japanese response. His search, which led him to Japan and the United States a number of times, had him believing that the fossils were interred in Japan. Though he had taken a position with the search committee, Zhou, like many in the field was worried the search would be a waste of time. The contention of skeptics was that many different scientists, historians, and explorers with much greater resources behind them had been looking for the fossils for years with no luck. How would a small municipal agency with limited resources do any better? The one thing the committee did was to set up a hotline which immediately filled with tips and clues from people across the country. Farmers, retired servicemen, and construction workers called in to tell of incidents where they had seen mysterious crates buried, hidden across China and shipped off to parts unknown. Grown children and relatives of people who had been around in those days called and related stories their elders had told them. A few promising leads were to be followed up and much enthusiasm filled the committee members' expectations. Jia Lanpo's son, Jia Yuzhang, commented that "searching [for Peking Man] could help the nation remember the loss as a pain China experienced during the war" (*People's Daily*, 2005).

JIA THE ELDER

Although religious dogmatism has always played a role in the study of human evolution, political dogmatism has as well. The Asia hypothesis in particular has been a magnet. Besides the Aryan-Nazi political connection to human evolution, there have been others. A fascinating example is the communist political theory that became part of the Asia hypothesis. Karl Marx and Friedrich Engels, two of the founding fathers of modern communism, embraced evolution because it seemed to fit into their belief that humans from lower social classes could pull themselves up to something better. Engels argued that humans evolved as a result of learning to work. Labor, he said, was the key factor in turning brutes into men. In 1949 the Communist Party under Mao Tse-tung had triumphed and taken control of China. With the country now following a socialist political path, all members of society were expected to do the same

It was during this period that a great upheaval hit China. The Cultural Revolution lasted roughly from 1966 through Mao's death in 1976. Although it was an attempt by Mao to consolidate his power, the Cultural Revolution was a popular revolt against government and party privilege. The worst thing a person in a leadership position could be accused of was using their position for personal gain and to undermine the party (regardless of whether they were actually doing so or not). In 1966 Chairman Mao unleashed hundreds of Red Guard groups upon China. Made up mostly of zealous teenagers who unwaveringly supported Mao, the Red Guards swept through society seeking to purge it of anything not of the proper political attitude. Thousands of people were imprisoned, put on show trials, or murdered in an increasing level of violence. The mob violence reached a point of spinning out of control by 1968, and so Mao disbanded the Red Guards. For Jia Lanpo this was an anxiety-ridden period. His position as a leader of the Chinese scientific establishment could have been easily used against him. When the Cultural Revolution hit, any Chinese citizen who had even the vaguest connection to foreigners could find themselves in deep trouble very quickly by being labeled "class enemies." The Peking Man site was accused of being a "stronghold of cultural aggression" because so many non-Chinese had worked there. Jia had begun to write a chronicle of the dig site so that the memory of what happened there and who had worked there and contributed so much would not be lost. At great peril to himself and those he wrote about he continued to work on the manuscript but carefully kept it hidden. He had to walk a careful line, as did thousands of scientists and intellectuals. If it fell into the wrong hands, his memoir would have become Jia's death warrant. This may account for some of his writing about the political aspects of human antiquity.

In *Early Man in China* (1980), Jia claims that socialism "has spurred the development of this branch [paleoanthropology] of science" (Lanpo, 1980, p. ii). After the communist victory in 1949 the government contacted Jia and lavished a large budget on him to get the excavations at Zhoukoudian going again. The old institutions of scientific study were reorganized as the Chinese Academy of Science, with Jia taking charge of the Institute of Vertebrate Paleontology and Paleoanthropology in 1953. In his books, Jia says that Chairman Mao's rule about intellectuals learning from the masses should be applied to the search for the first humans. Peasants could lead scholars to the sites where the fossils were located. Jia congratulates the Chinese people on dropping their old religious myths and superstitions about the divine creation of humans for more prosaic and scientific ideas. He cites Engels on his suggestion, made in the late nineteenth century, that Asia was the cradle of mankind. What is to be made of this? Jia may have been an ardent communist applying socialist theory to his work, or he may have felt obliged to pay lip service to it so that he would not feel the wrath of the party for "incorrect thought." He had watched powerlessly during the Cultural Revolution as many fine libraries were burned and destroyed in the name of socialist progress. He had kept exhaustive records of the Zhoukoudian project. In fact, it is through Jia's memoir of the events of those years that we know the details. He did not want to see his manuscript go the way of so many other books and papers during that dangerous time. At great personal risk he ferreted away the notes and papers from the academy (in the same way as he hid them from the Japanese) to his home for safekeeping and was just able to protect them from the ravages of the fanatical Red Guard. This fear may be what accounts for his fawning attitude to Chairman Mao and the principles of the Cultural Revolution. In his later works like *The Story of Peking Man* (1990), published after the death of Mao and a lessening of tensions inside China as well as between China and the West, there are no such party platitudes.

Though he did not find the very first Peking Man fossil (that was found by Pei Wenzhong) Jia Lanpo became the scientist most associated with the fossils. He has been called Peking Man's father, though he jokingly referred to himself as Peking Man's uncle as he only took care of him. Jia Lanpo was a central figure in Chinese paleoanthropology. Later in life he was often referred to as Jia Lau (Jia the Elder) as a sign of respect for a lifetime of work and important contributions. He had risen from an assistant to become the elder statesman of Chinese anthropology. His career began in 1931 when he entered the Central Institute of Chinese Geology and soon became part of the fledgling Zhoukoudian project. He worked alongside some of the great names of mid-twentieth-century Western paleoanthropology like Pierre Teilhard de Chardin, Henri Breuil, Davidson Black, and Franz Weidenreich. As a result he took an internationalist approach to science his entire career. Over the course of that career he worked for, ran, or started most of the important paleoanthropology institutions of China and was mentor to several generations of Chinese scientists and scholars. He published a wide range of books and articles for both the scholarly and popular press, including several in English. He also did extensive research on the ancient culture and environment of Peking Man. His work showed that the region Peking Man occupied at Zhoukoudian went through a series of climatic changes with alternating cold and warm spells which had an impact on *Homo erectus* evolution. He also argued that despite what Western scientists originally thought about Peking Man, they had a rather sophisticated culture and more modern morphology. He also believed Peking Man had the controlled use of fire. When Jia Lanpo passed away in 2001 his loss was deeply felt by colleagues around the world. His cremated ashes were interred at Zhoukoudian along with his friend Pei Wenzhong and Peking Man as well.

PEKING MAN'S PLACE IN NATURE

One of the problems with the disappearance of Peking Man is that many outside the paleoanthropology community seem to think that was the end of the story: The fossils are lost so nothing more can be done. Western creationists and anti-evolutionists in particular enjoy pointing to the loss of these fossils so they can say that either it was a hoax or a cover-up. They hold up Peking Man as just another straw man used by "Darwinists" to fool the faithful into believing the false idol of evolution. They argue that since Peking Man no longer exists—if it ever did at all—there is nothing to study, no proof of evolution. This would be a tough hurdle for evolutionists to get over except for one thing. While the original Peking Man fossils are lost an enormous amount of fossil material has been found at Zhoukoudian since. So much material has been found subsequently that even if the originals had not been lost they would constitute a small part of a much larger collection of evidence of the evolutionary paths the human line took in Asia. The overwhelming amount of evidence since collected and studied in China easily support all the contentions made for the original material found in the 1920s and 1930s.

The Peking Man site at Zhoukoudian is not one cave or dig site, but a complex of localities (twenty-six in all) grouped in the same general area. The caves are naturally occurring fissures in limestone deposits laid down about 450 million years ago. Running water slowly filled parts of the caves with sand making comfortable living spaces attractive for hominids well before any of them arrived on the site. Peking Man and his relatives probably moved into the cave valley about half a million years ago based on the fossil evidence found there. The youngest Peking Man fossils date to about 250,000 years old. This suggests that Peking Man/*Homo erectus* hominids were living at the site for about 260,000 years all told. By the time modern humans began finding fossils there and searching the site carefully, the water-borne sand deposition process had all but filled the caves in. This forced scientists to have to tediously remove the overburden first, but which was the primary factor in safeguarding the fossils from complete erosion and destruction. Chinese scientists saw the cultural life of Peking Man much like the traditional roles of Chinese society. The father led the family and did all the heavy work and providing—hunting game, butchering the kill, etc.—while the mother bore and raised children and saw to home life. The Chinese interpretation also saw Peking Man as the beginning of interfamily education with revered elders teaching the new generation about important skills to be passed along in turn by them.

In the late twentieth and early twenty-first centuries Peking Man again became a center of discussion. Noel Boaz and Russell Ciochan used the fossils as the basis of a new approach to understanding human origins in general. Boaz and Ciochan had done extensive work on Asiatic hominids including the peculiar *Gigantopithicus* (and no it wasn't Bigfoot). They came to a number of conclusions about the life of Peking Man and how these fossils relate to others and what they say about the human family line. In *Dragon Bone Hill: An Ice Age Saga of Homo Erectus* (2004) they addressed Peking Man/*Homo erectus*' skull, which is much thicker and harder than a modern human. They argue that Peking Man was using its flat, thick skull as a kind of head-butting devise for interspecies combat not unlike the way modern bighorn sheep use their thick horns to bash one another. They argue that marks on the fossil skulls suggested this sort of behavior. The thick skull was also useful for the attachment of tough chewing teeth, adapted for scavenging meat from already dead animals. From *Homo erectus* on the human line began a trend of increasingly thinner skulls as a result of a growing brain. Greater cranial capacity suggests greater brain power which was used to create more sophisticated tools and the ability to resolve disputes without the head banging. Also, a thinner skull allows for more blood vessels in the skull which help to cool the increasingly large and sophisticated brain: A thin skull with plenty of plumbing is a very good air conditioner for the brain. The thinner skull can also accommodate smaller more complex teeth allowing for a much wider diet including food cooked by fire.

The more controversial idea suggested to Boaz and Ciochan by the Peking Man material is a way of interpreting the human line and where and how it came to be. One of the great points of debate among scientists is just how and where humans appeared. The schools of thought which are vying for primacy in explaining where humans come from are "population replacement" and "multiregionalism." In other words, do humans have one point of origin or many? Population replacement, also known as the "Out of Africa" hypothesis (an idea first codified by British naturalist Christopher Stringer—see chapter 23), argues that a population of modern humans first appeared in Africa and radiated out around the world in a single wave, replacing those earlier more primitive hominids that had left Africa in previous waves. This scenario meant that as new species and variants of the *Homo* line appeared in Africa—especially *Homo erectus* and possibly *Homo ergaster*—they would eventually walk out of Africa and spread to Asia and Europe in a series of successive migrations or waves. As each group did this, they would replace the group that had gone before it. The last wave was modern humans (*Homo sapiens*) who managed to hang on while all other hominids died off. If correct this means that all humans alive today can trace their lineage back to that wave of people leaving Africa: making us all Africans regardless of the color of our skin.

Population replacement has the support of an allied theory sometimes known as the Eve hypothesis. Traditionally, human evolution theories were based on fossils. The late Allan Wilson (1934–1991) along with Rebecca Cann and Mark Stoneking turned to genetic evidence taken from living populations to work out human ancestry. Their groundbreaking work first appeared in 1987. Their argument was that modern people could be traced back to a relatively small group of females in Africa about 200,000 years ago. Moderns evolved from this group, then left Africa to populate the world. They claimed fossils alone were inadequate for determining species relationships and that it was genetics that would give the most information on the course of human evolution. Cann and the others said that their work showed that all living human mitochondrial DNA (mtDNA) came from a tiny population of females. They were quick to point out, however that even though many refer to this idea as the "Eve" hypothesis, there was no one individual responsible for all later humans. Genetic research supports the idea that all modern people sprang from the same original African *population*. Every time Cann and her colleagues attempted to construct a human family tree based on their data, it always led back to Africa. They used different populations of modern people from Papua New Guineans to Asians to African Americas as subjects. Always the trail stopped in Africa. Because of this they felt confident that Africa was the point of origin for human mitochondrial DNA, and thus the point of origin for all modern humans.

The main opposition to population replacement is Alan Thorne and Milford Wolpoff who developed the multiregional hypothesis. Wolpoff, of the University of Michigan, based his work on an exhaustive study of most of the human fossil material then known. He and Thorne, along with their Chinese colleague Wu Xin Zhi, argued that modern humans with their ethnic differences were the result of local evolution and interbreeding between different groups of hominids. Around 1.5 to 1.8 million years ago various modern and near-modern hominids left Africa and fanned out over the planet. These groups—versions of the Neandertals and *Homo erectus*—then evolved into modern *Homo sapiens* as a result of local climatic conditions as well as interbreeding. This would have mixed genetic material (a process known as gene flow) from diverse groups to form a final version. Modern human ethnic groups evolved in the places they are traditionally linked to in modern times. There was no replacement of one group by another, only a continuous series of linked and subtle transformations in place. So while the basic early hominid stocks formed in Africa, it was not until after they spread across Africa, Asia, and Europe that they evolved local adaptations and became modern humans. Wolpoff and his colleagues contended that archaic African populations did not have the same facial features as modern Africans. The ancient Africans had a mix of features that was both archaic and modern. In other words, not even modern African people looked like early humans. What accounts for today's African facial features are the same sort of processes that formed modern Europeans and Asians: evolutionary adaptations that occurred in their respective regions much later.

Although Wolpoff admired the work of the geneticists and accepted that it formed important new insights on human evolution, he rejected the idea that it gave conclusive evidence supporting population replacement theory. He claimed fossils and human artifacts gave the best evidence for the correct circumstances of human evolution. The mulitregionalists opposed replacement theory in part because they found it hard to believe one population could have completely replaced another without some interbreeding or that it happened in the relatively short period of 150,000 to 200,000 years ago. They found no archaeological or fossil record of such an event occurring and claimed that "only fossils provide the basis for refuting one idea or the other" (Wolpoff and Caspari, 1997, p. 65).

Some geneticists argued that fossils were prone to theoretical conclusions and interpretation, whereas some fossilists argued that genetics was dependent upon theoretical assumptions. Wolpoff said that accepting the Eve idea was to accept that modern African humans completely replaced all other existing hominids, that early humans must all have had African features, and that no interbreeding went on. The multiregional/continuity people could not accept these assumptions without fossil evidence. In fact, they said the archaeological evidence supported the opposite. They pointed to artifacts and fossils found at sites in Israel known as Skhul, Tabun, and Qafzeh. These sites, first discovered in the 1930s, were considered the oldest human evidence outside Africa. The fossils and artifacts there suggested the Tabun cave had been the home of Neandertals, whereas Skhul and Qafzeh had housed archaic moderns. This arrangement seemed to prove that the archaic moderns had appeared after the Neandertals, making them their descendants. The modern human artifacts, the multiregionalists argued, were identical to nearby Neandertal sites. If the population replacement theory was correct, there should be more advanced technologies that would be indicative of an outside invasion by more advanced people. Wolpoff argued there were none.

Replacement proponents continued to say genetic evidence ruled out the widespread interbreeding needed for continuity. It may have been disease that kept interbreeding from happening or the advanced language skills of the invaders effectively kept the groups apart. In 2000 scientists from the University of Uppsala, in Sweden, and the Max Planck Institute of Evolutionary Anthropology at Leipzig, Germany, did further mtDNA testing using fifty-eight subjects and a computer-modeling program to look at the gene sequences of current African and non-African populations. The computer created family trees based on the genetic material. The data showed that both Africans and non-Africans could be traced to a common ancestor. This work also pushed the African breakout date forward from 200,000 to 50,000 years ago. Eve proponents said this confirmed their view. Scientists of the Human Origins program at the Smithsonian Institution added that genetics alone could not answer everything, though the fossils and artifacts, in this case, supported the geneticists' study. They cautioned that modern human genetic material did not just appear overnight 60,000 years ago but was likely a gradual process. He also allowed for a certain amount of gene flow. Further muddying the genetic waters, a 1991 test of Neandertal mtDNA at first seemed to confirm that modern Europeans shared no Neandertal traits. A rival analysis of the same data, however, argued that the difference between Neandertal and modern European mtDNA was not as pronounced as was thought. There seemed no way to reconcile these two camps until Neal Boaz and Russell Ciochon took a long hard look at Peking Man.

Population Replacement has been popular in the late twentieth and early twenty-first centuries and not only because of the fossil and genetic evidence but because in part it reflects the modern liberal ideal of human relations. It suggests we really are one great big family despite the superficial differences. This supports the "why should we hate one another if we are the same" narrative as opposed to the "we are different and not the same so we are better than you" narrative so popular in the West in the nineteenth and twentieth centuries (and still quite popular in some quarters even now): a narrative which has caused so much strife and suffering in human history. Boaz and Ciochan propose a third, middle way, Clinal Replacement. A cline is a small population of individuals with certain genetic characters who live in a specific geographic area. At the outer edge of their reserve they come into contact with a neighboring cline. The boundary does not have a hard edge and so individuals mix their hereditary material between the clines. The genetic mixing going on at the overlap of clines breaks down the barrier even more and produces a relatively smooth genetic transition from one group to the next. These subtle changes occur relatively quickly and do not show up in the fossil record and as the *Homo erectus* populations encountered more advanced *Homo sapiens* populations any trace of the cline mixing would have been masked, giving the appearance of more abrupt population-wide replacement. This theory accounts then for the illusion of rapid population replacement and at the same time makes sense of the genetic anomalies pointed out by the multiregional theory, yet is neither completely. So Clinal Replacement suggests instead of massed waves of hominids replacing previous groups, or genetic mixing alone producing the modern human groups, what really happened was a combination of replacement and genetics but on a more circumspect and less easily detected scale. Boaz and Ciochon feel that Peking Man is key to clearing this up.

CONCLUSION

The Zhoukoudian site is so important to the story of human evolution, and so precious, that in 1961 local authorities declared it a protected cultural heritage site. The Beijing municipal government followed suit in 1983, and in 1987 the UNESCO world heritage committee placed the site on its "World Heritage List." This allowed for greater use by scholars from around the world and allowed for the site to be opened up to tourists, all of which bodes well for its long term protection.

Whether religiously, racially, or politically motivated, the search for the cradle of man in Asia did produce significant finds and discoveries. Although most scientists, at least in the West, believe Asia was not the home of the first humans, it certainly held important episodes in our history. Though the original cache of Peking Man fossils has disappeared, many more have been found subsequently at Locality I and elsewhere. Since the war, fossils from at least forty individuals have been found at Zhoukoudian along with a treasure trove of stone artifacts and traces of fire use. So while those original fossils are lost, Peking Man still lives and is well documented, taking its rightful place in the human family tree. Oh, and by the way, recent reevaluation of casts of the original lost Peking Man fossils show it was a female.

FURTHER READING

Boaz, Noel T., and Ciochan, Russell L. *Dragon Bone Hill: An Ice Age Saga of Homo Erectus* (New York: Oxford University Press, 2004).

Jaeger, J., Thein, Tin, Benammi, M., Chaimanee, Y., Soe, Aung Naing, Lwin, Thit, Wai, San, and Ducrocq, S. "A New Primate from the Middle Eocene of Myanmar and the Asian Early Origins of Anthropoids," *Science* 286 (15 October 1999): 528-530.

Lan-Po, Chia. *The Cave Home of Peking Man* (Beijing, China: Foreign Language Press, 1975).

Lanpo, Jia. *Early Man in China* (Beijing, China: Foreign Language Press, 1980).

Lanpo, Jia, and Weiwen, Huang. *The Story of Peking Man* (Beijing, China: Foreign Language Press, 1990).

Van Oosterzee, Penny. *Dragon Bones: The story of Peking Man* (Cambridge: Perseus Press, 2000).

Peppered Moths

David W. Rudge

H.B.D. Kettlewell's pioneering research in the early 1950s on the phenomenon of industrial melanism is widely regarded as the classic investigation of natural selection by scientists and the lay public. After briefly reviewing what the phenomenon of industrial melanism is, the context of Kettlewell's investigations and his specific work on the phenomenon as exemplified in the peppered moth, the chapter analyzes how and why it became an icon of evolution. A review of research since reveals the phenomenon of industrial melanism is far more complicated than textbook accounts would have us believe. Discrepancies between textbooks written for children and journal articles written by and for scientists who actually work on the phenomenon have led many commentators (most of whom have an obvious anti-evolution agenda) astray. The chapter concludes by discussing the import of scientific questions surrounding the phenomenon of industrial melanism and Kettlewell's work on it for the status of industrial melanism as an icon of evolution and its continued use in the teaching and learning of science.

THE PHENOMENON OF INDUSTRIAL MELANISM

Large-scale manufacturing associated with the Industrial Revolution (which started in Britain and Continental Europe during the mid-nineteenth century) led to a dramatic increase air pollution. This had profound effects on the surrounding environment. In manufacturing centers smog darkened the skies and buildings became visibly darker owing to the accumulation of grime and soot. Waste gases such as nitrogen and sulfur dioxide likewise adversely affected inhabitants, leading to widespread respiratory problems. Exposure to these contaminants also affected the flora and fauna of more rural areas downwind of industry, initially killing off the lichen cover and gradually over time darkening the surface of tree trunks. Naturalists throughout Britain and Continental Europe noticed, coincident with these changes, that rare dark forms of many moth species were becoming more common in the vicinity of manufacturing centers. The *phenomenon of industrial melanism* refers to this rapid rise in the frequency of dark (melanic) forms in many moth species that appeared to have occurred as a consequence of industrial air pollution.

The most famous and best-studied example of the phenomenon of industrial melanism is the peppered moth, *Biston betularia*, a common moth found throughout Britain and Continental Europe. Like other moth species, the adult form was believed to be nocturnal (active at night), spending most of the day motionless on trees, rocks and other resting sites. It was first described by Moses Harris, who in a brief description of the life cycle of the moth drew attention to the presence of color variation among larvae, but not among adults (Harris, 1766). The absence of any discussion of adult variation strongly suggests the melanic form was unknown at the time of his original description (Cook, 2003). And indeed, in a slightly later work, Harris specifically describes the adult as "white, and freckled as if peppered," with no mention of any known variants (Harris, 1775, p. 40).

The first published reference to the melanic form was by Robert Smith Edelston (1819–1892), who draws attention to multiple recent sightings near Manchester, a major manufacturing center, of the "negro" aberration (later known as *carbonaria*), which was "almost unknown" some sixteen years ago, i.e., 1848. His account also mentions how rapidly the dark form has increased in the area, concluding "if this goes on for a few years, the original type of A[mphydasis] betularia [later renamed *Biston betularia*] will be extinct in this locality" (Edelston, 1864). Over the next fifty years numerous additional sightings of the melanic form by naturalists and amateur lepidopterists led to widespread recognition that the range of the heretofore rare dark form was spreading throughout Britain and Continental Europe. These reports also drew attention to a dramatic increase in the frequency of the dark form in and around manufacturing centers. This widespread interest and curiosity surrounding a phenomenon occurring right before their eyes led the Evolution Committee of the Royal Society in 1900 to institute a collective inquiry in which they sent circulars to moth collectors throughout Britain and Continental Europe inviting them to survey local moth populations and reflect on any changes witnessed in their own lifetimes. Although the sporadic nature of the results of this inquiry prevented the committee from drawing any strong conclusions, it provided more systematic evidence that the change was indeed occurring.

As you might imagine, the phenomenon provoked a great deal of speculation regarding why it was occurring. Among the first was what might be referred to as a "Lamarckian" explanation. Nicholas Cooke (1877) drew attention to the fact that the dark form was becoming more common not only in industrial areas, but also areas where increased humidity likewise darkened tree trunks. He suggested that in industrial melanic species, like the peppered moth, the change to a darker form represented a physiological response to a changing environment, owing to changes in climate and the fact caterpillars were consuming soot deposited on their food plants.

J.W. Heslop Harrison alternatively suggested that the increasing frequency of melanic forms of moths in the vicinity of manufacturing areas was a direct consequence of industrial pollution. Harrison claimed lead and manganese salts contained in soot that covered food plants of moths had mutagenic properties that caused mutation of genes for melanin production, citing the results of breeding experiments he had conducted on caterpillars of *Selenia bilunaria* and *Tephrosia bistortata* fed on polluted foliage. Ford questioned the legitimacy of using species that did not exhibit a trend toward industrial melanism, pointing out further that in the vast majority of industrial melanic species, the melanic form was believed to be dominant in contrast to the two species Harrison used. Ford also drew attention to the fact that independent investigators were unable to repeat Harrison's results.

The first detailed published account of industrial melanism in terms of natural selection is generally attributed to James Tutt (1890), who, building off the work of Buchanan White (1876–1877) and others, drew attention to the role of selective elimination by birds. A brief comparison of the two forms of the moth when they rest against tree trunks in unpolluted and polluted settings reveals the intuition behind Tutt's theory. In unpolluted environments of the sort one finds in the rural countryside, trees are covered with lichen, which makes them a pale background against which the "typical" form of the moth would be very difficult to spot, but the dark form would be readily visible. This contrasts with the situation one finds in forests near manufacturing centers where years of exposure to air pollutants has led to the removal of lichen and a visible build up of soot, effectively darkening tree trunks. In these environments, it is the dark form that would be difficult to spot and the pale form would be easily seen. Thus, Cooke concluded, the spread of the dark form could be accounted for entirely in terms of selective predation by birds in the two environments. In unpolluted environments, whenever the dark form arose by mutation (or was introduced by migration) it would be quickly eliminated by birds; as such, in these environments the pale form is common. In polluted environments, in contrast, the pale form is the form most vulnerable to bird predation; as such, in these environs, it is the dark one that is prone to increase over time.

(a)   (b)

Biston betularia: one typical and one carbonaria resting on a lichen-covered tree
in unpolluted country (Dorset); and, one typical and one carbonaria resting on
blackened and lichen-free bark in an industrial area (the Birmingham district).
These photos originally appeared separately as Plates 14 and 15 in Ford (1975).

Interest in the phenomenon of industrial melanism began to spill over
from the naturalist community who were charting the spread of the dark
form and its rise in frequency near manufacturing centers to geneticists
who were as a group just beginning to recognize the numerous advantages of lepidoptera as model organisms for the study of heredity and
population genetics, such as their relatively short life spans and the fact
that their easily observed wing patterns had a genetic basis. Pioneering
genetics work by Col. W. Bowater established that the dark form in the
peppered moth, *carbonaria*, was the result of a single dominant gene.
Bowater also drew attention to the existence of intermediate melanic
forms, later referred to as *insularia*, which range in appearance from
individuals nearly as pale as the typical form to those that are nearly as
dark as *carbonaria*. Additional breeding experiments by E.B. Ford and
others strongly suggested that the gene responsible for dark coloration
might also have a physiological effect on the constitution of the moth, making it "hardier" than the pale form. Precisely what these investigators
meant by "hardier" seems to have varied with the investigator; some identified it with a tendency to emerge earlier in the year and at a lower temperature, others identified it in terms of higher than expected frequencies
in backcross broods. In view of these results, Ford offered an alternative

theory for industrial melanism in terms of natural selection that invoked two selective forces. Ford argued that the rapid spread of the melanic gene was primarily due to the physiological advantage it conferred. Ford explained why the spread was limited to industrial areas by drawing attention to the obvious handicap of dark coloration in unpolluted environments against visual predators such as birds.

The phenomenon of industrial melanism became of theoretical interest as well as a particularly striking example of natural selection. Charles Darwin's original presentation of the theory depicted natural selection as a slow process that led to the gradual accumulation of numerous slight variations over geological time. Indeed, Darwin himself went so far as to publicly doubt that natural selection could be directly observed during the brief span of a human lifetime. This consideration, coupled with numerous other apparent difficulties for his theory of natural selection, such as Lord Kelvin's 1868 theoretical estimate that the earth was about a hundred million years old (far below the amount of time required on Darwin's theory), briefly led to a period in the history of biology known as the "eclipse of Darwinism" in which Darwin's theory of natural selection was publicly doubted by many scientists during the turn of the twentieth century. Within this context, the phenomenon of industrial melanism provided a particularly striking example of natural selection for proponents of Darwin's theory, an example that J.B.S. Haldane in a very influential paper used to emphasize, in contrast to Darwin's portrayal, how very powerful the force of selection in nature could actually be. Haldane pointed out that if *carbonaria* was a simple Mendelian dominant and represented only one percent of the population in Manchester when it was first spotted in 1848 and completely ousted the typical form by 1901, the minimum selective advantage of the dominant gene would have to be roughly fifty percent greater than the recessive. This estimate was particularly important in that it was far in excess of what theoreticians had previously regarded as a realistic value for the force of selection in nature. Industrial melanism also figured prominently in theoretical debates between Haldane and Sir Ronald Fisher, two extremely influential figures in the development of mathematical theories of population genetics that provided an important basis for the "evolutionary synthesis," a dramatic period in the history of biology during the 1920s and 1930s during which a broad consensus on numerous fundamental issues in evolutionary biology was forged.

H.B.D. KETTLEWELL'S RESEARCH ON THE PEPPERED MOTH

With hindsight, it is easy to see how the foregoing developments mentioned above set the stage for someone to systematically study the phenomenon of industrial melanism. That someone was H.B.D. "Bernard" Kettlewell (1907–1979), a life-long amateur naturalist and entomologist, who at the age of forty-five left medical practice to pursue his hobby full time as a research worker under the supervision of E.B. Ford in 1952. Edmund Briscoe "Henry" Ford (1901–1988) was a pioneering researcher in the new field of "ecological genetics" he and others, including A.J. Cain and Philip Sheppard, were founding at Oxford University during the 1950s. Ecological genetics, broadly speaking, is the study of adjustments and adaptations of natural populations to changes in their environments by a combination of field and laboratory techniques. Cain and Sheppard originally worked on snail banding patterns in the grove snail, *Cepaea nemoralis*; Ford and others devoted themselves to the study of population fluctuations in several species of lepidoptera, including the scarlet tiger moth (*Panaxia dominula*) and the meadow brown butterfly (*Maniola jurtina*). As noted above, Ford was keenly interested in the phenomenon of industrial melanism, which he had independently studied from the standpoint of genetics and his own pilot field study. While it is fair to say that Kettlewell inherited the project of work on industrial melanism from Ford, the extent to which Ford actually mentored him is less clear.

Kettlewell initially pursued industrial melanism as one among several others, including some pioneering work on the use of radioactive tracers to track locust populations. Over time, however, his growing interest in all aspects of the phenomenon of industrial melanism led him to devote his entire career to the study of the subject. Kettlewell studied the direct effect of air pollution on local vegetation by means of monthly leaf washings that allowed him to quantify the amount of soot accumulating on the surfaces of leaves in the vicinity of manufacturing centers. Comparative studies of foliage and tree trunks in the two settings led him to recognize that aphids and lichen often die off in the presence of low levels of contaminants, presaging their use as bioindicator organisms. Kettlewell also orchestrated a comprehensive survey of industrial melanics throughout the whole of the British Isles. The latter ultimately led to the amassing of more that 100,000 records of melanic and typical frequencies in over fifty species of macrolepidoptera by one hundred part-time lepidopterists. With regard to the peppered moth, *Biston betularia*, these records amply documented the spread and rise in frequency of the *carbonaria* gene responsible for dark coloration in the vicinity of manufacturing centers. This work also documented a striking correlation between areas where the dark form was becoming more common and air pollution.

The studies for which Kettlewell is most famous, however, are a series of field investigations he conducted in the early 1950s. His initial investigation, conducted in the summer of 1953, involved three steps. First, he conducted what he referred to as a scoring experiment, in which he developed a method of objectively determining how conspicuous or inconspicuous pale and dark moths were when they rested against pale lichen-covered or soot-darkened pieces of bark. His specific technique involved placing moths representing the three forms (typical, *insularia*, and *carbonaria*) on representative pieces of bark from the two settings and then determining how far away one could walk from the bark and still spot the moth. As a result of these trials, Kettlewell and his associates determined that the typical (pale) form was regarded as inconspicuous when it rested on lichen-covered bark, but quite conspicuous when it rested on soot-darkened bark. The reverse was true for the *carbonaria* (dark) form, which was easily spotted when it rested against lichen-covered bark but much more difficult to spot when it rested on a soot-darkened piece of bark. Kettlewell laid great stress on the fact that multiple observers indepen-

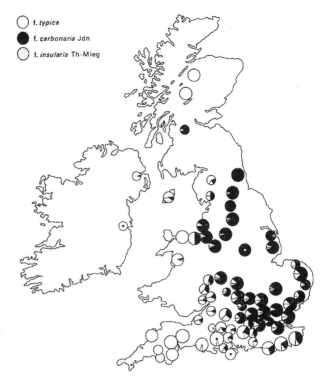

A frequency map of Biston betularia and its two melanics, f. carbonaria and f. insularia comprising more than 30, 000 records from 83 centres in Britain. (Reproduced from Kettlewell [1973], *The Evolution of Melanism*, from Figure 9.1, p. 135, by permission of Oxford University Press.)

dently reached the same conclusions as establishing the reliability of this scoring procedure.

The second step was to document that birds have the same difficulty humans do when it comes to spotting moths when they rest on matching backgrounds. Kettlewell had to consider the possibility birds would have keener powers of detection than humans; moreover, at the time of his studies, many naturalists publicly doubted that birds prey upon moths at all. To address this question, Kettlewell built a large cage within an aviary, which he subsequently divided by a large sheet into two sections, one of which contained two nesting great tits, *Parus major*. In the second section Kettlewell and his associates introduced numerous pieces of pale lichen-covered and soot-darkened pieces of bark, upon which sixteen moths representing the three forms were released. Kettlewell then removed the sheet, exposing the moths to predation by the birds, and monitored the experiment from a distance using binoculars and also by periodically checking to see which moths remained. Kettlewell found to his dismay that the birds ignored the moths for the first two hours, after which they ate all of the moths on their incorrect backgrounds as well as two on their correct backgrounds in the space of an hour. The high predation rate on both conspicuous and inconspicuous moths led Kettlewell to suspect the birds were becoming specialists on peppered moths. In a subsequent trial he introduced a broad spectrum of endemic insects. This alteration "proved successful" in that from this time onward, Kettlewell was able to document that the birds preyed primarily on the most conspicuous form when presented with a choice.

The third step was to assess whether Kettlewell could document these same results in nature. These are the classic field experiments he conducted in a heavily polluted wood in Birmingham (and later in an unpolluted wood near Dorset) using a technique known as mark-release-recapture previously developed by Fisher and Ford (1947). As the name suggests, the experiment involved three steps. First, Kettlewell raised a total of 630 male moths representing the three forms (137 typical, 46 *insularia*, and 447 *carbonaria*) marked with a dab of quick-drying cellulose paint on the undersurface of the wings. Second, Kettlewell released moths onto tree trunks, one per trunk, in a well-circumscribed forest with several natural boundaries to minimize migration from the test site. The third and final step was an attempt to recapture as many of the released moths as possible, which he did using a combination of a mercuryvapor light trap in the center of the wood and multiple assembling traps containing virgin females around the periphery of the wood. Kettlewell reasoned that, all things being equal, the recapture rates of the three forms should be the same. If, however, one form was better able to survive than another, for example the *carbonaria* form was better able to avoid avian predators than the typical form, more of the favored form should be recaptured owing to the fact that more of them would presumably survive during the interval between release and recapture. And this is exactly what Kettlewell found (see Table 1). Expressed as percentages, the recapture rate for *carbonaria* was 27.5 percent (123/447), over twice the recapture rate for typical, which was only 13 percent (18/137). Kettlewell clarified his interpretation of the results of this experiment by ruling out potential alternative explanations, such as the possibility that pale moths were more likely to migrate from the test site or that the dark form simply lived longer. Kettlewell also made numerous direct observations of bird predation when endemic birds were presented with a choice of moths representing the different forms. In each case he found they were much more likely to take the conspicuous moth first. In 1955, Kettlewell conducted a repeat smaller scale mark-release-recapture experiment in Birmingham and a slightly larger companion experiment in an unpolluted wood near Dean End, Dorset. The repeat experiment yielded similar results. In the unpolluted wood, he was able to document the reverse was true: In the Dorset wood it was typical that appeared to be at an advantage compared to *carbonaria*. The mark-release-recapture experiment in this setting resulted in a recapture rate for typical (13.7 percent [54/393]) that was nearly three times that found for *carbonaria* (4.7 percent [19/406]). Direct observations of bird predation in the unpolluted setting likewise suggested that birds were more likely to take the conspicuous form of the moth first, in this case, *carbonaria*. Kettlewell is widely regarded as clinching the argument by having his friend and colleague, the well known ethologist Dr. Niko Tinbergen, film the order of bird predation in the two settings. This latter piece of evidence was widely regarded as incontrovertible evidence that birds prey upon moths, and further that they do so selectively with reference to how conspicuous the moth is against its resting site.

Table 1. The result of Kettlewell's mark-releserecapture experiments. (Developed from figures reported in Kettlewell 1955, 1956.)

CENOZOIC ERA

Period	*Epoch*	*Years before present*
Quaternary	Recent	11,000 – today
	Pleistocene	1.5–2 million
Tertiary	Pliocene	13 million
	Miocene	25 million
	Oligocene	36 million
	Eocene	58 million
	Paleocene	63 million

MESOZOIC ERA

Cretaceous		135 million
Jurassic		181 million
Triassic		230 million

PALEOZOIC ERA

Permian		280 million
Carboniferous		345 million
Devonian		405 million
Silurian		425 million
Ordovician		500 million
Cambrian		600 million

PRE-CAMBRIAN ERA
Back to the formation of the Earth as a celestial body, possibly 3 billion plus years ago.

The results of Kettlewell's 1953 field experiments, conducted in the polluted wood near Birmingham, were initially published in E.B. Ford's *Moths* (1955), part of the New Naturalist series (a very popular series of scholarly books written for amateur entomologists). Kettlewell's results, as recounted in Ford's book, met with initial skepticism. Reviewers of the book publicly doubted that Kettlewell had actually observed large numbers of moths being preyed upon by birds as reported by Ford. It was this reaction, by amateur entomologists (clearly an important audience for Kettlewell) that led directly to his decision to make a film record of bird predation during the follow-up 1955 experiments. Kettlewell later published the results of the 1953 and 1955 experiments in the prestigious journal *Heredity*. The second paper included two plates featuring eight photographs, five of which illustrated different species of birds caught in the act of preying on the moths. These publications, as well as professional presentations and exhibits featuring a short silent movie made from excerpts of the film record collected during his investigations, effectively ended public doubt that birds prey on moths and do so selectively.

At this point it is important to distinguish the reactions of scientists and the public in general (including textbook writers) from those who actually work on the phenomenon. As discussed in detail below, the former learned about Kettlewell's work primarily from a series of popularizations, including a very influential article Kettlewell published in *Scientific American*, with the provocative title "Darwin's Missing Evidence" (1959). Scientists who actually studied the phenomenon (or took an interest in the techniques Kettlewell used for related research projects) consulted his professional publications with a much more critical eye. While none of them publicly doubted the plausibility of Kettlewell's basic conclusions, namely that differential bird predation was the primary selective agent that had led to *carbonaria* becoming more common in the vicinity of industrial areas, very early on several raised concerns about both the design of Kettlewell's experiments and the adequacy of his specific techniques. Kettlewell himself raised concerns about the large numbers of moths used in his field experiments. He intentionally used as many as possible to ensure that the results would be statistically significant, but fully recognized that the bird behavior he observed might simply be a response to greatly elevated densities of prey items. Both of his initial papers discuss how he continued to tinker with how he released moths onto tree trunks; moreover, he was aware that there wasn't a great deal of evidence to confirm what appeared to him to be a reasonable assumption, namely that moths rest in plain sight on the trunks of trees. Kettlewell also devoted considerable attention to the question of whether moths are capable of choosing their resting sites (e.g., dark moths that alight on birch trees might go out of their way to rest on dark scars in contrast to the rest of the tree-trunk surface, which is light). Others criticized Kettlewell's original experiments for using a combination of lab and wild-caught peppered moths, which potentially undermined his contention that there were no differences in longevity between the dark and pale forms (because we don't know how many of the dark, as opposed to the light, were caught in the wild). Questions were also raised about the precise role of lichen. Did lichen have to be present for the typical form to be at an advantage? Or was the typical form sufficiently camouflaged on the surface of trees with light bark surfaces (e.g., birch trees) in the absence of lichen? Much of the history of research on industrial melanism since Kettlewell's initial investigations was an attempt to remedy these perceived flaws in the design of his experiments. Continuing research has also raised questions about whether differential migration and sulfur dioxide concentrations might have played some role in the spread. In short, there are still many outstanding questions about the phenomenon of industrial melanism. It is also clear that, judged by contemporary standards, Kettlewell's original field experiments are problematic. This being said, it is important to recognize that none of these concerns has called into question the basic conclusion of Kettlewell's initial investigations. At least eight field-predation studies, several of which were specifically designed to remedy perceived deficiencies in Kettlewell's original experiments, have confirmed that differential bird predation is the primary selective agent. It is clear also that the phenomenon of industrial melanism is actually much more complicated than textbooks would have us believe.

Kettlewell continued to conduct research on all aspects of industrial melanism for the remainder of his professional career, from ongoing questions surrounding where moths actually rest during the day to questions concerning how the *carbonaria* gene became dominant over evolutionary time. He also extended his interest in industrial melanism to other species, and other types of melanism in nature. This culminated in the 1973 publication of his much awaited monograph *The Evolution of Melanism*, which attempted to provide a theoretical framework upon which to understand the origins and maintenance of melanic forms in nature.

THE SIGNIFICANCE OF THE PHENOMENON OF INDUSTRIAL MELANISM FOR OUR UNDERSTANDING OF EVOLUTION

As noted above, the phenomenon of industrial melanism and Kettlewell's specific work on it have turned out to be far more complicated than textbooks typically suggest. Many commentators have interpreted several well-known flaws in the design of Kettlewell's field experiments as somehow calling into question whether the phenomenon of industrial melanism really is an example of natural selection at all. Before turning to these issues, it is helpful to first clarify the importance of the phenomenon of industrial melanism and Kettlewell's work on it for evolutionary theory, and also why it became an icon of evolution.

Evolutionary biologists often distinguish between macro- and micro-evolutionary phenomena. Macro-evolutionary phenomena, such as speciation and extinction, are irreversible events that reflect processes that take place over hundreds of thousands, if not millions of years. Micro-evolutionary phenomena, such as an increase in the frequency of a gene within a species, are reversible phenomena that occur over much shorter periods of time. The phenomenon of industrial melanism is an example of a micro-evolutionary phenomenon, namely the local adaptation of populations to a changing environment over a very short period of time (decades). An obvious consequence of this observation is that Darwin's theory of common descent, namely life in all its diversity represents the descendants of a single or a few common ancestral organisms, is not at stake in debates about the merits of our current understanding of industrial melanism.

The phenomenon of industrial melanism is widely recognized by scientists to represent a particularly well-documented example of natural selection in nature. To see why this is the case, consider the following statement of Darwin's theory of natural selection, with reference to the phenomenon of industrial melanism as exemplified in the peppered moth:

1. If the peppered moth has both dark and light forms, and if these differences are correlated with survival differences in different environments; and
2. If the dark and light forms are heritable; and
3. If there is a competition in nature for resources, owing to the fact that the moths reproduce far in excess of those that can possibly survive; then
4. It follows that the form of the moth that is correlated with an increased chance of surviving in an environment will increase in frequency in the population inhabiting that environment over time (if it is not already in equilibrium).

We can then ask what evidence exists in favor of this explanation. Re. (1), Kettlewell (1958) documented a striking correlation between a rise in the frequencies of *carbonaria* and areas in the vicinity of manufacturing centers on the basis of extensive survey data provided by amateur collectors throughout the British Isles. Re. (2), as noted above, Bowater (1914) demonstrated that the dark coloration associated with *carbonaria* was the result of a single dominant gene. Re. (3), it is simply a fact in nature that organisms are capable of reproducing far in excess of those that can possibly survive; the peppered moth is no exception. In short, the evidence that the phenomenon of industrial melanism in the peppered moth is an example of natural selection is the unidirectional increase in the frequency of *carbonaria* in the vicinity of manufacturing centers during the Industrial Revolution (both in Europe and the United States) and its altogether predictable decline after the advent of clean air legislation. Again, at the risk of stating the obvious, the status of the phenomenon of industrial melanism as an example of natural selection is not at stake in debates about the merits of Kettlewell's field experiments. Industrial melanism is an example of natural selection regardless of whether and to what extent we understand the specific reason for why *carbonaria* rose in frequency in the vicinity of manufacturing centers. If this is the case, what did Kettlewell's much celebrated field experiments contribute to our understanding? Kettlewell's field experiments clarified the precise selective mechanism, namely differential bird predation, that was the primary cause for why the dark form was becoming more common.

WHY AND HOW DID INDUSTRIAL MELANISM BECOME AN ICON OF EVOLUTION?

Icons of evolution are entities whose fame extends well beyond the circle of scientists who do research on it, and even beyond the scientific community at large. They are so well known that they are generally recognized as symbols of evolution by laypeople with little or no background in evolutionary biology, as witnessed in the popular media, including books, encyclopedias, television, movies, and more recently the Internet. Perhaps the most reliable indication that an entity (or episode in the history of evolutionary biology) has achieved iconic status is when it is regularly used in biology textbooks to illustrate evolutionary concepts and processes. By this measure the phenomenon of industrial melanism stands as one of the preeminent icons of evolution. It is ubiquitous amongst U.S. biology textbooks, particularly in the 1960s–1980s, where it is commonly used as the only or one of a few examples of natural selection. Some textbooks refer to Kettlewell's investigations, often referring to them as *the* classic demonstration of natural selection.

The phenomenon of industrial melanism was not the only example of natural selection available to textbook writers in the early 1950s, and in the time since numerous other examples have come to light that are similarly well documented. Moreover, as noted above, despite its ubiquity among textbooks, the phenomenon of industrial melanism does not have any special theoretical importance unto itself. So why and how did the phenomenon of industrial melanism achieve its iconic status? The answer rests in three interrelated reasons: (1) specific advantages of this particular example for the teaching of natural selection, (2) the active role Kettlewell (and his colleagues) played in its dissemination, and (3) good timing.

PEDAGOGICAL ADVANTAGES

Much of the appeal of the phenomenon of industrial melanism for the teaching of science reflects several pedagogic advantages. From the standpoint of heuristics, one would be hard-pressed to find another example of natural selection that is as clear-cut as the classic textbook account of industrial melanism. The basic elements of the story (trees, soot, birds, moths) can be safely assumed to be familiar to all students regardless of their backgrounds in math and science. The mechanism of selection, selective elimination by birds, appeals to a widespread view of natural selection as a struggle for existence. Each part of the story from seeing how the surface of a tree trunk would darken as lichens died off and soot accumulates, to noticing how easy or difficult it is to spot typical and melanic moths against different backgrounds, to observing birds differentially prey upon the moths, lends itself to visual depiction. The same cannot be said for most other standard examples of natural selection (e.g., the evolution of pesticide and antibiotic resistance). The advantage of dark coloration in a soot-darkened environment (and pale coloration in an unpolluted environment) is also intuitive. (The contrast with A.J. Cain and Philip Sheppard's well-known work on snail banding patterns in this regard is striking, which also involved differential bird predation with reference to a visual characteristic. It isn't intuitively obvious why birds would more easily spot snails with certain shell-banding patterns over others.)

The phenomenon of industrial melanism is also particularly advantageous from the standpoint of content. The abbreviated descriptions one finds in textbooks discuss the example in terms of the replacement of one form by another with reference to the spread of a single gene and one readily observed effect of that gene on the organism. Students can readily appreciate why and how visual differences between the dark and light colored moths as they rest against soot-darkened and pale lichen-covered tree trunks might place a moth at risk of bird predation, depending on the environment. In this light, it is a particularly good vehicle for clarifying to students that fitness is a function of the fit of an organism with its environment, instead of some reified property (e.g., strength or intelligence) that the organism possesses. While there are other well-documented examples of natural selection available, several of which also involve readily obvious characteristics, most of these are far more complicated. Consider for instance Peter and Rosemary Grant's well-known studies of fourteen species of finch, collectively known as "Darwin's finches" in the Galápagos Islands. Over the years the Grants have amassed overwhelming evidence that variation in bird beak size is due to natural selection. To adequately even begin to explain these adaptations, however, would minimally involve students making sense of more complicated processes (e.g., disruptive selection) and the effects of multiple genes. Darwin's finches also raise questions surrounding what constitutes a species and how speciation takes place, important in their own right, but potentially distracting if one's immediate goal as an instructor is to teach the concept of natural selection.

Kettlewell's research on the phenomenon of industrial melanism, as recounted in textbooks, is also helpful from the standpoint of teaching science as a process. The design of Kettlewell's experiments, as described above, is nothing short of elegant. Students can readily understand how the data was collected, analyzed, and interpreted, with a minimum of mathematics. Kettlewell's results have the air of being definitive, providing students with a sense of closure. It has also been suggested that textbooks depict Kettlewell's field experiment in the unpolluted setting as a control for the field experiment in the polluted setting, thus illustrating how research in evolutionary biology can conform to a model of controlled experimentation. One can also use Kettlewell's investigations, questions about the merits of his specific experimental design and our current understanding of the phenomenon (discussed below) to illustrate several issues associated with the nature of science, such as how evidence is used in the evaluation of theories and the tentative nature of scientific conclusions.

KETTLEWELL'S ROLE AS A POPULARIZER

Insight into why the phenomenon of industrial melanism became an icon can also be gained from considering the very active role Kettlewell and his associates took in popularizing the example. Over the course of his remaining career, Kettlewell wrote twenty-three technical and nine review articles on the subject of industrial melanism and made numerous professional presentations. He created exhibits for annual meetings of the Royal Society (a prestigious British scientific society), the Darwin Centenary in London and World's Fair in Brussels (both in 1958), and the Royal Society Tercentenary in 1960. These elaborate exhibits featured numerous visual aids, including pieces of bark taken from polluted and unpolluted settings, live and dead moths, slides, graphs and maps, samples of pollution, and a color film developed in 1955 from footage taken during his 1955 field experiments. These temporary exhibits led to invitations for permanent exhibits for both the Birmingham Museum and the British Natural History Museum. Many of the attendees were biology textbook writers and teachers representing all grade levels, many of whom wrote him subsequently with requests for photos and use of his film. As early as 18 November 1955 he and Niko Tinbergen were already negotiating for the distribution of the film with the British Film Institute, and just a few years later in 1961 Kettlewell was commissioned to create a sound version for use in American classrooms.

Kettlewell also shared the results of his work by means of idealized retrospective accounts in the popular media, which were arguably much more influential among scientists than any of his professional writings and presentations. Kettlewell wrote at least five popular articles in newspapers and magazines devoted specifically to industrial melanism, and made numerous lectures to amateur entomologist societies and schools. He also made radio broadcasts and also occasional television appearances. Kettlewell's colleagues and their students, many of whom migrated from the University of Oxford to other prestigious colleges and universities throughout the British Isles, regularly cited the example in their professional presentations, lectures, textbooks, articles, and books. The collective effect of these numerous activities was to disseminate information about the phenomenon to a wide audience that went well beyond the British Isles.

Why did Kettlewell, and his associates, take such an active role in publicizing the phenomenon of industrial melanism? It would be misleading to say that Kettlewell, Ford, or anyone else associated with what has come to be known as Ford's "Oxford School of Ecological Genetics" set out with the intent of making it an icon, although it was probably soon obvious to all of them that public interest and acclaim by scientists and the lay public far outstripped any of their other numerous research projects. There are multiple reasons why Kettlewell took on the role of popularizer, not the least of which was his own ebullient personality and natural flair for explaining it. He genuinely enjoyed sharing his research with others, not that this sets him apart from other scientists. Part of the answer may lie in some personal insecurity surrounding the fact that he was in the process of making a midlife career change and needed to establish himself as a researcher (Hooper, 2002). This being said, it seems clear Kettlewell saw these activities as a way of thanking the Nuffield Foundation, who funded his work, as well as an opportunity to educate the public and solicit volunteers for his ongoing survey of melanism throughout Britain. In retrospect it seems clear that the active involvement of amateur entomologists from all walks of life (including members of the military, a bishop, an ironmonger, and a stockbroker) and all ages in his research increased public awareness of the phenomenon. It's also fair to say Kettlewell thought he had a special moral obligation to ensure that information about his work was not distorted by the media.

TIMING

A third and final set of considerations that account in part for why the phenomenon of industrial melanism became an icon can be ascribed to historical circumstances. As mentioned previously, the first published account of the finding of a melanic form occurred in 1864, just five years after the publication of Darwin's *Origin of Species* (1858). Darwin's epic work convinced the scientific community that life in all its diversity was the product of common descent from one or a few simple organisms. It also drew attention to the possibility that a process in nature, namely natural selection, could account for the ubiquitous adaptations one finds among living organisms. While it is clear that historically Darwin's theory of natural selection initially met with more resistance than his theory of common descent, *Origin* invited readers to consider the possibility that differential selection might account for how a species adapts to changing environmental conditions. In short, Darwin's *Origin* primed the public to look for examples of natural selection just as naturalists throughout Britain and Continental Europe were becoming aware of one of the most obvious examples of it, namely the phenomenon of industrial melanism.

Darwin's theory of natural selection was initially favored as one, but not the only, mechanism of evolution by the 1870s and 1880s among scientists. In successive editions of *Origin*, Darwin found himself relying increasingly on other mechanisms, such as Lamarckian inheritance (discussed above). He did so in an effort to address perceived problems for his theory of evolution by natural selection, such as the fact it relied upon a blending theory of inheritance, which posited that the characteristics of parents were mixed in their offspring. It also required vast amounts of time (hundreds of millions of years), in contrast to Lord Kelvin's generally accepted estimate on the basis of cooling rates that the earth could be at most about a hundred million years old. These two considerations (each of which is now regarded as erroneous) created what appeared to be insurmountable problems when it came to accounting for the origin of adaptations and species in terms of natural selection.

Indeed, the close of the nineteenth century is often referred to as the "Eclipse of Darwinism," when biologists increasingly preferred other mechanisms to account for evolutionary phenomena. In such a climate, defenders of Darwin's theory of natural selection searched for particularly good examples, such as the phenomenon of industrial melanism, to make their case. As noted above, J.B.S. Haldane's influential paper "A mathematical theory of natural and artificial selection" (1924) drew specific attention to the phenomenon of industrial melanism as one of several examples that illustrated how very rapid changes were possible by means of natural selection.

A series of scientific and mathematical discoveries during the next twenty years eventually led to the formation of a consensus view among biologists during the 1940s. This period of time, known as the evolutionary synthesis, was characterized by a renewed emphasis on natural selection as the mechanism of evolution and the exclusion of alternatives. Among these discoveries that made the evolutionary synthesis possible were a set of what are now regarded as definitive experiments by August Weisman among others that disproved the possibility of Lamarckian inheritance. The rediscovery of Mendel's work in 1900 led to the development of a viable alternative to blending inheritance. Marie Curie's discovery of radioactivity in 1903 led to recognition of a fundamental flaw in Lord Kelvin's estimate of the age of the earth, which failed to include this important additional source of heat. Particularly important in this regard was the formation of a comprehensive mathematical theory of population genetics by R.A. Fisher and J.B.S. Haldane (among others), which demonstrated how relatively low selective pressures could lead to dramatic changes in the frequencies of genes.

Biologists during this time period increasingly began to see evolutionary theory as central to the biological sciences, aptly captured by the geneticist Theodosius Dobzhansky's well-known dictum "nothing makes sense in biology except in the light of evolution" (Dobzhansky, 1973). One natural consequence of the development of this consensus view was the introduction of evolutionary topics into biology textbooks. Textbook writers, faced with the need for simple, clear-cut examples that made sense to introductory students, as well as practical space limitations and perhaps the need to use black-and-white photographs, naturally gravitated to the phenomenon of industrial melanism, which as noted above had multiple pedagogic advantages over other examples at the time.

The launching of *Sputnik* on 4 October 1957 was a catalytic event that ignited the so-called Space Race within the Cold War that existed between the United States and the Soviet Union. Among its many ramifications was a dramatic increasing in funding for science education in the United States, spearheaded by the active involvement of scientists. These developments directly led to the development of the BSCS (Biological Sciences Curriculum Study) headed jointly by Benjamin Glass and Arnold Grobman at the University of Colorado, Boulder in 1960. Among the emphases of the curriculum materials they created was a renewed focus on evolution and an emphasis on inquiry by observation and experiment. Kettlewell's investigations on industrial melanism lent themselves to idealization, and indeed demonstrated how it was possible to pursue research in evolution by experimental means.

In retrospect it seems almost inevitable that the phenomenon of industrial melanism would become an icon of evolution. As the preceding analysis has shown, there were multiple factors that collectively led to its widespread fame and use in textbooks. Some of these factors center around pedagogic features of the phenomenon itself that make it a particularly good illustration of natural selection, such as the fact that nearly every element in the story can be visually depicted and its intuitive character.

Particularly important in this regard is the apparently definitive character of the experiments that established selective bird predation was the cause. Kettlewell's active involvement as a publicizer of these experiments is also especially important. In numerous popular articles and other presentations, he provided brief, idealized descriptions of his investigations that not only made the phenomenon to scientists outside his field and the lay public, but could also themselves be simplified still further to meet the demands of textbook writers. The ubiquity of the phenomenon of industrial melanism is also due in part to a host of factors that fall under the heading of good timing. Kettlewell's classic investigations occurred just as textbook writers were introducing evolutionary topics into the science curriculum and were searching for simple examples of not only natural selection, but also examples of how questions in evolutionary biology (which heretofore was perceived to be pursued almost exclusively by observation) could be studied experimentally.

FURTHER RESEARCH ON THE PHENOMENON
OF INDUSTRIAL MELANISM

As mentioned above, most people (scientists and laypersons alike) have learned about the phenomenon of industrial melanism primarily from textbooks and the popular media, which for a multitude of reasons simplify science. It is well recognized that textbook descriptions of scientific phenomena focus on the sharing of results to the exclusion of any detail regarding how these results are obtained and interpretive problems that surround them. (This is sometimes referred to pejoratively as presenting science as a rhetoric of conclusions [Schwab, 1962].) The collective, albeit unintended, effect of such descriptions on the reader is to convince him or her that the facts being shared are known with certitude, and furthermore, that a decision to include a particular example must reflect its empirical or theoretical importance to the science.

Textbook discussions of the phenomenon of industrial melanism go well beyond the removal of superfluous detail. They often omit reference to the second melanic form (*insularia*) and any reference to Kettlewell's research (or if they do include it, only the details of the mark-release-recapture experiments). Indeed, the phenomenon and Kettlewell's work on it have become so simplified that one can seriously question the extent to which these textbooks are constrained by our understanding of the science at all. In a thoughtful review of textbook depictions, Michael Majerus, a well-known researcher on the phenomenon of industrial melanism, has analyzed standard textbook descriptions of the peppered moth story into seven components, most of which are either "wrong, inaccurate, or incomplete" (Majerus, 1998; see Table 2).

Table 2. Components of the "textbook" peppered moth story and some of the ways in which these claims are misleading. (Table created from Box 5.1 of Majerus, M.E.N. [1998], *Melanism: Evolution in Action.* Oxford: Oxford University Press, p. 98 and supporting text.)

Components of the "textbook" peppered moth story	Ways in which these components are misleading
1. The peppered moth has two distinct forms, the typical form being white with black speckling, the other f. *carbonaria,* being almost completely black.	There are intermediate forms, that vary in appearance between the pale typical and dark *carbonaria* forms, collectively known as f. insularia.
2. The two forms of the peppered moth are genetically controlled by a single gene, the *carbonaria* allele being completely dominant to the *typica* allele.	While there are two alleles at a single genetic locus, with the *carbonaria* allele being dominant to the *typica* allele, the insularia complex of forms involves at least three additional alleles.
3. Peppered moths fly at night and rest on tree trunks during the day.	Data on the actual resting site of the moth is scarce, there is some evidence that peppered moths rest in unexposed positions, e.g., on the undersides of the boughs of trees.
4. Birds find peppered moths on tree trunks and eat them.	The evidence that birds are significant predators of moths has increased since Kettlewell's investigations.
5. The ease with which birds find peppered moths on tree trunks depends on how well the peppered moths are camouflaged.	It's unclear that Kettlewell's experiments really do establish this fact; there is evidence birds have greater visual acuity than humans.
6. Typical peppered moths are better camouflaged than melanics on lichen-covered tree trunks in unpolluted regions, while melanics are better camouflaged in industrial regions where tree trunks have been denuded of lichens and blackened by atmospheric pollution.	This basic conclusion has been confirmed in at least 8 independent studies since.
7. The frequencies of melanic and typical moths in a particular area are a result of the relative levels of bird predation on the two forms in that area, and migration into the area of peppered moths from regions in which the form frequencies are different.	Differential migration may be a more important factor in accounting for relative frequencies of the two forms than has generally been recognized.

In his review of the discrepancies between the peppered moth story as discussed in textbooks and what is actually known about the phenomenon, Majerus stresses that in his expert opinion, "the huge wealth of additional data obtained since Kettlewell's initial predation papers . . . does not undermine the basic deductions from that work" (Majerus, 1998, p. 116). There is simply no doubt among scientists who work on the phenomenon that differential predation by birds is the primary cause that led *carbonaria* to become more common in areas affected by air pollution. This being said, there are outstanding questions to be addressed concerning some of the basic assumptions made by investigators, particularly with respect to development of theoretical models, and also the artificiality of some experimental work (e.g., the use of dead moths, and the continued positioning of moths on tree trunks in plain sight when it is clear they rarely choose to do so in the wild). And again, as mentioned above, none of these discrepancies calls into question the enormous body of survey data that is the basis for claims that the phenomenon of industrial melanism is an example of natural selection. Industrial melanism is still regarded as one of the best, if not the best, example of natural selection observed by humans in nature.

CHALLENGES TO INDUSTRIAL MELANISM'S STATUS AS AN ICON OF EVOLUTION

At the risk of understatement, not all commentators on this episode have interpreted the several discrepancies identified by Majerus as simply drawing attention to outstanding questions that surround an area of active research. Professor Jerry Coyne, a well-respected evolutionary biologist, in a review of Majerus's book interpreted these discrepancies as completely undermining the continued use of the peppered moth as a well-understood example of natural selection. Coyne's review was subsequently picked up by science commentators, many of whom with an obvious agenda seized upon the opportunity to raise fundamental questions concerning whether the evidence for evolution is as compelling as scientists would have us believe. The implicit rhetorical question underlying these attacks is transparent: If there are problems with the "best example" of natural selection, what faith should we have in other examples? It is precisely because of its iconic status that the phenomenon of industrial melanism as exemplified in the peppered moth has proven to be a lightening rod for debates surrounding the scientific status of evolutionary theory and the merits of intelligent design as an alternative theory of origins.

TEXTBOOKS SHOULD CARRY WARNING LABELS

It is instructive at this point to consider an example of one of the several commentaries on the phenomenon of industrial melanism which have criticized its continued use in science classrooms. In a recent book titled *Icons of Evolution* (2000), Jonathan Wells, a senior fellow at the Discovery Institute (a "non-partisan" think-tank, well known for its promotion of Intelligent Design Theory as an alternative to evolution), devotes an entire chapter to the peppered moth story. Many of the problems inherent in Wells's treatment reflect a fundamental misunderstanding of scientific issues associated with industrial melanism, evolutionary biology, and how field work is done. The following section will review more general problems in Wells's account involving the misuse of history and the logic of his argument.

Wells titles his chapter on industrial melanism "The Peppered Moth Story" to underscore his contention that it is just a story, symptomatic of a consistent pattern of distortions deliberately introduced by textbook writers to misinform the public about the evidence for evolution. His modest goal in this chapter is to reveal how poorly understood the phenomenon of industrial melanism is in order to raise questions about why textbooks continue to portray it as a particularly good example of natural selection when it is not. Wells starts the chapter by claiming that Kettlewell's investigations on the peppered moth was the first direct evidence in favor of natural selection, aside from Herman Bumpus's well-known statistical analysis of a fortuitous finding of English sparrows collected shortly after a severe snow storm. (This claim is historically inaccurate: See Dobzhansky (1951) for many examples known prior to Kettlewell's work.) Wells mistakenly identifies Kettlewell's experiments as a test of J.W. Tutt's theory (as noted above, Kettlewell's work was pursued with reference to E.B. Ford's theory in terms of two selective forces). Kettlewell's investigations are identified as three field experiments (ignoring additional research Kettlewell amassed before and after the field studies, as well as numerous studies since that have qualitatively confirmed the basic conclusions of Kettlewell's study). Wells then reviews what he portrays as numerous insurmountable interpretive problems associated with Kettlewell's original experiments (identified above), never acknowledging that many of the follow-up studies by Kettlewell and others were specifically designed to address these problems.

The conclusion Wells wants the reader to draw from this lopsided and biased review of research on industrial melanism has to do with how the phenomenon has and should be depicted in textbooks. He interprets discrepancies between textbook accounts and what researchers actually know about the phenomenon as evidence that textbook writers are engaged in an elaborate conspiracy to promote the "myth" of evolution. They do so by (intentionally or otherwise) systematically distorting how very weak the evidence is in favor of one of the so-called best examples of evolution. As he points out in an earlier chapter, "science is the search for truth," not myths. To begin with, we should recognize that even if one buys into Wells's interpretation of the questions surrounding the phenomenon of industrial melanism and Kettlewell's work on it, none of these concerns call into question either the fact that evolution has occurred, because, as noted above, descent with modification is not an issue in this example. Agreeing with Wells's analysis doesn't even call into question the fact that industrial melanism is an example of natural selection, because, as noted above, issues surrounding the assumptions and design of Kettlewell's experiments center around the issue of whether Kettlewell's results in fact establish bird predation as the cause. Here one can agree with Wells that if Kettlewell's original experiments were the only evidence in favor of his conclusion, this would be a problem. But they aren't. Scientists don't disagree about the fundamentals of this example nearly as much as Wells (and others) would have us believe. More to the point, a science textbook that consisted of only things that have been established as "true" would be a slim textbook indeed. Of course introductory textbooks written for children will differ from the accounts one finds in journal articles written by scientists primarily for the benefit of those actually doing research on the phenomenon. The fact scientists continue to pursue questions about the phenomenon of industrial melanism is indicative of an active area of research, not evidence of widespread doubts regarding whether it is an example of natural selection at all.

ALLEGATIONS OF FRAUD AND CONSPIRACY

Perhaps the most disturbing reaction to recognition that there are discrepancies between textbook accounts and what is actually known about the phenomenon is a book-length popularization by Judith Hooper (2002). *Of Moths and Men: An Evolution Tale: The Untold Story of Science and The Peppered Moth*, is an entertaining account of Kettlewell's research on industrial melanism that vividly depicts the very colorful personalities associated with E.B. Ford's Oxford School of Ecological Genetics. In it Hooper implies Kettlewell committed fraud and furthermore that E.B. Ford and his associates engaged in an elaborate cover-up to suppress dissension. These are particularly serious charges, not voiced previously by any scientist or historian of science. They are also completely baseless, reflecting poor historical scholarship, a lack of appreciation of the nature of field work, and fundamental misunderstandings about the nature of science. While the book has received rave reviews in some quarters, scientists familiar with the episode have universally condemned it. As with the discussion of Wells's chapter above, the following discussion will focus on how history is being misused by this writer and also several fundamental problems associated with the author's understanding of the nature of science.

Hooper's book-length account advances essentially a three-pronged attack in support of her allegations that Kettlewell committed fraud and he and his associates conspired to suppress this fact. Much of the book is devoted to character assassination in an effort to convince the reader that Kettlewell and his associates were of such a character that they would commit fraud and conspire to hide the details. The "evidence" Hooper provides in favor of these allegations are a series of previously published anecdotes centering around the personal lives of the members of the Oxford School of Ecological Genetics. There is no doubt that H.B.D. Kettlewell, E.B. Ford, J.B.S. Haldane, and others were characters having personalities that were larger than life. But Hooper's analysis never establishes the relevance of the details of Kettlewell's daughter's suicide, Ford's reputation as a misogynist, and the curious relationship between Ford and his mysterious "ward" (among others) for her allegations against them. One cannot help but conclude that the sharing of these details is simply a crass attempt to titillate the reader. To be fair, Hooper's account does share some details of Kettlewell's professional work that might actually be relevant to her arguments, but her argument on this score is entirely based upon innuendo. The account draws great attention to a comment by one of his former associates, R.J. Berry, who referred to Kettlewell as "the worst professional scientist I have ever known." By this, as Berry points out later in the same paragraph, he was referring to the fact that it was very difficult to write papers with Kettlewell, both because "writing papers with him was traumatic," and "as an experienced clinician he made rapid diagnoses and refused to be diverted by what he regarded as irrelevant evidence" (Berry, 1990, p. 322). Berry most certainly did not intend his candid appraisal of Kettlewell to be interpreted as suggesting he thought Kettlewell was capable of committing fraud, something he would have corrected her on if, as with other people she interviewed in the process of writing her book, she had made a point of letting him see and comment on her working manuscript before it went into publication.

A second general argument advanced by Hooper to establish her claims that Kettlewell committed fraud and his colleagues and associates conspired to hide this fact rests in repeated attempts to establish motive. Part of this involves drawing attention to Ford's pan-selectionist agenda and how the fate of this agenda rested on the success of Kettlewell's investigations. While the present chapter amply documents how very famous and influential Kettlewell's work ultimately became, it is simply not the case that Kettlewell's experiments were as important to Kettlewell's career (or that of his colleagues) as Hooper would have us believe. Kettlewell initially pursued several other lines of investigation (such as his pioneering research on the use of radioactive tracers to track locust populations), and at the time of his studies, the fame he would eventually receive for his work on industrial melanism was far from obvious. Like-wise, it is important to recognize that Ford and his associates had many other experimental studies that documented the importance of natural selection in support of their overall views. While it is certainly true that Kettlewell wanted his investigations to succeed, this consideration alone does not imply whenever a scientist's results conform to his predictions they should immediately be suspect.

The third and most specific argument Hooper raises is that of estab-lishing that Kettlewell did indeed commit fraud. The basis of her claim is a curious increase in Kettlewell's recapture frequencies that occurred on the day E.B. Ford wrote him a letter mentioning that he (Ford) shared Kettlewell's disappointment that some of the initial recapture figures were not as expected. Hooper considers and rejects one possible alterna-tive explanation (the increase might reflect a change in the local weather conditions). Scientists familiar with research on industrial melanism who have reviewed this book vehemently disagree with Hooper's interpreta-tion, which rests on fundamental misunderstandings regarding the nature of field work. A careful independent statistical analysis of Kettlewell's results likewise reveals there is no evidence that the figures were fudged. A careful review of surviving correspondence between Kettlewell and his associates, which Hooper consulted in the process of writing her book, has also revealed that her book rests upon shoddy historical scholarship. It ignores reasonable alternative explanations for passages which taken out of context might be interpreted as supporting her views, and multiple instances in which the evidence is actually contrary to her thesis.

Hooper's account also rests on several fundamental misunderstand-ings about the nature of science. It suggests, for instance, that proof in science only occurs in the context of experimental investigations, i.e., when one directly observes something under controlled conditions. Experiments do not establish facts with the certitude Hooper suggests (her own analysis of Kettlewell's experiments is ample testimony to the fact that our interpretations of the results of experiments may change over time). Moreover, the results of experiments are often used to make claims about other times and places that are not directly observed, e.g., claims about the effects of gravity on the moon. It is true that evolution-ary biology is not characterized by the use of experiments. This reflects the fact that it is a historical science, not its immaturity as a science. Short of a time machine, evolutionary biologists cannot directly observe many of the phenomena of most interest to them. But this does not imply that evolutionary biology, and historical sciences in general, can only provide a series of just-so stories. Once evolutionary biologists develop a scenario for what may have happened in the past, they test it by thinking through the implications of that scenario. Consider, for example, the phenomenon of industrial melanism. E.B. Ford developed a theory that would account for both why *carbonaria* was becoming more common and also why its rise in frequency was restricted to industrial areas. Kettlewell's field studies constituted a test of the second part of Ford's theory, namely that differential bird predation was responsible for the rise of *carbonaria* in soot-darkened environments.

CONCLUDING THOUGHTS

The preceding analysis has reviewed how and why the phenomenon of industrial melanism as exemplified in the peppered moth became an icon of evolution, its importance to evolutionary biology, and some of the controversies that have surrounded it since. Industrial melanism became the classic example of evolution by natural selection largely because of multiple pedagogic advantages unique to this example, such as its conduciveness to visual representation. Part of the answer also rests in how Kettlewell and his associates popularized the example. And at least part of the reason industrial melanism has become so widely known reflects good timing. Kettlewell's experimental demonstration that selective bird predation could account for why the dark form was becoming more common came at a time when textbook writers were searching for simple, clear-cut examples of natural selection that could demonstrate that it was possible to study evolution by experimental means.

We have also seen that much of the beauty of this example, and in particular Kettlewell's experimental demonstrations, reflects how it has been packaged and simplified by textbook writers and the popular media. The phenomenon of industrial melanism is much more complicated than textbooks would have us believe; Kettlewell's field experiments were not as definitive as they are often portrayed. Research by Kettlewell and others since has attempted to address some of the problematic features we now recognize in Kettlewell's initial field experiments, including the design of his experiments, his techniques, and what appeared to him at the time to be reasonable assumptions. Neither the fact that scientists who currently work on the phenomenon have not resolved some of these issues to their satisfaction, nor the fact they may disagree with one another about the relative importance of various factors, means they are ready to conclude that industrial melanism is no longer an example of natural selection. The existence of minor disagreements concerning the relative role of migration in explaining the spread of the melanic form (and its subsequent decline) doesn't preclude researchers from reaching broad consensus regarding the fact that differential predation by birds has been responsible for the spread. This isn't merely dogmatism on the part of scientists in the face of overwhelming evidence to the contrary. The continued presence of questions and disagreements among scientists is simply an indication that industrial melanism remains an active area of research.

It is important to recognize that none of the observations in the above paragraph are specific to the phenomenon of industrial melanism. All facts in science have a story behind them, one that if told would likewise draw attention to discrepancies between how textbooks depict the "facts" of science and what researchers actually know about those facts. A medical example makes this clear. Consider what textbooks typically tell high school students about the subject of alcohol. Textbook treatments typically warn students about the dangers of underage drinking, the importance of never drinking and driving, multiple adverse health effects, and also how and why a short-term consequence of drinking alcoholic beverages might impair judgment, leading teens to engage in risky behaviors such as sex and drug use. Readers are no doubt also aware of the occasional medical study that suggests drinking red wine in moderation might be beneficial to one's cardiovascular health. The presence of disagreements among medical researchers regarding whether drinking red wine is or is not beneficial does not undermine the wisdom captured in textbook accounts. Nor does the fact there is a consensus view about the dangers of excess alcohol drinking imply there is a conspiracy by prudent textbook writers to encourage abstinence.

It is, of course, entirely appropriate to be concerned about the simplistic way textbooks portray all topics in science, particularly when one considers the enormous role textbooks often play in setting the curriculum. As the foregoing analysis makes clear, neither labeling textbooks with disclaimers urging students to recognize that the information contained within is not as certain as the textbook writer suggests nor restricting textbook accounts to what is known to be true represent viable alternatives. While it is of course possible to write textbooks in a manner that more faithfully captures the actual process of science, warts and all, critics should recognize inherent limits to what can be done in the context of introductory accounts written for children. Textbook accounts are constrained by a host of pedagogic and practical constraints, constraints that favor brief, simple examples that can be summarized in terms that are familiar or easily understood by children. To address this dilemma, attention must instead be focused on the often uncritical stance teachers and students take to what they learn from such sources.

Fortunately there are ways of teaching that can provide introductory students with straightforward, accessible examples of unfamiliar scientific concepts that can also help them appreciate some of the complexities of the actual practice of science. One of the more obvious ways to promote reflection in the science classroom is by judicious use of history and philosophy of science. Discussing scientific debates in their original historical and philosophical contexts can promote active learning about how scientific claims are developed and evaluated. Hagen (1996) uses a vignette based upon the history of Kettlewell's classic investigations to foster student reflection and insight. Rudge (2004) provides lesson plans that illustrate how the phenomenon of industrial melanism can be introduced as a mystery for students to solve in a manner that directly addresses common misconceptions students have about evolutionary phenomena. The lesson invites students to share their sincerely held beliefs about the "mystery phenomena" and to consider evidence (observations and experiments by past scientists) that historically led investigators to conclude that differential bird predation was responsible for increasing frequencies of the dark form of the moth. Once students are comfortable with this explanation and see its advantages over others, the instructor shares the rest of the story—including details regarding problems with Kettlewell's original experiments and his assumptions. The lesson concludes by having students (all of whom are future elementary school teachers) compare how textbooks typically portray the phenomenon of industrial melanism with what is actually known about it and Kettlewell's investigations. The point of this discussion is not to disabuse them of the idea that industrial melanism is an example of natural selection, but rather to help them start to think, as future teachers, how they will help their students appreciate some of the complexities of science lost in textbook accounts. Counterintuitive as it may seem, the interpretive problems surrounding Kettlewell's investigations and our understanding of the phenomenon of industrial melanism actually enhance its continued value for the teaching of science, by inviting students to actively reflect on how very different the actual practice of science is from how it is typically depicted in textbooks.

Lamarck's Theory of Species Transmutation

Jean-Baptiste de Lamarck (1744–1829) was an influential biologist who became famous for a theory that attempted to account for the similarities and differences that exist among and between organisms using what is now regarded as an evolutionary point of view. Lamarck was initially an essentialist in that he believed, like other naturalists since Aristotle, that species do not change. At the age of fifty, he abandoned this view in favor of the possibility that species could transform over time. Lamarck's theory is perhaps best understood as the product of several Enlightenment themes common among eighteenth-century attempts to account for the origin and development of life. Lamarck believed that it was possible for the simplest forms of life to spontaneously arise from inanimate matter, and moreover that this had occurred throughout the history of the earth. He also believed there was a natural tendency of all organisms to become more complex, and further that this is reflected among living organisms. Relatively simple forms that exist today are the descendants of a recent act of spontaneous generation; more complex organisms represent descendants of earlier acts of spontaneous generation. Lamarck could not deny the existence of gaps in the fossil record, nor the fact that there does not appear to be a perfect continuum from simpler to more complex forms of life in nature. He did not believe in the possibility of extinction. To account for deviations, Lamarck instead drew attention to how use and disuse adapted organisms to changing local environmental conditions. One of his favorite examples of this was his explanation for why giraffes have long necks. Originally giraffes all had short necks and fed on low-lying leaves on trees. At some point in their history low-lying leaves became less available, and in response to this change, giraffes stretched their necks to reach higher leaves, which led to their necks becoming slightly longer. This change was inherited by their offspring, which in turn in response to similar conditions stretched their necks slightly more. Over multiple generations, the neck of the giraffe gradually became longer until it reached its present length. Lamarck's views are generally regarded as discredited. While no one doubts that the body of an organism may change as a result of changes in diet or habitat, numerous scientific experiments by August Weisman and others are generally regarded as disproving the possibility that such changes can be inherited by offspring.

FURTHER READING

To learn more about industrial melanism and the many outstanding scientific questions that continue to spark interest among scientists, see Majerus, M.E.N. *Melanism: Evolution in Action* (Oxford: Oxford University Press, 1998). The book includes a chapter that summarizes how textbooks have typically simplified the phenomenon of industrial melanism and a companion chapter that shares the rest of the story.

For an entertaining popularization of Kettlewell's work and the colorful personalities associated with E.B. Ford's so-called Oxford School of Ecological Genetics, see Hooper, J. *Of Moths and Men: An Evolution Tale: The Untold Story of Science and the Peppered Moth* (New York: Norton & Co., 2002). Read the text critically—it contains numerous historical errors and its main thesis alleging Kettlewell committed fraud is completely baseless. (Young and Musgrave, 2005: Rudge, 2005)

Those interested in learning more about fraud in science should consult Judson, H. F. *The Great Betrayal: Fraud in Science* (Orlando, FL: Harcourt, 2004). For a particularly good case study of accusations of scientific fraud against J.W. Heslop Harrison, one of the protagonists associated with early research on industrial melanism, see Sabbagh, K. *A Rum Affair: A True Story of Botanical Fraud* (New York: Farrar, Straus and Giroux, 1999).

For two examples of how to help students learn and appreciate a host of issues associated with the nature of science using Kettlewell's research on the phenomenon of industrial melanism see: (1) Hagen, J. "H.B.D. Kettlewell and the Peppered Moths," in Hagen, J., Allchin, D., and Singer, F. (eds.), *Doing Biology* (New York: Harper Collins, 1996), pp. 1–10; and (2) Rudge, D.W. "Using the history of research on industrial melanism to help students better appreciate the nature of science," and "The mystery phenomenon: lesson plans," in D. Metz (ed.), *Proceedings of the Seventh International History, Philosophy and Science Teaching Group Meeting* (Winnipeg, Canada), pp. 761–811. Available online at www.ihpst.org.Table 1.The results of Kettlewell's mark-release-recapture experiments. (Developed from figures reported in Kettlewell 1955, 1956).

The Horse Series

Christine Janis

The evolution of the horse (actually, of the genus *Equus*, the term "horse" here also including zebras and asses) has been a used as a classic example of evolution for over 150 years. Apart from human evolution, horse evolution represents the only "classic" example from the mammals, although note that, with the discovery of new fossils in the past couple of decades, the evolution of whales is fast assuming iconic status.

Horses are ungulates (hoofed mammals), belonging to the order Perissodactyla—odd-toed ungulates (i.e., with one toe or three toes), also including tapirs and rhinos. The other living order of ungulates, closely related to the perissodactyls, is the Artiodactyla—even-toed ungulates (i.e., with two or four toes), including pigs, camels, and the "cloven-hooved" mammals like giraffe, deer, antelope, and cattle. (There are also animals called "subungulates," including elephants and hyraxes, but they are not closely related to perissodactyls and artiodactyls.) Today artiodactyls are more numerous (i.e., have many more species) than perissodactyls, and the traditional view is that perissodactyls are some-how "evolutionary failures" that lost out to competition with artiodac-tyls. However, this predominance of artiodactyls is actually rather recent (within the past 5 million years or so of 55 million years of history), and can largely be explained by accidents of geography and climatic change.

Steven J. Gould (1996) discusses the history of ideas in horse evolu-tion, from the original perception of a straight line to the subsequent perception of a "bushy" pattern, but points out that the "iconic" nature of this evolutionary series largely relates to the fact that only a single genus of horse, *Equus*, remains today. Thus, in his view, the story of the success of horse evolution is actually a story of evolutionary *failure* (this

"Evolutionary" view of horses (and their potential predators) through time. From left to right: Mesohippus, being chased by the creodont Hyaenodon (Oligocene); Merychippus, being chased by the bear dog Amphicyon (Miocene); Equus, being chased by the dog Canis (Pleistocene—Recent).

being the "little joke" in the title of Gould's article). I will later contend that this apparent failure has a strong temporal perspective to it: It has been estimated that 99 percent of all species that have ever lived are now extinct, and thus every lineage is ultimately an "evolutionary failure." However, I would agree with Bruce MacFadden (1992) and Stephen Budiansky (1997) that the evolution of the horse is also held as iconic because of the incredible importance of horses to human history, in agriculture, commerce, transport, warfare, and sport, and their countless depictions in art and culture, from cave paintings to My Little Pony. Only in the past half century has the use of horses been primarily for sport and recreation. MacFadden (1992) notes that horses were used in warfare as late as World War II, and I am old enough to remember horses pulling commercial carts in England in the 1950s (and seeing the signs, reminiscent of *Black Beauty*, saying "loosen bearing rein on hill"). Thus the "noble horse" is an appropriate animal to inspire the popular imagination in accounts of evolutionary processes.

Horses are also appropriate for study because of the exceedingly complete and well-preserved nature of their fossil remains. This stands in sharp contrast with the handful of fossils known for at least the early stages of human evolution. The main collection of North American fossil horses is in the Frick Collection in the American Museum of Natural History in New York. Each floor of the Frick building has an area of 5,234 square feet; one entire floor is taken up with fossil horses with an estimated 50,000 specimens (figures from Christopher Norris, AMNH fossil mammal curator). There are also extensive collections in the Smithsonian Institution, Washington, DC, in the Museum of the University of Florida, Gainesville, and in the Museum of the University of Nebraska, Lincoln, as well as numerous other smaller museum collections, and many collections in Europe. Although I am not a primary researcher on horse evolution, unlike colleagues such as Bruce MacFadden and Richard Hulbert, I have visited the majority of these collections, and studied and measured countless fossil horse specimens in the course of my research into ungulate evolution, and consider myself to be very familiar with the actual physical evidence for horse evolution.

The iconic status of horse evolution has attracted much attention from anti-evolutionists. The Web site www.bible.ca calls the horse evolution series "the second Piltdown man of paleontology," and charges the authors of various textbooks that discuss horse evolution with deliberate fraud. Where appropriate in this chapter I shall point out how our actual knowledge of the fossil record counters various claims of anti-evolutionists. My actual examples are limited by space and appropriateness; this does not mean that I am unable to counter other published claims. I also refer the reader to the excellent article by Monroe (1985). My basic tenet here can be expressed by the following quote (attributed to the former U.S. Senator, Daniel Patrick Moynihan): "Everyone is entitled to his own opinion, but not to his own facts."

EARLY IDEAS: STRAIGHT LINE EVOLUTION

The earliest recognized horse fossil is the early Eocene *Hyracotherium* (still the earliest known horse), excavated in England in 1838 and described in 1840 by Sir Richard Owen (superintendent of the Natural History Department of the British Museum, and famous as an antagonist of Darwin's ideas). (Note that *Hyracotherium* is essentially the same animal as the North American early Eocene "dawn horse" "Eohippus" described by Marsh in 1876: Because *Hyracotherium* was the first name applied to this genus, it has taxonomic priority.) A series of European fossil horses, interpreted as representing an evolutionary lineage, was recognized by Owen as early as 1851, and described by Gaudry (1867) as a progression from the middle Eocene *Pachynolophus*, to the late Eocene *Palaeotherium*, to the middle Miocene *Anchitherium*, to the late

Miocene *Hipparion*, and finally to the Pliocene *Equus*. (*Pachynolophus* and *Palaeotherium* are now recognized as belonging to the horse-related family Palaeotheriidae, rather than being true horses.) Note that the reason why this series, and all subsequent fossil horse series, was recognized as an "evolutionary" series (or at least, as a "progressive" one) is as follows: The general mammalian condition, seen in over 50 percent of mammals today, and the general norm for two-thirds of mammalian history (i.e., during the Mesozoic), was to be small, with relatively short limbs and the full (five) complement of fingers and toes, and to have low-crowned cheek teeth. Thus departures from this condition, such as those seen in horses over time, are interpreted as the acquisition of more advanced features ("derived features" is the less value-loaded term now usually applied, as there is nothing inherently maladaptive about being a small, short-legged mammal).

The Issue of Body Size

The issue of the body size of the horse increasing through time is a favorite issue of the anti-evolutionists. They claim that evolutionists follow Cope's Rule, lining up the fossil horses from smallest to largest, then claiming that this represents an evolutionary sequence (see www.bible.ca). However, this is a conflation of pattern with process. Cope's Rule is merely an explanation for the evolutionary pattern seen in many mammals, that they tend to increase in size over time (this is largely true because mammals were initially all fairly small, so increase in size is one obvious evolutionary trend). This "rule" is not a mandate of evolutionary process, that body size *must* increase over time in a linear sequence. The ordering of the fossil horses was originally done based on their stratigraphic sequence, and it was then noted that horses in this ordering proceeded from smaller to larger. Nowadays taxonomists usually ignore stratigraphy, and base their phylogenies on the acquisition of derived features. The sequence of horse phylogeny derived by this method correlates well with the stratigraphic sequence. With the discovery of more fossils we can see that there have been instances of dwarfing in genera such as *Archaeohippus*, *Nannippus*, and *Pseudhipparion*. This is an interesting evolutionary pattern, not a challenge to the veracity of horse evolution.

One anti-evolutionist argument here is that "modern horses come in a variety of sizes," noting that they range in size from the "tiny Falabella to the massive Clydesdale" (www.bible.ca). However, horses as small as Falabellas (Argentinian dwarfed horses) or as large as Clydesdales (British draft horses) are a product of artificial selection, and so are of no relevance to the evolutionary debate. Even if Falabellas are as small as some ancient horses, they still resemble modern horses in every other detail, i.e., they are not "re-creations" of *Hyracaotherium* or *Mesohippus*. But this argument is in general irrelevant as increase in body size is an evolutionary *description*, not a *prescription*.

Additionally, the Web site www.bible.ca accuses textbooks of fraud.

if they fail to tell you that "Moropus" that lived in the Miocene Age, but is not included in the fossil series although it resembles the horse in great deal [*sic*]. If [*sic*] was not included in the horse sequence because it does not serve to the purpose of the evolutionists, since Moropus was two meters high and is larger than both Merychippuston [*sic*] "horses" of the same age and the horses of today.

Quite apart from the irrelevance of size as discussed above, *Moropus* is a chalicothere (family Chalicotheriidae), not a horse. It does belong in the same order (Perissodactyla) as horses, but is no more closely related to a horse than to a rhino. Also, although its head is vaguely "horse-like" in that it has a long face, it doesn't look anything like a horse. In the skull the upper incisors have been lost, and the hooves on the feet have been replaced by claws. Thus there are other reasons, besides that of deliberate fraud, that this animal does not serve the purpose of evolutionists in a discussion of horse evolution.

Thomas Henry Huxley (also known as "Darwin's bulldog") was the next important player in this evolutionary tale. While visiting O.C. Marsh in 1876, a noted fossil collector and professor of paleontology at the Yale Peabody Museum, Huxley saw the extensive collections that Marsh and his crew had made of North American fossil horses. He became convinced that the fossil horse sequence represented in that collection showed that the main lineage of horse evolution was in North America, with the European sequence representing phases of dispersal to the Old World. He presented these ideas, touting the horse as a classic demonstration of evolution, in a famous lecture in New York. Marsh (1876) then published an article noting the gradual evolution of the horse through time, noting progressive trends in the evolution of the teeth (from low-crowned to high-crowned) and the limbs (from three toes to one toe). The figure on page 256 shows the classic picture from around this time by William Diller Matthew (1903), a professor of paleontology at the American Museum of Natural History, showing the change in body size over time as well as the changes in the feet and the limbs. This sequence was considered to represent the evolutionary progression from a small forest-dwelling browsing animal to a larger one adapted to live on the plains, the high-crowned teeth of later forms now allowing for

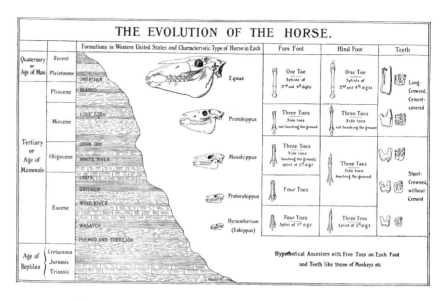

Depiction of horse evolution by Matthew (1902).

the consumption of abrasive grass, and the single-toed, longer legs of later forms allowing for faster running and better escape from predators.

This classic issue of increasing body size through time has been a favorite item for attack by anti-evolutionists. Note also that the number of ribs has never been used as part of this evolutionary sequence, thus claims from anti-evolutionists that rib number is highly variable between fossil horse genera, and thus does not "show progression," has no bearing on the evolutionary argument.

At this time in human history, evolution was interpreted as representing a march of increasing progress, with the notion of "striving" toward some ideal of perfection. This notion of increasing progress, or *ortho-genesis*, dates back to the Greek philosopher Aristotle in the third century b.c.e., with the accompanying notion of the *Scala Naturae*, the organized, hierarchical ladder of life (or the "Great Chain of Being"). These ideas were certainly prevalent among nineteenth-century European evolutionists, such as Geoffroy Saint-Hilaire and Jean-Baptiste Lamarck, and championed in the late nineteenth century by the American paleontologist Henry Fairfield Osborn, and so it is not surprising that the fossil record of horses was interpreted to represent some straight line of progress. Unfortunately, this simplistic portrayal of horse evolution remains in many textbooks and museums today, even though the changes in evolutionary thought and the discovery of more fossils greatly changed the way horse evolution was perceived in the twentieth century.

LATER IDEAS: THE NOTION OF "BUSHINESS"

As time went on, evolutionists started to see the process of evolution as being more the result of random processes rather than directional. The phylogeny ("family tree") produced by Matthew in 1926 now showed a complex branching pattern, with the co-occurrence of many types of horses during the Miocene and Pliocene. This complex branching pattern was further promoted by Matthew's student, Ruben Stirton, and George Gaylord Simpson (chairman of the Department of Paleontology at the American Museum of Natural History, and one of the most prominent evolutionists of the twentieth century) synthesized these ideas and his own research on fossil horses in his 1951 book on horse evolution, with the resulting often-reproduced figure shown on page 258.

The issue of co-occurring horse genera seems to be of considerable concern to anti-evolutionists. Nevins (1974) seems to think that the co-occurrence of "the three-toed horse *Neohipparion*" with "the one-toed horse *Pliohippus*" somehow refutes the claims of the evolutionists. However, the current view of evolution is of a branching pattern (i.e., "cladogenesis"), not of the overall transformation of one genus into another (i.e., "anagenesis"). (Notwithstanding the fact that *Neohipparion* and *Pliohippus* belong to different tribes, and so are not in any type of ancestor/descendent relationship.)

Gould (1996) makes the extensive point that this now bushy, branching phylogeny negates the old notion of directed evolution. Anti-evolutionists have noted that this branching tree pattern does not provide definitive evidence *against* directed evolution, but merely fails to refute it. However, as Gould notes, any of the terminal branches of the horse tree can be traced back to *Hyracotherium*: This branching tree shows that *Equus* was not the orthogenetic goal of the evolutionary processes that produced the equid radiation. He also notes that, had the diminutive, three-toed *Nannippus* survived to the present rather than *Equus*, we might tell a very different picture of horse evolution today.

THE CURRENT PICTURE

The work of Bruce MacFadden provides a comprehensive view of our current knowledge and interpretation of horse evolution. Page 259 shows a current phylogeny of North American horses, which basically follows MacFadden. This figure shows the horse sequence as a strict cladistic branching pattern, with no genus being directly ancestral to any other. In practice, we know that this is not strictly true, and that various species of any one genus may be more or less derived: For example, *Parahippus*

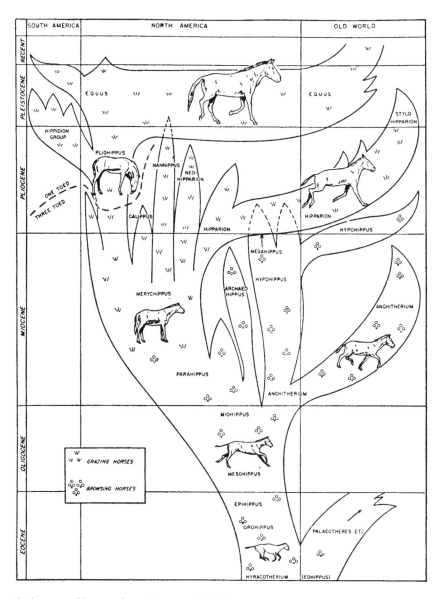

Phylogeny of horses from Simpson (1951).

leonensis is regarded as the species closest to *Merychippus*. However, this mode of representation is the standard one these days, and serves here as an albeit simplistic interpretation.

The updates from the phylogeny of Simpson 1951 are as follows: (1) redating of the Eocene/Oligocene boundary (now older) and the Miocene/Pliocene boundary (now younger); (2) the recognition of various new genera—*Kalobatippus* (previously included with *Anchitherium*), *Desmatippus* (previously included with *Parahippus*), *Pseudhipparion*

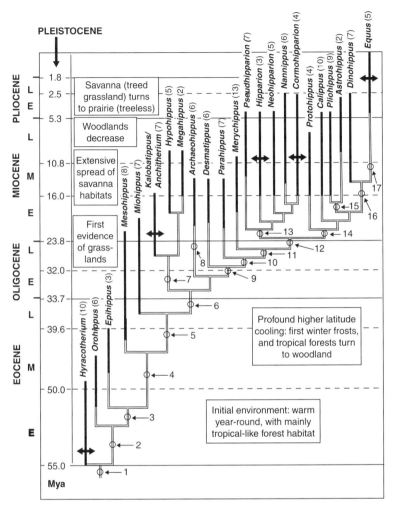

Current phylogeny of North American horse genera, also showing major environmental changes in North America during this time period. Solid lines show the temporal duration of the genus; open lines show the phylogenetic (evolutionary) links between genera. Numbers in parentheses after each genus indicate the number of valid described (North America) species. Double arrows indicate periods of dispersal to the Old World. The numbers (arrows pointing to circles, indicating the point on the phylogeny in question) refer to the various anatomical characters, listed below. Mya = millions of years ago (adapted from MacFadden, 1992, 1998).

(previously included with *Nannippus*), *Cormohipparion*, *Protohippus* (both previously included with *Merychippus*), *Astrohippus* and *Dinohippus* (both previously included with *Pliohippus*) (other more recent additions and changes are noted in the text); (3) a change in the perceived patterns of dispersal to the Old World—two lineages of hipparionines

(*Hipparion* and *Cormohipparion*) are now considered to have dispersed, and the supposed Old World *Hypohippus* (*Sinohippus*) is now considered to have evolved *in situ* from the original dispersal of *Anchitherium*, so there was only a single, early Miocene dispersal of anchitheriine horses; (4) the recognition that *Pliohippus* itself is too specialized to have given rise to *Equus* (see later discussion), and that the *Dinohippus* is the more likely ancestor.

Note that, despite various changes and additions, the basic overall shape and ordering of the horse phylogeny has remained constant, contra the assertion "more than 20 horse bushes have been invented—this proves that the bushes are guesswork and speculation" (Bowden, 1982). The following claims are also without foundation. (1) "A complete series of fossils in the correct evolutionary order does not exist anywhere in the world" (Kofahl, 1997). On the contrary, all of the genera depicted on page 259 can be found within a single area, the Great Plains region of North America (essentially the states of Nebraska and Wyoming), even if some genera are also known from elsewhere in North America or from the Old World. (2) "The 'evidence' has been garnered spanning geologically isolated regions. The series starts in North America, jumps suddenly to Europe, and then back to America" (Wysong, 1981). As previously discussed, we can now document that the main evolution of the horse took place in North America, with several successive dispersals to the Old World (discussed in more detail below). Wysong also makes the bizarre, and totally unsubstantiated, claim that *Hyracotherium* and *Equus* co-occur in the same rock formation in Nebraska. (3) "The fossils of these horses are found widely scattered in Europe and North America. There is no place where they occur in rock layers, one above the other" (Moore and Slusher, 1974). There are many places in North America where horses can be found superimposed within a rock sequence, for example the late Miocene to Pliocene Ash Hollow formation (spanning around 7 million years) in Nebraska. MacFadden provides a comprehensive discussion of issues relating to fossil occurrences and biostratigraphy.

HOW DO WE KNOW WHAT WE KNOW ABOUT HORSE EVOLUTION?

As previously discussed, the transformation of anatomical characters in the horse has been linked to adaptive changes: high-crowned cheek teeth for eating grass, longer legs with reduction of side toes for faster running. How do evolutionists know what these anatomies mean in functional terms? First, there is the issue of biomechanics. More highly-crowned teeth will take longer to wear down (because there is more crown to wear away), and thus are better-adapted to abrasive food, such as silica-containing grasses. Longer legs allow for longer strides, and the reduction of side toes makes the lower limb lighter for more efficient limb movement. However, we can determine the correlation in a more quantitative fashion between mammalian anatomy and function because, even though only one type of horse is around today, there is a great variety of artiodactyls (especially the ruminants: giraffe, deer, antelope, and cattle), with a wide range of morphologies, that have evolved in a convergent fashion with horses. (Note that derived features such as adaptations for grazing and cursorial locomotion have evolved convergently several times among ruminants, enabling a better determination of the correlation between form and function.)

An additional source of information is something called "dental microwear" (Solounias and Semprebon, 2002). Studies on living mammals show that different types of food produce different characteristic types of microscopic scratches (or pits) on the teeth. Such microwear features can be an additional source of evidence of the diets of fossil mammals.

THE NATURE OF THE EARLIEST HORSE

Hyracotherium, as is well known, was a smallish animal, with various species weighing between 2 kg (cat-sized) and 10 kg (corgi-sized). Note that there has been some debate as to whether *Hyracotherium* is actually a true horse, or merely a primitive perissodactyl that could also be ancestral to rhinos and tapirs. MacFadden summarizes this debate, and notes that *Hyracotherium* in fact possesses certain derived features seen only in equids.

Although *Hyracotherium* is commonly described as "unspecialized" for its time, and in comparison with the contemporaneous or slightly older archaic ungulates ("condylarths") it was highly specialized. *Hyracotherium* was one of the earliest mammals to show modifications of the limb characteristic of mammals that today we term "cursorial" (i.e., running-adapted). Debate exists, however, as to whether "cursorial" limb modifications were initially evolved specifically for high-speed locomotion, rather than for efficiency at slower gaits. These include the lengthening of the forearm (ulna and radius bones) and the hand bones (metacarpals), and the lengthening of the shank (tibia and fibula bones) and the foot bones (metatarsals). The foot posture was also of the derived subunguligrade form: Rather than standing with the foot flat on the ground (like us), or even on tiptoes (like dogs and cats), early horses bore their weight primarily on the tips of the phalanges (fingers or toes), but with the retention of a rather dog-like footpad, and they had only small hooves. Modern tapirs have a similar type of foot. More highly cursorial mammals, such as later horses, lost the footpad and acquired a fully unguligrade foot posture, complete with larger hooves. Highly cursorial mammals also have relatively longer limbs (especially the metapodials, the bones of the hands and feet), and show modifications of the limbs that restrict the motion to the fore and aft plane, including the reduction of the ulna and fibula, more hinge-like joints between the major limb bones, and the alignment of the bones of the wrist and the ankle into a more "blocky" form. The beginnings of some of these anatomical changes are indeed seen in *Hyracotherium*.

In terms of the skull and teeth, the skull of *Hyracotherium* was longer than that of earlier ungulates, with a distinct gap (diastema) between the cheek teeth (i.e., molars plus premolars) and the incisors (this is the place where the bit goes in modern horses). The brain was also relatively larger than contemporaneous mammals, although not as large as that of a modern horse. The molars, although still low crowned (brachydont) and fairly simple in form (bunodont, with low, bumpy cusps, like those of most condylarths, and also seen in us humans), nevertheless showed signs of forming high ridges between the cusps. Bunodont teeth mainly pulp and crush, while ridges between the cusps are used to shear flat, fibrous vegetation such as leaves. Most hoofed mammals today (except the omnivorous pigs) have teeth with these types of ridges, or "lophed" teeth (termed lophodont in perissodactyls and selenodont in artiodactyls, the difference in detailed form reflecting independent instances of evolution of lophed teeth from bunodont forms in the two groups). The teeth of *Hyracotherium* should properly be termed "bunolophodont," and the tooth wear is indicative of a "folivorous-frugivorous" diet of buds, berries, and young leaves (similar to that of living mouse deer).

The reason for this rather detailed description is as follows. Most texts on horse evolution stress the primitive nature of *Hyracotherium*, but this is "looking down" on this animal from the perspective of modern horses. "Looking up" at *Hyracotherium* from the perspective of other early Eocene ungulates provides a different picture: In comparison with the more archaic ungulates *Hyracotherium* had long, slender legs modified for more cursorial and strictly terrestrial locomotion, and had a diet with a much greater percentage of leaves. For its time, *Hyracotherium* was a highly derived, specialized animal in its tropical forest habitat, and although it was not especially large for an ungulate (hoofed mammal), it was large in comparison to contemporaneous non-ungulate mammals.

Anti-evolutionists often point out that *Hyracotherium* was named for the modern hyraxes, with the implication that the existence of hyraxes today somehow refutes horse evolution. In fact, although the cheek teeth are somewhat similar, hyraxes are quite different from early horses: They have a shorter skull (although note that on the Web site www.bible.ca

the skull of the hyrax has been mysteriously elongated so that it now appears to be as long-snouted as *Hyracotherium*); their front teeth are completely different (with the incisors transformed into canine-like teeth); and despite the fact that they have the same number of digits (four in front, three behind) as *Hyracotherium*, so do many other mammals (e.g., tapirs), and their legs are much shorter. Molecular studies confirm that hyraxes are not at all closely related to horses, nor even to other ungulates. However, if the hyrax actually was proved to be a surviving hyracothere, would this be a problem for evolutionists? Absolutely not! Evolutionists would be ecstatic to discover such an evolutionary relict. Relict forms (see later discussion about relict artiodactyls) have absolutely no bearing on whether or not evolution actually occurred.

THE RELEVANCE OF THE FEATURES CHARACTERIZING THE HORSE LINEAGE

MacFadden notes detailed anatomical changes (derived features) that happened at every step of horse evolution. What do these tell us about evolutionary changes, specifically about how horses might have been changing their way of life over time? As shown on page 259, horses can be divided into three subfamilies: the Eocene Hyracotheriinae, the Eocene to Miocene Anchitheriinae, and the Miocene to Recent Equinae. Anti-evolutionists have made the claim that these three subfamilies represent three different types of animals, and that the different genera within each subfamily are basically all the same genus (see www.bible .ca). As will be noted below, it is certainly possible to trace intermediate steps between these different subfamilies, and to see distinct differences between different genera within each subfamily. I will also provide a refutation of a specific claim, that *Mesohippus* and *Miohippus* are the same animal.

THE HYRACOTHERIINAE

The traditional North American hyracotheriine genera *Hyracotherium, Orohippus*, and *Epihippus* are shown in the figure on page 259. Earlier reports of Paleocene specimens of *Hyracotherium* have now been shown to be incorrect. Excluded from this figure are two other genera known only from fragmentary material, *Xenicohippus* (placed between *Hyracotherium* and *Orohippus*) and *Haplohippus* (placed between *Orohippus* and *Epihippus*). Recently Froehlich has suggested that the genus *Hyracotherium* can be subdivided into a number of separate genera. He reserves the name *Hyracotherium* for Eurasian Eocene horses, and divides the species previously ascribed to *Hyracotherium* among the genera *Sifrihippus*, *Minippus*, *Arenahippus*, *Eohippus*, and *Protorohippus*. Whether or not this proliferation of genera becomes adopted by the majority of paleontologists remains to be seen.

Although hyracotheres in general were small (for an ungulate) animals, with a general appearance suggestive of a forest-dwelling form (arched back, front legs shorter than hind legs), a progression of characters indicating the adoption of a more fibrous diet can be seen through time. Temperatures started to decline during the middle Eocene, when *Orohippus* and *Epihippus* were around, which would have decreased the availability of non-fibrous food like buds and berries, at least seasonally. Dental changes in these genera, with the progressive molarization of the premolars, reflect this change in vegetation, indicating a shift to processing more leaves. Unmolarized premolars are simple, triangular teeth, used mainly for crushing. Molarized premolars now have the high ridges of the molars, enabling them to be used for shearing leaves. Thus molarizing the premolars results in a greater tooth area for leaf processing. The dental wear on the teeth also shows an increase in shearing wear, and a decrease in crushing wear, from *Hyracotherium* to *Epihippus*, reflecting a shift to a diet containing more leaves and fewer buds and berries. The possession of relatively larger third toes in *Epihippus* likely reflects a shift to ranging more widely in search of food each day, consistent with the environmental changes.

Hyracotherium was also known from Europe (and possibly also from Asia) during the early Eocene. In Europe the equid radiation also gave rise to a separate family, the Palaeotheriidae, which was known from both Europe and Asia. Some palaeotheres were much like the early horses, while others were larger (some as large as modern horses) and had teeth suggestive of more specialized browsing. However neither the palaeotheres nor the European equids survived past the early Oligocene. Note that for most of the Cenozoic there was no connection between North and South America. In the Old World, although Asia retained a tropical zone, there was no connection to Africa until the early Miocene. Thus there was no easily accessible tropical refuge for horses in the changing higher-latitude environments of the early Cenozoic.

An interesting aspect of horse evolution, and one that is rarely noted, is the decline in numbers of hyracothere specimens during the middle Eocene in North America. *Hyracotherium* fossils are among the most common in early Eocene fossil localities, while *Orohippus* is rare, and *Epihippus* known only from a handful of specimens. As fossils of other ungulates of this body size (e.g., tapiroids and early artiodactyls) remain abundant during this time period, it is likely that this change in number of specimens reflects a real decline in actual numbers of horses, as their original preferred habitat became more restricted. North America was an island continent at this point in time, and the horses could not follow the tropical forests back toward the equator. Thus the North American horses almost succumbed to the fate of the Eurasian ones. However, in the colder, more seasonal climate of the late Eocene, a new type of horse emerged to again become common in North America; this was *Mesohippus*, the first of the subfamily Ancitheriinae.

THE ANCHITHERIINAE (EARLY MEMBERS)

Mesohippus was not much bigger than the hyracotheres (about the size of a goat), but its feet and teeth were somewhat different. Specifically, the teeth were now more fully lophed, both this morphology and the dental wear indicative of a diet more strictly consisting of leaves, with minimal inclusion of buds and berries (this proposed dietary shift is also borne out by studies of dental microwear). Going along with this, the premolars were fully molarized (apart from the first one, that was eventually lost, although a minute one can persist even in modern horses as a "wolf tooth"); the face was somewhat longer (as seen today in browsers compared with more omnivorous forms); and the posterior part of the lower jaw was enlarged, suggesting larger chewing muscles (masseter) that would befit a more fibrous diet. The limbs were further elongated (although less so in *Mesohippus* than in other anchitheres), and the forefeet were now functionally tridactyl (i.e., three-toed, rather than four-toed as in the anchitheres, although remnants of the fifth digit has been found in an individual specimen of *Mesohippus*). The limbs were more derived than the hyracotheres in other ways: The third metapodial was enlarged and elongated, with a larger hoof than those seen on the side toes; the ulna and fibula were reduced in size; and the footpad was apparently reduced, although still present.

Despite these differences, there is no doubt that *Mesohippus* was descended from the hyracotheres. It possessed all the derived features of the skull and teeth that characterize *Hyracotherium*, and the overall form of the limbs (e.g., details of the articulations of the wrist and ankle bones) was similar. Although the teeth were now more highly lophed, they still had the same detailed arrangement of the cusps and other features as in the hyracotheres. In comparison to hyracotheres, anchitheres in general were more committed leaf eaters, with limbs somewhat more adapted for efficient long-distance locomotion. This goes along with the dominant late Eocene to Oligocene habitat of North America being subtropical to temperate woodland, rather than tropical forest.

Mesohippus, like all later horses except for *Equus*, now possessed a facial fossa, a depressed area in front of the orbit (the dorsal pre-orbital fossa). The function of this fossa is not known, but Gregory (1920) proposed that it housed an extension of the nasal diverticulum (a chamber extending up from the nose), similar to the condition seen in tapirs, who possess a similar facial depression. The form of this fossa (and the presence of an additional, lower fossa, the malar fossa, in some later horses) differs between closely related genera, and researchers have speculated that it was somehow related to taxon recognition, much as the way that different genera of deer and antelope have characteristically different horns or antlers. In modern horses, the diverticulum "flutes" when the horse snorts, possibly acting as a sound resonator. In a "backyard experiment" with my own horse, I determined that if this diverticulum is packed with cotton balls the horse is unable to snort (or, rather, it tries to snort in vain, and then finally expels the cotton balls with a loud trumpet). My preferred, admittedly speculative, explanation is that the fossa is indeed related to an expansion of the nasal diverticulum, and that different fossa morphologies are indicative of different types of snorting. Modern horses snort as a greeting, and thus these differences may relate to different sound production associated with species recognition. Gregory proposed that the malar fossa of pliohippines housed the insertion of a muscle that elevates the upper lip. This muscle is large in modern horses, and the lips are used extensively in food prehension. Perhaps an enlarged malar fossa was related to different ways of obtaining food, or to raising the lip in the "flehmen" type of display seen when stallions sniff mares' urine, or when domestic horses are presented with some unusual-smelling food, such as a peppermint. It is interesting that the absence of facial fossae in *Equus* is associated with the extinction of all of the other genera of horses. In an environment with only one or two species at any one site, perhaps species recognition was not so crucial.

Miohippus, the descendent of *Mesohippus*, was primarily an Oligocene horse, only a few specimens being known from the latest Eocene. *Mesohippus* and *Miohippus* are found together in many fossil faunas, and as they are rather difficult to tell apart from their teeth alone they have often been confused. However, *Miohippus* does tend to be a little larger (as indicated by bigger teeth), and had an occlusal (tooth surface) morphology that was more complicated, indicative of a more fibrous diet. However, if the entire skeleton is available for study, the differences become more apparent. *Miohippus* had a longer face than *Mesohippus*, and the ankle joint showed an enlargement of the bone that articulates with the middle toe. Thus further progression toward a strictly leaf-eating diet, and more efficient locomotion, is seen in this transition.

The Web site www.bible.ca has the following contention:

> Miohippus and Mesohippus (and Parahippus) are really the same animal. They are classified differently because different skeletons of the same animal can exhibit minute differences. They may be two closely related cross-fertile species like Mule deer and White Tail deer, for example. The

differences are also easily accounted for by if adult and juvenile specimens had been found. Teeth and leg differences between these three animals are no different that [*sic*] we find within men or dogs given "race or age" differences.

First, adults and juvenile mammals are easy to distinguish: Juveniles have skulls where the teeth have not completely erupted, and limbs where the ends of the bones (epiphyses) are not fully fused to the shafts. No competent scientist would make the mistake of describing a juvenile as a different species to the adult.

Second, there really are distinct differences between these horse genera, as noted above. These differences are much greater than those seen between mule deer and white-tailed deer (both species of the genus *Odocoileus*), which are largely distinguished by coat color and antler form, and are very hard to tell apart from the bones alone. The difference between *Mesohippus* and *Miohippus* is more like the degree of anatomical difference between a white-tailed deer and a wapiti (*Cervus canadensis*), different genera that most certainly cannot interbreed. *Parahippus* is even more distinct, with its somewhat hypsodont cheek teeth, complete postorbital bar, and more reduced side toes.

Domestic dogs have been bred by artificial selection to have longer or shorter faces, or longer or shorter legs. However, anatomical features as profound as the presence or absence of a postorbital bar in the skull, or changes in the ways the ankle bones articulate, show absolutely no variation between different dog breeds, and certainly no variation between different races of humans.

THE ANCHITHERIINAE (LATER MEMBERS)

Mesohippus and *Miohippus* were primarily late Eocene and Oligocene horses (although a few species persisted into the earliest Miocene). They showed a continuation of the earlier Eocene trend of only one, or at most two, horse genera being around at any one time, with no more than five sympatric species. During this time period the numbers of origins and extinctions among horses was fairly equally balanced. However, in the early Miocene, a change was apparent in the pattern of horse diversification, with many more originations than extinctions. This is the time when the horse lineage can first be described as "bushy." By the late early Miocene, around 20 Mya (millions of years ago), three distinctly different lineages of horses (all anchitheriines at this point) were present.

The first was the specialized large browsers of the tribe Anchitheriini. This group of horses was largely around in the mid-Miocene, and represent a specialized radiation of larger browsing forms that flourished in the woodland environments associated with the Miocene savannas. They were extinct by the latest Miocene, along with a number of other browsing-adapted ungulates (e.g., the deer-related dromomerycids) and predators that seem to have been woodland adapted (e.g., the bear dogs). The first member of this tribe, *Kalobatippus*, was apparent in the late Oligocene, and was not greatly dissimilar from *Miohippus*. It differed in a few dental details, and in having long slender legs, a feature not apparent in the other anchitherines. Miocene species ascribed to *Kalobatippus* have now been assigned to the genus *Anchitherium*, a generally more robustly built, shorter-legged genus previously only described from Eurasia. The *Anchitherium* species in Eurasia represent a migration from North America in the early Miocene, via Beringia (which was the probable route for all horse migrations subsequent to the early Eocene). The anchitherines had a moderately successful Miocene radiation in the Old World, but did not reach Africa.

Later North American anchitheres *Hypohippus* and *Megahippus*, known from the late early Miocene to the early late Miocene, were large-sized, highly specialized browsers. Most species were pony-sized, but some reached the size of a large modern horse. They had heavily built limbs, large forward-projecting incisors, and relatively short faces, as in earlier horses, with the orbit still in the primitive position over the posterior molars. Their cheek teeth were relatively large for their body size, with high-relief ridges that resembled the teeth of present-day specialized browsers (e.g., browsing rhinos) rather than those of more mixed-diet feeders that have ridges of lower relief (e.g., most deer).

The second Miocene anchithere radiation is represented by the genus *Archaeohippus*, known from the early to middle Miocene, primarily (but not exclusively) from the more subtropical environments of Florida. *Archaeohippus* was distinguished by its small size; other contemporaneous anchitheres, including later species of *Miohippus*, were around the size of a white-tailed deer, while *Archaeohippus* was around the size of a roe deer or a goat. MacFadden studied this genus as an example of evolutionary dwarfing.

The third Miocene anchithere radiation was that of the parahippines, *Desmatippus* and *Parahippus*. The taxon *Desmatippus* was coined for the more primitive, less hypsodont parahippines, although the systematics of this group has yet to be fully resolved. Most recently, the species *Parahippus coloradensis* has been considered as belonging to *Desmatippus*, and the species *Desmatippus texanus* as belonging to a separate genus, *Anchippus*. *Parahippus* is a more highly derived form, and the later species in particular (e.g., *P. leonensis*) provide a gradation of features intermediate between anchitheriines and equines. *Parahippus* had a complete postorbital bar, a feature of the skull seen in most living ungulates and thought to be related to reducing the forces on the skull during chewing. The more hypsodont (high-crowned) cheek teeth indicated a more abrasive diet than that of earlier horses, probably including some grass (see discussion below). The greater reduction of the side toes also foreshadowed the equine condition to a more derived, more cursorial type of foot.

THE EQUINAE (EARLY MEMBERS)

By the late early Miocene the first members of the subfamily Equinae appeared, with the first occurrence of the genus *Merychippus*. Equines can be distinguished by their moderately hypsodont cheek teeth (more hypsodont than in *Parahippus*), with the addition of cement (a supportive bony infilling). Going along with the hypsodont cheek teeth were several modifications of the skull related to the adoption of a more fibrous diet, such as the greater elongation of the face, the shifting of the orbits posteriorly to behind the tooth row (thus accommodating the unerupted high crowns of the upper cheek teeth), and the deepening of the lower jaw (to accommodate a larger masseter muscle). The limbs showed further cursorial modifications, including the loss of the footpad (with the foot posture now fully unguligrade, like a ballerina en pointe), and the development of a system of foot suspensory ligaments, as indicated by scars left from muscles and ligaments on the bones.

Merychippus is shown as a single lineage on page 259. However, this does not represent the true diversity at the time. Some species of *Merychippus* were generally primitive, but others were more closely related to members of the tribe Hipparionini or to the tribe Equini. Previously, the species of *Protohippus* and *Cormohipparion* were considered to belong to *Merychippus*. More recently several more of the former species of *Merychippus* have now been given separate generic status (the new hipparionine genus *Acritohippus*, and the new equine genera *Scaphohippus* and *Parapliohippus*). As noted by Stirton, "The genus *Merychippus* is extremely difficult to diagnose, since its species intergrade almost imperceptibly with those of *Parahippus* on the one hand and those of *Pliohippus*, *Calippus*, *Nannippus*, *Hipparion*, and *Neohipparion* on the other."

Merychippus is usually recognized as the "first grazing horse" due to its hypsodont cheek teeth, and the appearance of *Merychippus* was thought to herald the appearance of the North American savanna habitat. Recent research has modified this view considerably. First, hypsodonty is no longer unequivocally associated with grazing; while grass may be the prime abrasive element, other factors (such as grit on the food) can also abrade the teeth. Most living mixed-feeding ungulates (eating both browse and grass, often switching seasonally) have hypsodont teeth, and although there is a correlation between the degree of hypsodonty and the amount of grass in the diet, it is a weak one. Other features of the skull (e.g., a broad muzzle) can also be used to characterize grazers, and patterns of dental wear can also distinguish diet. Studies of the microwear of *Merychippus* teeth show them to have been mixed feeders, which goes along well with their relatively moderate degree of hypsodonty. These features refute the claims of anti-evolutionists that there are only browsing and grazing types of horses known, and that no intermediate types of teeth have been found.

Second, paleobotanical evidence has now shown that grasslands were present in North America prior to the appearance of *Merychippus,* by the earliest Miocene at least. There has been some speculation as to why horses did not seem to "respond" to the environmental change until several million years later. The notion that horses such as *Parahippus* were grazing to a large extent without having highly hypsodont teeth seems unlikely on biological grounds: Grazers have rapid rates of tooth wear, and a low-crowned animal subsisting on a grass diet would wear its teeth down before reproductive maturity. Whatever the true story behind the evolution of grasslands and grazing horses, the late early Miocene saw the great proliferation of many species of *Merychippus,* and this was coincident with the diversification of other groups of ungulates, suggesting the appearance of a savanna-adapted large mammal fauna.

THE EQUINAE (LATER MEMBERS)

By the start of the middle Miocene, 16 Mya, genera traditionally ascribed to equines more derived than *Merychippus* made their first appearance. These horses were now highly hypsodont and were the size of a largish pony (*Merychippus* species were generally the size of a small pony). There is a distinctive split in the subfamily at this point in time, into the tribes Hipparionini and Equini, which are distinguished on the basis of detailed dental features. Later members of these tribes are also distinguished by foot morphology: Hipparionines remained persistently tridactyl whereas later equines became monodactyl.

The Case of Tridactyly to Monodactyly

Anti-evolutionist Dwayne Gish argues that the claim that horses evolved from three toes to one toe is confounded by the "rather astounding and revealing fact" that South American ungulates apparently evolved from a single hoof to three toes at the same time. He says, "I don't know any evolutionist who suggests such an evolutionary sequence of events, but why not? Perhaps it is because the three-toed to one-toed sequence for North American horses became so popularized in evolutionary circles that no one dare suggest the reverse condition."

The detailed evidence here involves lining up three members of the extinct South American order Litopterna: *Thoatherium* (one toe; Miocene), *Diadiaphorus* (one main toe and two small side toes; Miocene), and *Macrauchenia* (three large toes; Pleistocene). This "sequence" is then compared to the sequence of "Eohippus," *Merychippus*, and *Equus*. Here is the truth of the matter (leaving aside the unlikely possibility that an animal with the derived condition of a single toe could easily regain the missing digits). *Diadiaphorus* and *Thoatherium* belong to one litoptern family (Prototheriidae), while *Macrauchenia* is in a completely different family (Macraucheniidae) that did not undergo toe reduction. I could present a similar "astounding" sequence by lining up members of the order Perissodactyla as follows: *Pliohippus* (one toe, Miocene), *Cormohipparion* (one main toe and two small side toes, Miocene), and *Rhinoceros* (three large toes, Recent). And, apart from anything else, there would be no problem for evolutionists if a completely different group of ungulates had been shown to follow a different evolutionary trajectory to the horses. There is no "rule" that evolution has to proceed in an identical fashion in different lineages.

Despite the detailed differences between the middle- to late-Miocene horse genera that allow us to place them in different tribes, they were basically all rather similar animals, with dental wear indicative of a mixed-feeding diet rather than an exclusively grazing one. A diversity of evolutionary trends were apparent: Dwarfing within lineages was seen in both *Pseudhipparion* and *Nannippus*, and *Calippus* possessed an exaggeratedly broad muzzle. Ten sympatric genera, and dozens of sympatric species, found during this time (not including the continuing presence of the large anchitheriines). (Note that the figure on page 259 does not include the South American genera, *Hippidion* and *Onohippidium*, that made a brief appearance in the Pliocene and latest Miocene of North America, respectively, before their migration across the isthmus of Panama, nor a new, recently described pliohippine, *Heteropliohippus*, from the early late Miocene.)

This later Miocene equid diversity certainly rivaled the diversity of present-day African antelope. Individual fossil localities frequently contained up to a dozen different types of horses. Basically, the middle to late Miocene of North America was "horse heyday," with no indication that only one lineage would survive from this radiation. Between 12 and 10 Mya there were 12–15 sympatric species of horses at any one fossil site. However, by the start of the Pliocene general cooling and drying in the higher latitudes had turned the more productive savanna grassland habitat into the less productive prairie. Only four genera of horses survived past the earliest Pliocene (many other mammals also became extinct at this time), and only *Equus*, first appearing in the early Pliocene, survived into the Pleistocene.

In the late Miocene (by 10 Mya) *Hipparion* and *Cormohipparion* migrated over to Eurasia with a resulting diversification of Old World hipparionines (now including Africa) that blossomed during the late Miocene and lingered on until the early Pleistocene. However, hipparonines were never as diverse or as individually numerous in the Old World as they were in North America, perhaps because they there faced the indigenous radiation of the bovids (antelope, cattle, etc.). Note that bovids did not migrate in the opposite direction, from the Old World to North America, until the Pleistocene. Old World fossil localities rarely contained more than one or two different horse species, similar to the condition with the different species of *Equus* today.

The details of the precise relationships among the Equini are not well resolved, but it is clear that the more derived forms can be united into two groupings: *Pliohippus* and *Astrohippus* (the pliohippines) on the one hand, and *Dinohippus* and *Equus* on the other. These horses are usually described as monodactyl, and *Pliohippus* was traditionally considered the ancestor of *Equus*. However, more recent discoveries have shown that early *Pliohippus* species were actually tridactyl. Additionally, in the Ashfall fossil beds from the late Miocene of Nebraska, where volcanic ash has resulted in the fine preservation of many completely articulated skeletons, Voorhies notes that the development of the toes is variable among different individuals of *Dinohippus*, with some individuals definitively tridactyl, and others with more reduced side toes. *Pliohippus* (and the related *Astrohippus*) were also rather different from *Equus* in a number of ways. In addition to the usual facial fossa they had a second, lower depression in front of the eye (the malar fossa), their cheek teeth were more curved than in *Equus*, and they were rather slenderly built animals, more like a modern Arabian horse than like the stocky form seen in most wild species of equids today.

In contrast, the more heavily built *Dinohippus* (once confused with *Pliohippus* when known only from dental remains) resembled *Equus* in the reduction of the facial fossa (lost completely in *Equus*), and also had details of the tooth occlusal morphology that ally it with *Equus*. *Equus* itself was larger than earlier horses (most species about the size of a modern Przewalksi's horse, around fourteen to fifteen hands), with teeth that were more highly hypsodont and with a greater complication of the enamel on the occlusal surface, befitting of an animal that by this time was probably a specialized grazer, as are all the modern species of *Equus*.

THE EVOLUTION OF MONODACTYLY

The fossil record thus shows us that monodactyly evolved twice in horses, once in in the *Pliohippus/Astrohippus* lineage and once in the *Dinohippus/Equus* lineage, probably appearing (as best as can be determined, because fossilized limbs are rare) in the late Miocene. In traditional views of evolution, monodactyly is usually considered to be superior to tridactyly, largely because it is a feature of modern horses. Both limb anatomy and footprint evidence shows that hipparionids were functionally monodactyl: That is, the side toes did not touch the ground during normal locomotion, but might act to protect the fetlock joint during occasional hyperextension. Thus what might be the reason for the loss of the side toes?

Note that the top portions, and part of the shafts, of the side toes are retained in modern horses: This is likely because the side toes were originally involved in the articulation of the feet with the wrist or ankle, and so had a function independent of the lower portions. (In a similar fashion, the ulna bone may be reduced in cursorial ungulates, but it is never completely lost because the top portion, the olecranon process, is used for the insertion of the triceps muscle.) Anti-evolutionists claim that these side toes can not be "vestigial" if they can be seen to have some existing function today, but this claim has no relevance to the issue of whether the "splint bones" of modern horses were originally ever fully developed toes. Note that these "side toes" occasionally appear as more fully developed toes in modern horses.

Monodactyly is not necessarily superior to tridactyly overall, but there must be some functional reason for it, or else it certainly would not have evolved *twice* (a single instance could represent some sort of developmental accident that had no negative consequences). Losing the side toes could reduce the mass at the end of the limb, possibly bestowing greater locomotor efficiency. However, the single toe of monodactyl equines is bigger and more robust than the middle toe of hipparionines, and it is not clear that much mass is reduced. Was there some difference in hipparionine and equinine paleobiology? Hipparionines remained as mixed feeders, while derived equinines showed a trend towards more specialized grazing, with more highly hypsodont teeth. A specialized grazer would need to cover more ground per day to search for food, perhaps migrating with the rains, as do present-day zebra. Shotwell (1961) also thought that tridactyl and monodactyl horses favored different habitats, with monodactyl horses being less woodland and more open grassland in their habitat preference (as would befit the proposed differences in diet).

It is also likely that the locomotory gaits of the two types of equines were different: All extant equids use the trot as an efficient gait for distance locomotion, while Renders (1984) showed that the footprints of

Hipparion from Laetoli (a Pliocene site in Tanzania that also preserves the hominid footprints) indicate a "running walk," a type of gait seen today only in some breeds of horses, such as Paso Finos and Icelandic horses. Finally, study of muscle scars on the limbs shows that the full suspensory ligament system of the lower leg, which acts to conserve elastic energy in a spring-loaded foot, is only seen in monodactyl horses. One possible explanation for monodactyly, then, is that as derived equinines became more grazing specialists, thus needing to cover more ground per day, they adopted the trot gait as their main mode of locomotion. The trot is a more stiff-legged gait than the running walk or the gallop, and perhaps the suspensory ligament system worked better with this gait to conserve energy, while the side toes did not touch the ground at any point during this gait. Thus the side toes might have been lost in relation to the evolution of a more extensive system of foot suspensory ligaments. (More research would be needed to confirm this evolutionary scenario.)

A final specialization first seen in rudimentary form in *Dinohippus*, and elaborated in *Equus*, is bony evidence for the "stay apparatus" of ligaments that allow modern horses to lock their legs and doze standing upright. This system also reduces the amount of energy that is required to maintain a standing posture.

THE EVOLUTION OF THE GENUS *EQUUS*

Equus first appeared in the Pliocene, and survived through the Pleistocene in North America: Note that Winans (1989) has synonymized some 230 proposed species names for North American *Equus* to 5. For most of the Pliocene only one North American species, *E. simplicidens*, was known, but further speciation occurred at around 2 Mya, following the extinction of the hipparionines (a similar pattern was observed in the Old World). There were two to three sympatric species of *Equus* present during the North American Pleistocene. *Equus* became extinct in North America at the end of the Pleistocene, around 10,000 years ago, along with many other large ungulates (mammoths, camels, etc.). Whether or not this "megafaunal" extinction was caused by climate change or by human action is a matter of debate that is not relevant here. However, it is clear that present-day North America contains habitat suitable for *Equus*, as evidenced by the proliferation of feral horses and donkeys over the past couple of centuries.

The genus *Equus* migrated to the Old World by middle of the Pliocene: 3.7 Mya was the first recorded appearance, and the genus became common in both Eurasia and Africa by around 2.5 Mya. *Equus* had a moderately successful Old World radiation, resulting in the variety of horses (one species, *E. caballus* [the domestic horse, inc. *E. przewalskii* the Mongolian wild horse]), asses (three species; *E. asinus* [domestic donkey, inc. *E. africanus*, the African donkey], the Asian *E. hemionus* [the onager], and *E. kiang* [the kiang]), and zebras (four species; *E. burchelli* [Burchell's zebra], *E. greyvi* [Greyvi's zebra], *E. zebra* [the mountain zebra], and *E. quagga* [the recently extinct quagga]), found in Eurasia and Africa today. These three main groupings can be distinguished by cranial and dental features, and the groupings are also upheld by molecular studies. Mitochondrial DNA indicates that zebras and asses may be more closely related to each either than either is to the horse. Note that although there is only a single genus of modern equid, and the species rarely co-occur (a notable exception is the co-occurrence of Burchell's zebra and Greyvi's zebra in Samburu, northern Kenya), equids are always highly numerous as individuals when present. A similar pattern can be seen in the fossil record.

AN ALTERNATIVE VIEW: THE EVOLUTION OF THE COW

As previously discussed, one reason why horse evolution has become such an icon is that only one genus remains today, just as only one species of hominid, *Homo sapiens*, remains today. Hence there is the temptation to ascribe to horse evolution the "Hero's tale" (Landau, 1991), whereby the horse is seen as having undergone a victorious journey through the vagrancies of time, from the Eocene to the Pleistocene, to finally arrive at the final state of magnificence of the modern horse. One might wonder why other familiar domestic animals, such as the cow, are not seen as having undergone such a splendid evolutionary journey. To be sure, the cow is somewhat less glamorous than the horse and has not directly aided humans in warfare, etc., but it is still a vital animal to human civilization.

There is no lack of fossils of potential cow ancestors, although these fossils are from the Old World, and so not among the spectacular fossil deposits of the American West that were well known in the late nineteenth century. However, I submit that one reason that the cow is not exonerated is that there are many other cow-like animals (i.e., bovids and other ruminant artiodactyls) around today, and so the cow does not stand out as some sort of sole survivor. I will elaborate later about how the place of horse evolution, on high latitude island continent of North America, likely led to a very different pattern of evolution and survival to that of ungulates in the Old World, with interconnected continents and the presence of tropical refugia.

The cow belongs to the family Bovidae. Unlike the family Equidae, which appeared 55 Mya, the Bovidae appeared around 20 Mya, about the same time as the subfamily Equinae. However ruminant artiodactyls, of which the bovids are one of a number of families, first appeared in the early Eocene, around 55 Mya, along with the first horses. Higher level taxonomic ranks, such as family, order, etc., may be somewhat arbitrary, and the great diversity of living ruminants may have resulted in the more finer subdivision of this group than of the horse group. Perhaps if a greater horse diversity had survived, we would perceive different types of horses as being very different, and the present day subfamilies might be considered to be families. Thus I shall treat the suborder Ruminantia (which also includes deer, giraffes, etc.) as if it was broadly equivalent to the family Equidae. I will also primarily consider the radiation of Eurasian ruminants as a more appropriate comparison with temperate North America (but note that Asia contains a tropical zone).

The earliest ruminants, first becoming prominent in the late Eocene, were forms broadly referred to as "traguloids." These represented a number of families, but some families were more closely related to the horned ruminants than were others. These animals were in generally small forms, without any horns, and with teeth indicative of a diet of buds, berries, and soft leaves, and most of them lived in what was probably tropical forest at the time in the middle Eurasian latitudes. These animals were probably ecologically similar to the hyracotheres, but unlike the Eocene equid subfamily, this basic type did not go extinct. A few members of the family Tragulidae (mouse deer and chevrotains), that retain this type of diet and general body form, survive today in both tropical Africa and Southeast Asia.

The Oligocene to early Miocene radiation of ruminants was of forms a little bigger than the traguloids, with teeth indicative of a more leaf-eating diet. These were the "gelocids," a term that encompasses many early forms including some closely related to the living horned ruminants. This type of ruminant is broadly equivalent, in terms of both anatomy and ecology, to the evolutionary grade represented by early anchitheres such as *Mesohippus*. A ruminant that survives today of this general evolutionary type (although probably more closely related to true deer than these Oligocene forms) is the musk deer (family Moschidae, including several species of the genus *Moschus*) of eastern Asia.

In the early Miocene, various different families of horned ruminants (pecorans) made their first appearance. For simplicity, I will only consider the families found in Europe today, the Cervidae (deer) and the Bovidae (antelope and cattle). Cervids generally remain as persistent woodland-dwelling browsers with low-crowned cheek teeth (although many deer do take at least some grass their diet). In contrast, bovids rapidly became hypsodont, and from their early evolution seem to represent forms that were largely mixed feeders or grazers in more open habitats (savanna and prairie, although note that these types of habitats appeared much later in the Old World than in North America). Note that in the absence of cervids, some African bovids become woodland browsers, such as members of the Tragelephini (kudu, bushbuck, etc.). This split into browsing cervids and more mixed-feeding (at least initially) bovids parallels the early Miocene split within the derived anchitheres, one group giving rise to the specialized browsing Anchitherini, and the other to forms such as *Parahippus* that gave rise to the Equinae.

Finally, specialized grazers evolved within the Bovidae in the Pliocene, a little later than in the North American horses (probably because climatic changes in Europe lagged behind those in Eurasia). Perhaps the bovid lineage that continued as mixed feeders, such as the one containing gazelles and goats, can be seen as analogous to the hipparionines, while the linage giving rise to the cattle that became more specialized grazers can be seen as being analogous to the equines. This is of course not a perfect analogy, but one that is at least a provoking comparison. Now, if one were to imagine that Eurasia never conjoined with Africa (as in fact happened around 25 Mya), and did not contain a tropical zone, and that the effect of late Miocene to Pleistocene cooling and aridification on the vegetation was as profound as in North America, then it's quite possible that only large specialized grazers like cattle would have survived out of the initial diversity of bovids and deer.

If this evolutionary scenario had actually happened, then we might be tracing a lineage of cow evolution from *Archaeomeryx* (an Eocene traguloid) to *Gelocus* (an Oligocene gelocid) to *Eotragus* (an early Miocene bovid) to *Bos* (the genus that the cow belongs to, first appearing in the Pleistocene), much in the way that horse evolution is traced from *Hyracotherium* to *Mesohippus* to *Merychippus* to *Equus*. Of course, one could still make that straight line connection through time if one chose to, but we usually don't perceive events that way because of the large current diversity of living ruminants, and the survival of a few of the more primitive forms in tropical refugia.

The point of this thought experiment is to illustrate that the notion of "straight line" evolution versus "bushiness" is really irrelevant. It is an accident of time and geography that only one genus of horse survives today. The real issue is the fact that the family Equidae was a diverse radiation on the North American continent that paralleled the Old World radiation of ruminants. Had the survivors Miocene diversity of horses survived to the present (which is basically what happened to the Old World ruminants, given their luck in their geographic setting), then our interpretation of horse evolution would have been different, but the actual processes would have been the same.

WAS THE EVOLUTION OF THE HORSE DIRECTED?

I will contend that one *can* perceive the evolution of the horse as being directed, but only in the context of the forcing of environmental change, rather than by any drive toward "perfection" as once believed. Armed with the thought experiment about the evolution of the cow, perhaps it can be seen that the evolution of both horses and ruminants was directed by changes in climate and environment in the higher latitudes. The Eocene of higher latitudes was covered in tropical-like forest, a good environment for both early horses (hyracotheres) and ruminants (traguloids). Late Eocene higher-latitude cooling resulted in the extinction of these groups (at least outside of the tropical zones in the case of the ruminants), and the temperate woodlands of the Oligocene resulted in the evolution of both horses (early anchitheres) and ruminants (gelocids) that were more specialized leaf-eaters that did not need a year-round supply of buds and berries. The warming and aridification of the Miocene resulted in the spread of grassland savannas and savanna woodlands (more profound in North America at this time than in the Old World, which remained wooded until at least the Pliocene). Now larger types of horses and ruminants appeared, specialized either for browsing in the woodland (later anchitheres and deer, respectively) or for more mixed feeding in more open habitats (equines and bovids, respectively). Finally, by the Pliocene the spread of grasslands resulted in the evolution of specialized grazers such as *Equus* and *Bos*.

The differences between horses and ruminants is not so much in the patterns of *evolution* as in the patterns of *extinction*, and this can be related to an accident of geography. Who could have known in advance that the tropical world of the early Eocene would give rise to the colder, more temperate world that prevailed from the Oligocene onward? The ruminants had the good fortune to have their evolutionary radiation on a land mass (Eurasia) that retained a broad tropical zone, and which later connected up to a large, largely tropical land mass (Africa). Thus more tropically-adapted animals at least had the opportunity to survive in tropical refugia. In contrast, horses evolved on an isolated, largely temperate land mass, with no opportunity to retreat to the tropics. Thus with each major change in higher-latitude climate and environment, the previous radiation of horses went extinct. (Note that although Central America is subtropical today, it is a narrow area of land in contrast to the broad swath of Old World tropics; for most of the time of horse evolution it was not a solid land mass, but divided up into a number of small islands, so it did not represent an easily attainable refuge.)

A brief caveat here. One might critique this scenario by noting that there were various other horses in the Old World, immigrants in different waves from North America, that did not survive, so it can't just all be about geography and climate. However, I will note that, at the risk of appearing to be doing special pleading, there might be a good biological explanation for this pattern. First, at every point in time that Old World horses went extinct (hyracotheres in the late Eocene, anchitheres in the early Pliocene, hipparionines in the early Pleistocene) many other ungulates (including ruminants) also went extinct at that time. The relatively low diversity of the Old World horses, as interlopers into an existing ecosystem (at least in the case of anchitheres and hipparionines), may mean that it was just happenstance that they all went extinct at those points along with numerous ruminants. Alternatively (or in addition), in times of food shortage ruminants can survive better than horses because of differences in digestive physiology, and they may have had the competitive edge over horses during periods when a number of different types of ungulates would be trying to find environmental refuges (Janis et al., 1994, discuss these differences in physiology, and provide a description of this type of competition between North American browsing horses and browsing ruminants).

SUMMARY AND CONCLUSIONS

The evolution of the horse shows a progressive change in anatomical adaptations, from forms adapted to tropical forests, to forms adapted to temperate woodland, to animals adapted to woodland/savanna condition, and finally to forms initially adapted to seasonal, arid grasslands, now spread into more tropical conditions. This apparently "directed" evolution can be perceived as a response to environmental change on the island continent of North America, which did not make contact with South America until a couple of million years ago. The pattern of evolution of horses closely parallels the evolution of ruminants (deer, antelope, etc.) in the Old World. The main difference between the evolutionary radiations of these two different types of hoofed animals is in the patterns of extinctions. The essentially one-way environmental change in North America resulted in the relatively recent survival only a single genus of horse. In contrast, the easy access to tropical refugia in the Old World resulted in the retention of a great diversity of ruminants today.

FURTHER READING

Budiansky, S. *The Nature of Horses: Exploring Equine Evolution, Intelligence, and Behavior* (New York: Free Press, 1997).

Gould, S.J. "Case Two: Life's Little Joke," in *Full House* (New York: Three Rivers Press, 1996), 57–73.

MacFadden, B. J. *Fossil Horses: Systematics, Paleobiology, and Evolution of the Family Equidae* (Cambridge: Cambridge University Press, 1992).

Monroe, J. S. "Basic Created Kinds and the Fossil Record of Perissodactyls," *National Center for Science Education Journal* 5(2) (1988). Available online at www.natcenscied.org/resources/articles/4661_issue_16.

Rose, K. D. *The Beginning of the Age of Mammals* (Baltimore, MD: Johns Hopkins University Press, 2006).

Simpson, G. G. *Horses: The Story of the Horse Family in the Modern World and through Sixty Million Years of History* (Oxford: Oxford University Press, 1951).

Voorhies, M. R. "Dwarfing the St. Helens Eruption: Ancient Ashfall Creates a Pompeii of Prehistoric Animals," *National Geographic* 159 (1981): 66–75.

Taung Child

Anne Katrine Gjerløff

If the skull of a three-year-old child can be conceived as beautiful, the fossil of Taung Child is indeed beautiful. Described as a "jewel of a fossil" (Lewin, 1997, p. 48), this small uniquely well-preserved face and petrified cast of the child's brain is impressive to see, but even more impressive is the story of the discovery—the number of theories, interpretations, and scientific implications of the little toddler's head. The skull was blasted out of a limestone cave in Taung in South Africa in 1924 and was shortly thereafter presented to the world, under the name *Australopithecus africanus*, by the anatomist Raymond A. Dart in the scientific journal *Nature*.

The Taung Child was the first fossil of the hominid genus *Australopithecus* to be found and described, and after initial rejection by the international scientific community and further discoveries of *Australopithecus* remains, it completely changed the way human evolution was interpreted and explained—both when, where, and how. As such the fossil is not only an icon of evolution, but is also a symbol and a poignant case study of the fact that science and society are intimately connected and that the willingness to risk scientific mistakes and ridicule is a prerequisite for scientific progress.

DISCOVERY

After millions of years in the dark, Taung Child saw the light of the day again in the summer of 1924. The young Australian-born anatomist Raymond A. Dart (1893–1988) had studied in England and been a part of the growing milieu of physical anthropologists in Britain, but had

The Taung child. (J. Reader / Photo Researchers, Inc.)

more or less reluctantly accepted the position as a professor in anatomy at the University of Witwatersrand in Johannesburg, South Africa. The following tale of how he found the Taung Child is based on his memoir *Adventures with the Missing Link*.

Dart had brought an interest in fossils with him from Britain and he encouraged his students to find and bring him fossils. In the early summer of 1924 his only female student, Josephine Salmons, showed him a fossilized baboon skull, a rarity at that time. She also brought his attention to the fact that the skull and several other fossils were regularly found at the limestone mines at Taungs (now Taung) in what was then the Bechuanaland Protectorate (not to be confused with what is today the state of Botswana). The baboon skull was in itself a rare fossil, and Dart immediately arranged for the mine manager to ship any fossils in the limestone caves to Dart. The fossils arrived in two large boxes at Dart's house just as he and his wife were expecting guests for a wedding they hosted for a close friend. The arrival of the crates caused a bit of panic, but Dart could not resist opening the boxes and among other interesting fossils, to his surprise, he found a brain—or rather a cast of the interior of a skull, called an endocast. As a neuroanatomist, Dart recognized the brain as not belonging to a baboon or chimpanzee. It seemed too large and too complex in its organization to be anything but human. In the same box Dart found the face of the owner of the brain, barely visible and embedded in stone.

The wedding ceremony tore Dart away from the fossils, but the same evening and for the next thirty-seven days he painstakingly studied and cleaned the fossil until the face was finally freed from its stone matrix. To his surprise "the creature which had contained this massive brain was no giant anthropoid such as a gorilla. What emerged was a baby's face, an infant with a full set of milk (or deciduous) teeth and its first permanent molars just in the process of erupting. I doubt if there was any parent prouder of his offspring than I was of my 'Taungs baby' on that Christmas of 1924" (Dart, 1959, p. 9). As any proud father Dart didn't waste time in showing his baby to the world. On 7 February 1925 the journal *Nature* included a short article by Dart presenting the Taung Child. The content of the article will be quoted here in quite some detail, as it is important to know Dart's claims to understand the reactions of the scientific community. In four and a half pages and with a few small illustrations Dart both described the fossil and argued for his interpretation for it as a creature "intermediate between living anthropoids and man" (Dart, 1925, p. 195). The anatomical features that separated Taung Child from the apes were mainly the face, the dentition, the brain organization, and the inferred way of movement. Regarding the facial anatomy, the very non-protruding eyebrows especially pointed in a human direction, as well as the fact that "the facial prognatism is relatively slight"—which means that there was no protruding ape-like muzzle. The dentition was described as human-like in both the size of the teeth, as well as the lack of big canines and hence no diastema (the gap between the teeth that gives room for the opposite canines). Also the mandible had no trace of the so-called simian shelf—a ridge found in other apes.

Who Killed the Baby?

There is no doubt that the Taung Child did not die a natural death. Many of the adult Australopithecines have damages that Dart interpreted as meaning they were slain by their own kind, as he believed was the case of the many baboons found in the caves. In one instance the holes in an Australopithecine skull have been proved to fit perfectly with the canines from a big cat, like a leopard. *Australopithecus* was not the only carnivore around. In 1995 Lee Berger and Ron Clarke of University of Witwatersrand suggested a new solution to the fate of Taung Child (Berger and Clarke, 1995). They wondered why other hominid fossils had not been found at Taung if the cave was the living space for a whole group, and why there were no remains of big game animals found at the place, but only small prey. Berger and Clarke found that the remains at Taungs matched the remains from big birds of prey, and they suggested that an eagle had been responsible for the death of the child, or at least responsible for the child ending up in the Taung cave after its death. In trying to imagine this event—the fear and the cries of the child, the desperation of its mother, and the hunger of the eagle—prehistory suddenly comes to life, and becomes emotionally important to us. With such interpretations—as with Dart's theory of Killer Apes—the past and its individuals turn into more than just bones, and that is one of the reasons why bold interpretations are not only important but necessary.

Dart also made a bold and indirect interpretation of the child's way of moving. It was based on the position of the foramen magnum—the hole in the base of the skull where the spinal cord and the brain are connected. In quadruped animals the foramen magnum is placed in the back of the skull, in bipeds such as humans it is based more in the middle since the head "balances" on the vertebral column. The position of the foramen magnum is expressed as an index number, which for Taung Child made Dart conclude:

> It is significant that this index, which indicates in a measure the poise of the skull upon the vertebral column, point to the assumption by this fossil group of an attitude appreciably more erect than that of modern anthropoids. . . . It means that a greater reliance was being placed by this group upon the feet as organs of progression, and that the hands were being freed from their more primitive functions of accessory organs of locomotion. (Dart, 1925, p. 197)

From this Dart could even infer something about the behavior of the child: It probably could use its hands to manipulate and carry objects, as well as for "offence and defence" which was necessary since it lacked the large canines.

The braincase of the skull was not preserved, but on the basis of the cranial endocast Dart figured that the brain of an adult creature would be much larger than that of the gorilla who has the biggest brain among the anthropoid apes. As a neuroanatomist, Dart also believed he could observe a human-like pattern in the anatomical organization of the brain:

> Whatever the total dimensions of the adult brain may have been, there are not lacking evidences that the brain in this group of fossil forms was distinctive in type and was an instrument of greater intelligence than that of living anthropoids . . . It is manifest that we are in the presence here of a pre-human stock, neither chimpanzee nor gorilla, which possesses a series of differential characters not encountered hitherto in any anthropoid stock. (Dart, 1925, p. 198)

After his description Dart consequently announced that Taung Child represented not only a new species, but a group which he proposed to call *Homo-simiadæ*. This name never caught on, but the new name he proposed for the fossil did: *Australopithecus africanus*—the southern ape of Africa. In the end of the article Dart expressed surprise that such a species could be found far away from the usual habitats of anthropoids, and he figured that the dry harsh climate of South Africa had given the evolutionary pressure by which the Australopithecines had evolved their big brain, and a behavior different from the forest-living apes.

The fossil was in every way a remarkable discovery. It was well preserved, showed characters hitherto unseen and was found in Africa which had yielded only very few and very *Homo sapiens*–like fossils of early man. One would suspect that the anthropological scientists all over the world would congratulate Dart and rewrite their textbooks in light of this extraordinary fossil. Congratulate they did, but almost nobody accepted Dart's interpretations of the child as an intermediate between man and ape.

REACTIONS

Only a week later in the next volume of *Nature* came the first reactions from Dart's English colleagues, all well-known and respected anthropologists. First and foremost was the extremely influential anatomist Sir Arthur Keith, then the anthropologists Prof. E. Elliot Smith, Sir Arthur Smith Woodward, and Dr. W.L.H. Duckworth. They expressed their sympathy for Dart and his important discovery, but they all concluded that the importance was due to the rarity and completeness of the fossil, rather that its position in man's evolution. Actually all of the commentators concluded that *Australopithecus* was an unknown and extinct ape. Elliot Smith wrote, "The simian infant discovered by [Dart] is an unmistakable anthropoid ape that seems to be much on the same grade of development as the gorilla and the chimpanzee without being identical with either" (Smith, 1925, p. 235). Woodward continued: "The new fossil from Taungs is of special interest as being the first-discovered skull of an extinct anthropoid ape . . . I see nothing in the orbits, nasal bones, and canine teeth definitely nearer to the human condition than the corresponding parts of the skull of a modern young chimpanzee" (Woodward, 1925, p. 235), and Keith echoed the verdict: "Of humanity there is no trace except in one respect . . . we have made known to us a high anthropoid, one which is clearly related to chimpanzees and gorillas but differs from both" (Keith, 1925, p. 326).

They all emphasized the biggest problem: the juvenile status of the fossil. Young apes have much more human-like features than adult apes. The big canines, the prognatism, and the large eyebrows do not develop until a certain age. The human-like anatomy of Taung Child was thus easily explained away by its young age: "I feel fairly certain that some of the other characters mentioned are related preponderantly to the youthfulness of the specimen" (Duckworth, 1925, p. 236). The authorities agreed: Taung Child was an ape, and the human features were caused by its immaturity. Later Elliot Smith even suggested that the only reason Dart had made such a mistake to believe Taung Child was human-like, was because Dart did not have access to skulls of gorilla and chimpanzee infants to compare his fossil with. Had this been the case, Dart would certainly have interpreted the skull in another fashion.

Not only British anthropologists greeted the child in this way. The American anthropologists Earnest Hooton and Henry F. Osborn also rejected the fossil's importance for human evolution. Alfred Romer was only a bit more enthusiastic, since he dismissed the idea that *Australopithecus* was a chimpanzee, and he concluded (after having studied the fossil himself) that "*Australopithecus* is not a chimpanzee, but a new and separate type of anthropoid ape, worthy of careful consideration in any discussion of higher primate phylogeny" (Romer, 1930, p. 483). Romer also gave his opinion about why so few scientists acknowledged Taung Child: "Unfortunately, few who have discussed it have seen the original, and have based their opinions on photographs or somewhat inadequate casts" (Romer, 1930, p. 482). South Africa was far away from the scientific centers of Europe and the United States, and the sparse material Dart had been able to present was not enough to convince anyone. Only a few researchers had the wish or possibility to study the fossil themselves.

To illustrate just how coldly the new member of the human family was welcomed, it should be mentioned that even the name *Australopithecus* was criticized, since it combined Latin (austral- meaning "south") and Greek (pithecus meaning "ape") vocabulary. This is allowed in the international rules for biological nomenclature, but didn't fit the traditional scholars' ideals at that time (Barther, 1925, p. 947). The anthropological community agreed that further information was necessary, and they expressed the wish that a monograph or lengthy article would soon be published by Dart. This never happened and in 1935 Woodward wrote:

> The only fossil hitherto discovered in Africa, which suggests that the continent may have produced man, is the immature skull from a deposit of uncertain age (probably Pleistocene) at Taungs in Bechuanaland, which was named *Australopithecus* by Professor Raymond A. Dart in 1925. It belongs to an ape and seems to exhibit more human characters than the skull of any of the existing apes; but professor Dart's complete account of the fossil has unfortunately not yet been published. (Woodward, 1935, p. 401)

It seems like Dart gave up the defense of his baby after receiving such negative responses to his interpretation, but the truth is more complicated. Dart did write a monograph about the fossil in 1930 but for some reason it was rejected by the Royal Society and never published. According to science writer Roger Lewin this was perhaps due to the influence of the South African anthropologist Solly Zuckerman, one of Dart's students and a later protégé of Elliot Smith (Lewin, 1997, p. 82). Zuckerman was, like many of his contemporaries, convinced that the common ancestor of apes and humans had lived at such an early time in evolution that *Australopithecus* was way too young to be placed at the human branch.

Anyway, Dart's book was never published and for a period of time he turned away from study of the Australopithecines.

In all this misery it should be mentioned that Dart had some supporters. One of them, Robert Broom, will be mentioned later; another one, William King Gregory, was one of the few anthropologists who at that time defended the theory that humans were closely related to the African great apes, and he became only more convinced that Dart was right after having studied the Taung Child himself during a visit to South Africa in 1938. Even before that, Gregory engaged in many debates about the interpretations and rejections of *Australopithecus* and it caused him in 1930 to ask the desperate question: "If *Australopithecus* is not literally a missing link between an older dryopithecoid group and primitive man, what conceivable combination of ape and human characters would ever be admitted as such?" (Gregory, 1930, p. 650). Gregory had a point: Why on earth wasn't the combination of human and ape-like anatomy taken as proof of Dart's assertion that Taung could be an intermediate between man and his ape-like ancestors? To explain this, it is necessary to take a closer look at the scientific paradigm for paleoanthropology in the 1920s.

PALEOANTHROPOLOGY IN THE EARLY TWENTIETH CENTURY

At the time when Dart published his baby only a handful of other early human fossils were known to science. It can therefore seem surprising that the anthropological community didn't react with more enthusiasm, but responded to the discovery with rejection of the fossils humanlike status. If one consider the whole context of paleoanthropology in the early twentieth century though, the reaction was actually hardly a surprise at all. In few words: The Taung Child was at odds with almost every accepted idea of human evolution at that time; it just didn't fit into the scientific paradigm of where, when, and how humans had evolved. Considering the "where" of human evolution, Taung Child simply came from the wrong continent. In the beginning of the twentieth century it was generally believed that the earliest humans first had seen the light in Asia—ex oriente lux. Even though most of the known fossils were found in Europe—Neandertals and early *Homo sapiens*—the earliest forms of humans had been found in the 1890s in Indonesia, on Java, which gave the fossil the nickname Java Man besides the scientifically *Pithecanthropus erectus* (now known as *Homo erectus*). Java Man had suffered the same initial ridicule and rejection as *Australopithecus*, but was by the 1920s generally accepted as a very early human, and was commonly known as "the ape-man"—a literal translation of the name *Pithecanthropus*. That the fossil was genuine was further supported in the late 1920s when a population of rather similar fossils were unearthed in

China and became famous as the Peking Man fossils. Asia was also preferred because some anthropologists found that the gibbon was that of the living apes whose anatomy was most human. Others believed that the split between the human and the anthropoid line in evolution was so ancient that the common ancestor was something like the small Asian tarsier or lemur.

Pithecanthropus and prehistoric monkeys were not the only reason Asia was preferred as the cradle of humankind. Also more emotional and esoteric reasons played a part. Paleoanthropology was affected by a kind of Orientalism, that found Asia not only climatically more fitted for evolution, but also found it simply more appropriate that man should come from the cooler, nobler, enlightened Asia, than from that dark uncivilized continent, Africa. Science was most certainly not unaffected by the Eurocentric and racist views of that time. In this Orientalism the American anthropologist Henry Fairfield Osborn played a major role, but the reasons among anthropologists to prefer an Asian origin were many and it seems that the dislike for an African connection and a close kinship with the great African apes was not the least influential of these reasons. It has been said that paleoanthropology in those years suffered from acute pithecophobia—fear of apes. A lot of time was used to explain why humans were not closely related to the African great apes whom we resemble so much. A part of the problem was that the African apes were commonly believed to be too specialized and to live too easy a life in the lush green jungle, whereas humans were believed to be resourceful, flexible, and to have sharpened their wit by the fight for survival. In the same vein it was not uncommon to have the same attitude toward the people of Africa, who were in the race theories of that time described as lazy, non-progressive, and rather unintelligent. In light of this Orientalism, anti-Africanism, and pithecophobia Taung Child couldn't have been found in a worse place.

Connected to the where, the when, and the how of human evolution was also the notorious fossil who carried the scientific label *Eoanthropus dawsoni*. This enigmatic species exemplifies a lot about why Taung Child received such a cold welcome in the human family, and is worth spending a few words on.

A central question in paleoanthropology is how to know that the fossil you have found is from a human ancestor and not that of an ape. The answer is simply that you look for those characters that to you and in your historical context define humanness. In our days early humans are normally recognized by their bipedality—their ability to walk on two legs—and a lot of discussion about newly discovered early hominid fossils centers on if and how they walkedbipedal.

In the beginning of the twentieth century this was not the case. Humans were defined not by their legs, but by their heads: the mind, the intellect, the big brain of humans was what characterized us. This was an apparently logical conclusion that explained how the hairless, clawless, canine-less, slow, and vulnerable human being had survived in the evolutionary race. In the idea of survival of the fittest, humans had been made fit for fight by their intelligence that made them outsmart other creatures and made it possible to create tools, weapons, fire, clothes, etc. The brain was how humans had survived in the first place, and the reason they had been able to conquer every climatic niche on earth. Thus it was believed that the enlarged human brain preceded the evolution of the human anatomy, and earliest humans were envisioned by most scientists as ape-like creatures with big brains.

Actually one defender of Taung used this argument to explain why he believed that *Australopithecus* was definitely on its evolutionary way to man. The German anthropologist Dr. Paul Alsberg believed that man, like apes, lived by a "fighting principle," but that, contrary to apes, man had to rely on "'extra-physical' means (tools)." According to Alsberg the absence of large canine teeth proved that *Australopithecus* was already evolving in the human direction: "If we were to paint a theoretical picture of the first stages of man, we should necessarily arrive at a form such as the Taungs child represent: the jaws are beginning to recede, the brain is about to increase" (Alsberg, 1934, p. 158). Just like William King Gregory, Alsberg raises the question of how to determine where "the animal ends, and man begins" (Alsberg, 1934, p. 155). The problem was that many anthropologists then (and now) had a quite specific idea of how to define human versus animal, and this determined how they defined the fossils, and not the other way around.

In the first years of the twentieth century an amateur archaeologist Charles Dawson found fossil pieces of a human skull in Sussex, England. He published the fossils in 1912, gave it the name *Eoanthropus dawsoni*, and the British anthropologists were delighted. Not only did the geology imply that these were the oldest human fossils ever found, but British anthropologists had also longed for their own human fossil in the light of the many French and German Neandertals and early *Homo sapiens*. *Eoanthropus* (literally dawn-man) struck a cord of national pride, but it was also the perfect fossil by the combination of its anatomical characters. The fossil was a partial skull and a jaw, and while the jaw was very ape-like, the skull had dimensions that showed that the brain had been a lot larger than those of both apes and Java Man. *Eoanthropus* lived up to all expectations: It demonstrated that the foundation of human evolution was a British, big-brained male, and thus also served as a support for paternalistic imperialism which at that time was believed to be the natural order of the world (at least if you yourself were a British big-brained male). To put it bluntly: *Eoanthropus* was such a perfect ancestor, that if an anthropologist should have created the most likely forefather, this would be how he would have made him. The only problem was—as it became apparent in the 1950s—that somebody had actually created *Eoanthropus*, or Piltdown Man as he was commonly known. Chemical analyses showed that the jaw and the skull were not of the same age, and that the fossil pieces were colored to make them appear old. Also the jaw had been tampered with to make the wear of the teeth look human. The conclusion was clear: Piltdown Man was a forgery, intentionally made by the combination and preparation of a human skull and the jaw of an ape. It has never been proved to satisfaction who did it.

The story of the Piltdown forgery and the ready acceptance by the scientific community explains a lot about why Taung Child wasn't accepted by the same community. Taung had none of the characteristics of Piltdown Man. It had small brain, small jaw, small canine teeth. It was too late in date to be the ancestor of *Eoanthropus*, and was thus perceived as an extinct side branch, or perhaps the ancestor of anthropoid apes. And not least: It came from Africa, from where—as it was commonly believed at the time—nothing new or progressive could evolve. Piltdown and Taung were complete opposites: an African small-brained, human-like child versus a European big-brained, ape-like man. The latter simply fit the expectations and the scientific paradigm, while the former failed completely.

NEW PROOFS

One of the very few scientists to support Dart's claims for human-like status for Taung Child was the Scottish-born paleontologist Robert Broom. Broom had made a name for himself by several fossil discoveries of extinct animals, and the idea of *Australopithecus* apparently pleased him. Dart described their first meeting as follows:

> Broom immediately sent a letter of congratulations and two weeks later burst into my laboratory unannounced. Ignoring me and my staff, he strode over to the bench on which the skull reposed and dropped on his knees "in adoration of our ancestor" as he put it. He stayed with us over the week-end and spent almost the entire time studying the skull. Having satisfied himself that my claims were correct, he never wavered. (Dart, 1959, p. 37f)

Following this meeting Broom wrote a report to *Nature* to argue further for the true nature of Dart's claims for Taung Child. Here he compared Taung Child with Piltdown Man which he accepted as genuine:

Eoanthropus has a human brain with still the chimpanzee jaw. In *Australopithecus* we have a being also with a chimpanzee-like jaw, but with a sub-human brain. We seem justified in concluding that in this new form discovered by Prof. Dart we have a connecting link between the higher apes and one of the lowest human types. . . . While nearer to the anthropoid apes than man, it seems to be the forerunner of such a type as *Eoanthropus*, which may be regarded as the earliest human varieties. (Broom, 1925, p. 571)

Unlike other paleoanthropologists at that time Broom did not let *Eoanthropus* stand in the way for *Australopithecus*, but rather saw in the latter the ancestor of the former.

Just as the young Dutch doctor Eugéne Dubois had set out for Indonesia in the 1890s to find "the missing link," so did Broom decide to find the fossils that would prove Dart right and make Taung Child an only child no more. Most amazingly he succeeded just as Dubois had done. Dubois found his *Pithecanthropus*, Broom found what in his opinion were several species related to Taung Child. He reported his first findings in *Nature* in September 1936 and concluded:

This discovery shows that we had in South Africa during Pleistocene times large non-forest living anthropoids—not very closely allied to either chimpanzee or the gorilla. . . . They also show a number of typical human characters not met with in any of the living anthropoids. (Broom, 1936, p. 488)

From the 1930s until his death in 1951, Broom excavated several South African caves and found a number of fossils, some very complete and well preserved, of several individuals, all of which is now ascribed to the genus *Australopithecus*. These fossils were published in a number of articles in *Nature* and other journals and slowly other scientists began to understand that Taung Child was not what they originally thought, and that Africa had yielded a number of early human species. Broom did not himself wholly support the idea that *Australopithecus africanus* was the direct ancestor of modern man, but he believed that it belonged to a group to which this forefather also belonged. And Broom most certainly did not believe that Taung was just an immature chimpanzee.

Broom himself belonged to the category of anthropologist who are known as "splitters." "Splitters" are (contra "lumpers") quite inclined to name new species, and put more emphasis on differences between fossils than on similarities. The splitting tendency was prevalent in anthropology until the 1950s where a general sorting out of names resulted in the recognition of fewer, but more diverse, species, for example, the *Homo erectus* that included many of the Asian species like Java Man and Peking Man, and the inclusion of nearly all of the South African fossils in the one genus *Australopithecus*.

Before this however, Broom, in collaboration with Dart, J.T. Robinson, and G.W.H. Schepers, was the man behind the discovery and naming of the species *Plesianthropus transvaalensis* (in Sterkfontein), *Australopithecus prometheus* (in Makapansgat), *Paranthropus robustus* (in Kromdraai), *Paranthropus crassidens* and the more modern-looking type *Telanthropus capensis* (both found in Swartkrans). Except for *Telanthropus* that today is recognized as a species of *Homo*, all these names represent two different forms of Australopithecines: a small, gracile form like Taung Child, and a more robust large-teethed form now known as *Australopithecus robustus*. It should be noted however, that paleoanthropology is currently on its way back to a more splitting paradigm, and a number of scientists think that the robust Australopithecines represent not only a different species, but another genus. Therefore Broom's name *Paranthropus* is often used as genus name for all the robust Australopithecines.

The main achievement of Broom and Dart's further fossil discoveries was that it on the one hand gave access to a much larger collection of material than the single skull of Taung, and thus created possibility for studying the post-cranial anatomy of *Australopithecus*, that proved Dart's interpretation that they were bipedal. On the other hand it was important to get fossil material from adult individuals, as most critics had explained Taung Child's human features by its young age. The many new fossils that the discoverers published internationally in several articles and big monographs now proved that there had existed a prehistoric population of quite human-like creatures in Africa, and this forced the anthropological community to rethink the evolutionary position of Taung Child and its many new relatives. *Australopithecus* could no longer just be ignored or explained away as a distant relative of apes. It was bipedal, had a distinct human set of features in the skull, jaw, and post-cranial anatomy, it was relatively large-brained compared to apes, and some of the excavations even hinted that the species had possessed some kind of material culture.

ACCEPTANCE AND BEYOND

After the many discoveries of adult Australopithecines, the opinions slowly started to change. In the 1940s many of the arguments against *Australopithecus* as a possible human ancestor was beginning to fade. This was also due to the fact that many anthropologists had traveled to Africa to study Taung and the other fossils, and thus had been able to compare and measure, without having to rely on drawings, descriptions, and casts. The South African fossils had also been published in major monographs by Dart, Broom, and their colleagues and had thus been made accessible to those who had not been able to hold the fossils in their own hands. Especially the monograph by Broom and G.W.H. Schepers, *The South African Fossil Ape-Men The Australopithecinae* from 1946 was important in this respect. So was the thorough anatomical study by the very well-respected British anthropologist Wilfred le Gros Clark that convinced many of the non-ape status of *Australopithecus*.

The 1947 Pan-African Congress of Prehistory stands as a turning point for the acceptance of the Australopithecines. Arranged by Louis S.B. Leakey this was the first gathering of scientists studying African prehistory from all the fields of archaeology, paleontology, physical anthropology, ethnography, etc.

In a special session on human paleontology on Monday, 20 January 1947, the participants could listen to Dart, Broom, Le Gros Clark, and the French archaeologist Camille Arambourg, who recapitulated all that was known about the Australopithecines until then. Dart described the stratigraphy and the geology of the limestone caves, and argued that the inhabitants of Makapansgat had been able to use fire—hence the name *Australopithecus prometheus* that was soon to be dropped again when the ashes proved to be of non-human origin. Dart also claimed that the Australopithecines had possessed a Bone Culture: "Lacking manufactured implement of stone, these creatures were able to destroy big game animals by virtue of their superior intelligence, their dexterity in wielding clubs of wood, bone or horn, and their management of fire" (Dart, 1952, p. 106). After these quite provoking claims, Broom and Le Gros Clark described the anatomy and the characters of the fossil types found in the preceding ten years—*Paranthropus*, *Plesianthropus*, *Australopithecus*—and Le Gros Clark concluded: "there must be a relatively close zoological relationship between the *Australopithecinae* and the *Hominidae*" (Clark, 1952, p. 112). To support this he enlisted both many of the anatomical features that Dart had already found in Taung Child twenty years before, and several other which the post-cranial fossils had revealed, not least the proof of bipedal locomotion.

The turn of the tides had been coming for some time, but after Le Gros Clark's firm conclusion almost the entire anthropological community accepted the place of the Australopithecines as an important phase in human evolution. As an example of this—and of the humble acceptance of a great scientist—is the letter Sir Arthur Keith a few months later published in *Nature* where he twenty years earlier had dismissed Taung Child's human status:

When Prof. Raymond Dart, of the University of the Witwatersrand, Johannesburg, announced in nature, the discovery of a juvenile *Australopithecus* and claimed for it a human kinship, I was one of those who took the point of view that when the adult form was discovered it would prove to be near akin to the living African anthropoids—the gorilla and chimpanzee. Like Prof. Le Gros Clark, I am now convinced, on the evidence submitted by Dr. Robert Broom, that Prof. Dart was right and that I was wrong; the Australopithecinæ are in or near the line which culminated in the human form. (Keith 1947, p. 377)

This is one of the clearest admittances of mistakes ever to be published in the history of science and it illustrates an important lesson that can be learned from the story of Taung Child: What characterizes true science is not just the quest to prove that you are right, but the ability to accept change and new ideas. Science should not be rigid or conservative, but flexible and ready to try new theories than might be right or wrong. In this important way only can science differ from belief.

One should not be led to believe that everything was now fast forward for the Australopithecines. Most of the anthropological establishment doubt that it is possible to draw a simple family tree with *Australopithecus* leading to *Homo sapiens*. The consensus then (and now) is that humans passed through a phase like the one that is represented by the Australopithecines, but that it is extremely difficult to judge which species of Australopithecine is the one and only ancestor, and which are side branches on the family tree. This has not been made easier in the last thirty years where several other species have been discovered and named.

Neither should one forget that not everybody was as convinced as Arthur Keith. Some anthropologists still found that evidence pointed to the possibility that human evolution had followed its own course for some time before the evolution of the Australopithecines, and that the human line had started in an even earlier period with more generalized apes as ancestor, e.g., the famous *Proconsul* which the Leakey family had found some superb fossils of in the 1940s. Among the researchers that still dismissed *Australopithecus* were Louis Leakey and Solly Zuckerman.

One stumbling block was still left, but was removed in the 1950s. The Piltdown Man was exposed as a forgery in 1953, and thus *Australopithecus* was not only the most primitive candidate, but also the oldest human ancestor. Still, dating only relied on relative chronology, and not absolute dating of the geological layers or the fossils themselves, but in the 1950s and 1960s the natural sciences provided new methods for dating fossils. It now became clear that the South African Australopithecines were 1–3 million years old, but also that early members of the genus *Homo* were up to 2 million years old. It was then evident that not only had some Australopithecine species and *Homo* coexisted, but the possible ancestor of *Homo* should be found among the oldest of the Australopithecine forms. Until the late 1970s *Australopithecus africanus* was the oldest and thus still a likely candidate for human ancestry. Parallel with this Louis Leakey announced the earliest species of *Homo*: *Homo habilis* from East Africa. Attention shifted to the East African

Australopithecines though Leakey denied their relevance for the evolution of *Homo*. The *Telanthropus* found by Broom in Swartskrans was now taken to be a South African *Homo habilis*, and this made it possible that South African *Australopithecus africanus* could have played a part in the evolution of *Homo* after all. But the main attention of the anthropological community was—also for political reasons—not directed toward South Africa but firmly fixed on East Africa. Then something happened that shows—like the case of Taung Child—that science is prone to change and debate. Donald C. Johansson and co-workers published the discovery of Lucy—a female gracile *Australopithecus* of approximately 3.7 millions years of age that they assigned to a new species *Australopithecus afarensis*. They claimed that this species was the common ancestor of *Homo* and the other Australopithecines. This thesis has sparked plenty of debate, and after discovery in the new millennium of several other and considerably older species (*Orrorin tugenensis*, *Sahelanthropus tchadensis*, *Kenyanthropus platyops*, etc.) that are all claimed to be *Homo*-like in some way or another, the confusion is complete. Paleoanthropology stands at a point that can be compared to the 1920s and it will be most interesting to watch which theories will prevail and which of the fossils that will eventually be deemed most human-like. Or it might be the case that science has leaned from the Taung case and will resist in making too-firm conclusions on very little evidence. It is said that there exists an old Chinese curse: "May you live in interesting times!"—this is certainly a curse that is true for paleoanthropology in the twenty-first century.

But from the 1950s the road was open to *Australopithecus* to travel his way to scientific importance as a human ancestor. And while this position became even further accepted, ideas of what it means to be human followed him along the road.

MAN THE TOOLMAKER?

In his 1959 memoir *Adventures with the Missing Link*, Dart claims that from his very first encounter with Taung Child he suspected that Australopithecine behavior was that of a tool-carrying carnivore. His suspicion was caused by holes in the forehead of the baboon skulls found among the other fossils from Taung. Dart figured that the holes had been caused by a blow by a weapon, held by an *Australopithecus*. Also, Dart had wondered what an ape-like creature would eat in the dry South African climate, nothing like the lush forest where anthropoid apes usually live. His answer was that the Australopithecines had mainly been carnivores or omnivores and that the harsh conditions had placed an evolutionary pressure on their intellect:

> For the production of man a different apprenticeship [than that of the apes] was needed to sharpen his wits and quicken the higher manifestations of intellect . . . a more open veldt country where competition was keener between swiftness and stealth, and where adroitness of thinking and movement played a preponderating role in the preservation of the species. (Dart, 1925, p. 199)

The ability to hunt, think, and create tools were thus a necessity to survive for the earliest human-like creatures; need was most certainly the mother of inventions. And what inventions! At the same time as the general acceptance of the Australopithecines, Dart launched his idea of an Australopithecine tool culture with the almost unpronounceable name the osteodontokeratic culture. It was primarily based on his excavations in the Makapansgat limestone caves.

The name in itself explains a lot. It simply means "bone-tooth-horn culture" which implicates that it pre-dates a stone-tool culture, which presumably demands more intellect and is normally only ascribed to members of the genus *Homo*—ourselves and our immediate ancestors. Dart imagined that Australopithecines had an extensive use of the bones, teeth, claws, and horns from the animals they ate and killed, and that this use was a prerequisite for they success as hunters, and thus for their survival. The tools were in most instances not deliberately made, but consisted of bones with the right shape for the job:

> Thus the chief cultural tools of the Australopithecinae were clubs formed by the long limb bones of antelopes; but it is clear that small punctured, round and triangular holes in skulls have been formed by thrusting home the sharp ends of broken long bones, or the horns of antelopes, employed as daggers, or by smashing blows with the upper canine teeth of carnivores when using their jaws as picks. The lower canine teeth of carnivores and pigs have usually been wrenched out of the mandible as though they have been used as natural curved kukri blades. (Dart, 1953, p. 204)

Some of the tools Dart described were more straightforward and easy to accept for most researchers than others, such as a femur and humerus (the long bones) used as clubs, and jaws or sharp teeth used as knives. Others were more problematic and carried further implications for Australopithecine social behavior, such as a bone used as spoon to feed young or elderly individuals or tails used as flags to signal victory or group identity. While it is very likely, almost logical, that the Australopithecines used some bones as natural tools or weapons, the more sophisticated uses are less likely and have been criticized by other scientists.

Dart's theory fit like a glove on the more mainstream ideas of human evolution in the 1950s and 1960s. In this period the first humans were defined as Man the Toolmaker, and even scientists that did not believe in the osteodontokeratic culture agreed that the ability to make tools had been one of the main steps in human evolution. Tools were connected to the growing intellect and to the possibility to control nature by making fire, clothes, cooking, cutting and storing food, and making weapons for hunting and defense. Tools were from this point of view the reason why man had succeeded, had been able to migrate all over the earth, and now ruled it. This emphasis on tools and man's prerequisite for evolutionary success also became apparent when the paleoanthropologist Louis S.B. Leakey in the early 1960s named the new species *Homo habilis*—the handy man, and thus used the ability to create tools as one of the definitions of humanity. The osteodontokeratic culture was intimately connected with Dart's description of Australopithecines as hunters or killers. That idea is the subject of the following section. The point in this section is that the osteodontokeratic culture has been and still is highly debated, and that no firm consensus has been reached regarding its existence or its degree of sophistication.

Several studies conclude that the accumulations of bones in the South African caves can easily be explained by natural agents such as animals (hyenas, leopard, etc.) bringing bones to their dens and other animals rearranging and gnawing on the bones. Especially C.K. Brains's study (called *The Hunters or the Hunted?*) point to this fact. Dart has eagerly protested against this idea which he called the Hyena Myth (Dart, 1959, p. 120). At the same time there are also plenty of arguments for the use of bones as tools by the Australopithecines, and one must conclude that this is one of the questions about the past that awaits its answer—if it will ever be possible to find. The conclusion must be that the osteodontokeratic culture is a very appealing idea, but one that has been very hard to prove, as witnessed by this last sentence from an anthropologist observer of the debate: "One must perforce admire the ingenuity and imagination displayed by Dart in his reconstruction of Australopithecine society. Yet the data upon which his deductions are based, and hence his conclusions themselves, have proved somewhat short of convincing to at least some students of human evolution" (Strauss, 1957, p. 1,108).

THE KILLER APE: THE BLOODTHIRSTY HUMAN?

While Dart received no general acceptance of his osteodontokeratic culture, his description of *Australopithecus* as hunters and even as aggressive killers became extremely popular and made a strong impact on the general idea of how prehistoric humans had behaved. Dart's discovery of the osteodontokeratic culture was directly connected to his wish for explaining the many cases of broken baboon skulls from Taung and later Makapansgat. And by explaining the head traumas as caused by

Australopithecines with bone clubs, Dart made the inevitable connection between tool and weapon, hunting and killing. While animals who kill for food are rarely described as murderers, this was the not the case for *Australopithecus*. Their possible human intellect, and the imaginations of researchers, the press, and the public immediately turned the Australopithecines into the Killer Apes. Nowhere is this clearer than in the vocabulary of Dart himself: "They were murderers and flesh hunters; their favourite tool was a bludgeon of bone, usually the thighbone or arm bone of an antelope. . . . The fragile porcelain-thin skulls of infant baboons were emptied of their brains and then crushed in the hand and thrown aside as a human child might throw away a breakfast eggshell" (1959, p. 114f), and

> The australopithecine deposits of Taungs, Sterkfontein and Makapansgat tell us . . . a consistent story not of fruit-eating, forest-loving apes, but of the sanguinary pursuits and carnivorous habits of proto-men . . . man's predecessors differed from living apes in being confirmed killers; carnivorous creature, that seized living quarries by violence, battered them to death, tore apart their broken bodies, dismembered them limb from limb, slaking their ravenous thirst with the hot blood of victims and greedily devouring livid writhing flesh. (Dart, 1953, p. 204 and 209)

Early human history in Dart's words reads like a gruesome horror story!

To highlight his messages Dart further evoked analogies to ancient, biblical, and imperial myths and spoke of the club of Heracles, the fire of Prometheus, the jaw of Samson, and the fire of Mowgli, and the *Australopithecus* was compared to Cain who killed his brother—in this instance the baboons and other Australopithecines. This rhetoric helped get Dart's message out—everybody could understand the implication that "the South African man-apes were hunters of large game in terrifying possession of Herculean club, Samsonian jawbone and Mowglian firebrand; unless speech was also a barrier, they are separable from man only by the intellectual wall that was finally breached by implement of stone" (Dart, 1952, p. 103).

Presented with these theories and the powerful images they evoked the press could not resist publishing headlines such as "The First Murder Solved" or "The Ape Who Killed Its Brother," and the writer Robert Ardrey popularized Dart's theories in the book *African Genesis*, which begins with the impressive words, "Not in innocence, and not in Asia, was mankind born" (Cartmill, 1993, p. 13). The idea of the murderous ape was simply too good a story to dismiss, and even more important: It fit man's idea of his own inner demons. From the 1940s on a certain misanthropic feeling evolved about man's place in the world, probably caused by the horrors of the two world wars.

That not only early man, but mankind as such were considered aggressive is illustrated by this quote from Dart himself:

> The loathsome cruelty of mankind to man forms one of his inescapable characteristics and differentiative features; and it is explicable only in terms of his carnivorous and cannibalistic origins ... the blood-bespattered, slaughter-gutted archives of human history from the earliest Egyptian and Sumerian records to the most recent atrocities of the Second World War accord with early cannibalism, with animal and human sacrificial practices or their substitutes in formalized religions and with the world-wide scalping, head-hunting, body-mutilating and necrophiliac practices of mankind in proclaiming this common bloodlust differentiator, this pre-dacious habit, this mark of Cain that separates man dietically from his anthropoid relatives and allies him rather with the deadliest of the carnivore. (Dart, 1953, p. 207)

In taking up hunting as a necessary practice to survive, humans have, according to Dart, opened a Pandora's box of horrors.

In popular media and fiction the killer ape became a frequent guest. He is found in the dystopic poetry of Robinson Jeffers, in William Golding's *Lord of the Flies* and *The Inheritors*, and several stone-age novels. Jeffers even dedicated a whole poem (*Original Sin*) to this "man-brained and man-handed ground-ape, physically the most repulsive of all hot-blooded animals up to that time of the world" (Jeffers, *The Double Axe*, 1946).

The function of the Killer Ape is also highlighted in the beginning of Stanley Kubrick's movie *2001: A Space Odyssey*, where a band of Australopithecines discovers that bones can be used as weapons. After committing the first murder, the *Australopithecus* throws the bone-club into the air where it whirls around and transforms into a spaceship. The interpretation is straightforward: All human technology stems from this initial murder weapon; it is our technology that made us masters of other creatures as well as of space itself. In the 1960s the idea of inborn aggression became subjected to study by the science of ethology—the study of behaviour—and especially in two popular books, Desmond Morris's *The Naked Ape* (1967) and Konrad Lorenz's *On Aggression* (1963). The latter tried to explain human evilness by our use of technology which had put evolution out of the game. Lorenz explained that all animals have an aggressive behavior that makes it possible to survive, but that animals have also evolved biological mechanisms to tame this aggression. For example can a wolf show his throat to another aggressive wolf, and this will make the aggressor stop. The problem for humans, according to Lorenz, is that we have evolved weapons too fast for our biological evolution to keep up with. We do not have any natural and automatic mechanisms that will stop an aggressor in killing us if he is determined to do so. Our psyche, our biology, and our technology have not followed the same path nor walked at the same pace. Lorenz vividly draws an analogy of a creature as aggressive as a chimpanzee and with a sharp ax in its hand—this is humankind, which in another sentence is described as a creature with a nuclear bomb in its hand and the aggression of an ape in its heart. Lorenz's conclusion is positive though: Humankind must use the intellect that made us dangerous, in order to survive together rather than to fight. In all these works the conclusion was simple: Man is aggressive and bloodthirsty because of his past; our aggressiveness and our ability to create murderous and effective weapons are the reason why we have survived as a species, and it is therefore not altogether an evil thing, but in our society it must be considered a moral problem which has to be solved.

An interesting place to find reflections of the Killer Ape theory is in one of the most benevolent places of international history: UNESCO, the United Nation's organization for education, science, and culture. After World War II physical anthropology tried to come to terms with the fact that scientific theories had been misused as instrument in the German genocide. The international scientific community decided to make an announcement on the concept of race to prevent any such horrors in the name of science again. The result was the UNESCO Statement on Race from 1950–1951. The statement had a first draft issued that received wide criticism, mainly for not acknowledging that races even existed—quite a few traditional anthropologists were opposed to regarding races as cultural and ethnic groups. One of the changes from the first draft to the second and accepted statement is interesting with regard to the human aggression theories. In the first draft the anthropologists tried to underscore the unity and cooperative nature of humankind with the following words: "Indeed the whole of human history shows that co-operative spirit is not only natural to men, but more deeply rooted than any self-seeking tendencies. If this were not so we should not see the growth of integration and organization of his communities which the centuries and millennia plainly exhibit" (Anonymous, 1950, p. 138). This sentence was criticized, and it was argued that co-operative spirit was not natural, but could only exist in spite of the real human nature that was perceived as a battle for survival, and loaded with aggressiveness, which were clearly witnessed by the "blood-bespattered, slaughter-gutted archives of human history" that Dart had referred to. In the second and final statement the sentence was deleted and the committees work was described as follows:

> We also tried hard, but again we failed, to reach some general statement about the inborn nature of man with respect to his behaviour towards his fellows. It is obvious that members of a group show co-operative or associative behaviour towards each other, while members of different groups may show aggressive behaviour toward each other—and both of these attitudes may occur in the same individual. (Anonymous, 1951, p. 155)

The conclusion was clear; man was most likely aggressive, at best unpredictable and only truly faithful to his own group.

The moral problem was for all the mentioned writers and scientists that the intellect and the ability to kill in an effective way was intimately linked in human evolution to our ability to create tools, and hence to the wonders of modern technology and scientific progress. Many popular writers saw a direct line from the first bone or stone tool to the cold cruelty of Nazi *endlösung* technology and the atomic bomb horrors of World War II. If intellect was founded on the ability to kill, how was man supposed to praise his intellect at the same time as he tried to end war and punish killers? If aggression is what made us human, should we then embrace it, and let biology and evolution forgive us all our sins? These are only some of the questions that Dart's theory of hunting Australopithecines caused, and of course, even through they were debated, no firm answers were suggested. Luckily this was a period where times literally were a-changing, and the problems disappeared. Not because the questions were answered, but because the scientists began to ask new questions and get new notions of what it meant to be human. During the late 1960s and the 1970s ideas about human evolution and behavior shifted. Not so much because of new discoveries, but rather because the old dystopic picture of both human past and future began to look to grim and dark for the present.

THE APE IN THE MIRROR

It is scarcely surprising that a period in history where technological progress and war was ever present fostered a theory of human evolution that depended on aggression and weapons. It is a common feature of science that concerns humans, that scientific progress and new discoveries walk hand in hand with common notions and popular ideas of what it is to be human. Exactly as the belief in the original big brain was a stumbling block for the acceptance of Taung in the 1920s, the belief in human aggression became a source of acceptance of the Killer Ape in the 1950s. And as expected things changed again, this time in a more social and less male-centered direction. The 1970s was the period of strange theories such as Elaine Morgan's Aquatic Ape Theory that suggested a phase of aquatic life for early humans, and the much less believable Eric Von Dänikens theory of aliens that mated with apes and created humans. These theories lived their own life in popular media and did not have much to do with the paleoanthropological discoveries in this decade, but some scientific theories turned things upside down. Among these were a group of women anthropologists that addressed the idea that hunting had been central in human evolution, and that tools were originally designed for killing. This theory that Dart had been important in creating, and which was popularized by Desmond Morris and Robert Ardrey, was now criticized for focusing on male activities and completely ignoring the rest of the Australopithecine population. In a number of provocative essays it was suggested that early humans should be described not as Man the Hunter, but rather as Woman the Gatherer. Most important, it was suggested that big-game hunting was not the main activity, but rather small-game hunting and collecting of nuts, berries, eggs, tubers, etc. which could easily have supported an Australopithecine community. Also the first tools should not be imagined as killing weapons, but were more likely baskets for collecting, sticks for digging, and slings for carrying babies. To put it briefly: The focus was turned around. You did not have to debate what to do with inborn aggression if it did not exist, and if the first tools were not clubs to hit with and stones to kill with, but rather sticks and stones to dig roots and crack nuts with. The focus on female activities was further supported by ethnographic studies of modern hunter-gatherers, and the female ancestor became personified by the discovery of a very well-preserved skeleton of *Australopithecus aferensis*, which was nicknamed Lucy. Another discovery that paved the way for new questions were the many lengthy studies of chimpanzees that revealed that they not only hunted and ate meat, but that they also used tools. Hunting, killing, and using tools were thus not solely a human activity, and new theories were needed to explain what was so special about humans and their evolution if their nearest family possessed the same behavior as the earliest humans.

In the early 1980s it was suggested that what was so special about humans was not our inborn ability to fight, kill, and create weapons, but rather our ability to talk, cooperate, and socialize. To be human was now conceived as exactly the opposite of what was believed after World War II, and it is tempting to believe that it was not only science that had progressed, but also human self-understanding that had changed dramatically.

If we—laypeople and scientists alike—look at our distant past, we look in a mirror as well. We can only explain how humans became human by using our self-understanding of what it is to be human. That means that what we see in the past depend on what we see when we look at ourselves in the mirror. And a mirror can be many things. In Oscar Wilde's novel *The Picture of Dorian Gray* (1891), the man stays beautiful and young, but his picture becomes grotesque, just like in the Killer Ape theories, where the mirror showed us bloodthirsty killers as the foundation of our glorious present. Later anthropological theories have more in common with the magic mirror of Snow White's stepmother. We ask who is the most beautiful, and we get the answer we wish for: ourselves. It can perhaps be considered progress that paleoanthropological theories in recent years has focused more on the positive features in human evolution, but when it comes to reconstructing past behavior, mentality, and social life, I doubt we have seen the last theory yet. One should not forget that parallel with all these more or less plausible ideas about Australopithecine behavior, paleoanthropology has since the discovery of Taung Child reached remarkably new insights on this phase of human evolution. The last section of this chapter will try to summarize the generally accepted knowledge about the Taung Child.

THE MANY FACES OF TAUNG CHILD

In 1984 the University in Johannesburg celebrated the sixtieth anniversary of the discovery of Taung Child. Dart participated and was applauded for his insight and integrity in his original interpretation of the fossil. As described above much has changed in the world, in science, and in the way *Australopithecus* behavior is understood, and I doubt what is status quo now will remain so for long. In this final section I will anyway try to recapitulate some of the changing opinions on Taung Child since that sixtieth anniversary with regard to its age, its taxonomic status, and its death.

The Look of Childhood Innocence

One way to come closer to the prehistoric people we only know from fragments of bone is to reconstruct their looks with flesh and skin, hair and eyes. Most reconstructions have been made by artists as illustrations to more or less serious accounts of life in the Stone Age, and the art has often been made with more imagination than knowledge of the fossils. The small child from Taungs is one of the earliest examples of reconstructions made by the scientist himself, since Broom made a very charming portrait of the child. With its slightly protruding lower face it looks surprised, and not much different than a small modern child.

Most reconstructions of Australopithecines reflect the belief in inborn aggression, and this tendency has survived much longer than the Killer Ape theory itself. Most pictures of Australopithecines show them fighting each other, killing prey, or scaring away predators. Amidst all this aggression most of the portraits of Taung Child served as an antidote. Simply because of its young age Taung Child has always been portrayed as more human—not only in look, but also in temper. There is something touching about a portrait of a dead child—whether it is 200 or 2 million years old and the child gives the image of brutal Australopithecines a softness and something we can identify with.

In the late 1940s the Danish ornithologist, paleontologist, and artist Gerhard Heilman wrote a furious defense of the evolutionary theory and an attack on human cruelty, a book called *The Universe and the Tradition*. In this he argues that *Australopithecus* must be much closer to the childhood of humanity than, for example, Peking man, since Taung Child's face expresses an innocence that humankind has long lost. For Heilman Taung symbolizes the beginning of humankind—the Eden outset from where all went wrong, and he sounds a lot like Dart in his descriptions of the cannibalistic and violent practices of humans through history. Only the small baby face gives us a glimpse of an earlier and more peaceful humanity.

One of the questions to which we have wanted an answer is how old the child was when it died. It should be noted that anthropology can not tell any gender differences in the skeleton of children before puberty and therefore it will still be spoken of as an "it," not he or she. Dart has—because of the newly erupted molars—judged the child to be six years old. But this is founded upon the notion that Australopithecines has a growth pattern like humans, not like apes, and not something completely different or inbetween. Later analysis in the growth lines of other Australopithecines has indeed shown that they have their own pace of growing and that the child probably was 2.7 to 3.7 years old when it died (Lewin, 1985, p. 42). Then there is the geological date—how old is the fossil itself? This has also been a point of some debate, and everything depends on what you really measure the date from. Is it the geological layers (and how do we know what they were precisely), is it on comparison with other well-dated fossils or something different? That this question still needs solving, and might never be solved, is illustrated by the fact that the dating of Taung Child ranges between 1 and 3 million years depending on which method is used. This gives the odd result that Taung is either the oldest *Australopithecus africanus* of all or might not even be an *Australopithecus africanus*, since they are believed to have been extinct more than 1 million years ago. The dating and other analysis has actually suggested that the child might even be robbed of its species-related identity. Comparisons between the face and that of other fossils have also led scientists to suggest that it is not an *Australopithecus africanus*, but rather a child of one of the robust Australopithecine species—those Broom called *Paranthropus*. This is a most interesting idea, that has been challenged by other comparisons and lacks scientific consensus (Olson, 1985, p. 540). Most datings point to an age of about 2.5 million years.

All these new interpretations, many of which are possible because of new scientific techniques and technology, thus threaten Taung Child as Dart described it, from being the oldest human ancestor, to being quite young and not even the species we thought. The common belief—or consensus—is that Taung is an *Australopithecus africanus* of indeterminable date and about three years at the time of its death. What can never be questioned is the fact that the fossil was the first to make scientists look to Africa for our origin, and to consider that locomotion and use of tools might be the first human characters to evolve, and not the big brain that had been worshiped for years. *Australopithecus* has shown us that the earliest humans were African and bipedal.

That one fossil should be the source of so many different interpretations as Taung Child has can serve as a warning, not only to science itself, but also to the interested public, of not being too easily seduced by new technology and new opinions. This might seem to contradict my argument that science needs to take the chance of being proved wrong, but science is defined by both characteristics: Dare to be wrong, dare to accept that mistakes have been made, but also be careful that no one idea or method shall overthrow everything we thought we knew. This is one of the problems in the theory of evolution itself, and that is why Taung Child is indeed an icon of evolution. Not only of evolution itself, but also of the way we study the proofs of evolution.

One of Dart's successors at Witwatersrand University, the paleoanthropologist Phillip V. Tobias, put it this way at the sixtieth anniversary symposium:

> Here we shall encounter arguments, controversies, paradoxes. It should never be forgotten that that is a sign of health in a discipline. A branch of science which is not generating new hypotheses, contradictory viewpoint and dissentient voices—has ceased to grow, has passed its prime and is slipping into senescence prior to fossilization . . . if we argue with one another over detailed interpretations, these are arguments abut patterns of evolutionary change: they are not arguments about whether evolution has occurred. (Tobias, 1985, p. xxiiif)

The history of the changing interpretations of Taung Child illustrate that science isn't simple. Science is an act of balance between right and wrong, truth and belief, and last, but not least: Science is made by humans, and paleoanthropology is not only made by humans but also tries to explain what the definition of humanity is. Taung Child's face not only reflects our deep history but also serves as a reminder of why we should be wary when we see such a face in the mirror—do we see the past as it was, or do we see ourselves, as we think we are?

FURTHER READING

Broom, R., and G. W. H. Schepers. *The South African Fossil Ape-Men, The Austalopithecinae*. Transvaa Museum Memoir no. 2. Pretoria, 1946.

Cartmill, Matt. *A View to a Death in the Morning: Hunting and Nature through History* (Cambridge, MA: Harvard University Press, 1993).

Dart, Raymond A., and Craig, Dennis. *Adventures with the Missing Link* (London: Hamish Hamilton, 1959).

Lewin, Roger. *Bones of Contention: Controversies in the Search for Human Origins* (Chicago: University of Chicago Press, 1997).

Tobias, P. V. ed. *Hominid Evolution—Past, Present, and Future.* Proceedings of the Taung Diamond Jubilee International Symposium (New York: Alan R. Liss, 1985).

The Monkey Trials

Brian Regal

It was so hot that July day in 1925 that the judge presiding over the trial moved the proceedings at the Rhea County Courthouse outside. The accused, John T. Scopes, was on trial for teaching evolution to schoolchildren in contravention of Tennessee state law. On the stand was the prosecuting attorney William Jennings Bryan. The defense attorney, Clarence Darrow, had called him to the stand in an unusual move. Jennings was called as an expert witness on the content of the Bible (the judge would not allow Darrow to call experts on evolution to testify, at least not before the jury). Darrow peppered Bryan with irksome questions concerning contradictory details of Holy Scripture. Growing increasingly agitated at what seemed like insulting and intentionally annoying questions Bryan burst out, "I am simply trying to protect the word of God against the greatest atheist or agnostic in the United States!" The crowd of Christians making up the audience cheered loudly in approval.

Superficially the trial was about whether Scopes had taught evolution to his students. Some observers felt the trial was more about which was true, evolution or revealed religion. Acute observers saw that the trial was about something far more culturally significant. The questions were what role would evolution play in American society and could you teach unpopular ideas? Would factual authority—the underpinning of law, commerce, and industry—be with science or religion, and how really free was speech?

The work on evolution by paleontologists, geneticists, paleoanthropologists, and others is not done in a vacuum. Their work has implications for and impact on wider society. Their search for facts about how

William Jennings Bryan and Clarence Darrow in courtroom during Scopes trial, 1925. (Library of Congress, Prints and Photographs Collection.)

life began and then spread around the planet has fascinated and inspired; it has also frightened. In the study of evolution there is what scientists know about it and what it is perceived to be by the public. Unfortunately, these two are not always the same. There have always been, for various reasons, those who refuse to believe evolution a fact of the natural world and some of them have actively fought against it (as if keeping evolution from being taught would make it go away). The most overt and public battles over evolution waged by anti-evolutionists have come in the form of a series of court cases where complainants have tried to control evolution through legislation. These trials—which appeared in a series of phases—have become icons of the antagonism that has sometimes bubbled up between religion and evolution.

To be sure, evolution cannot be willed away or legislated away. As a biological process it can never be stopped any more than gravity can be. What anti-evolutionists have tried to do is suppress knowledge of it by explaining it away as nonsense, or to squelch it being taught in the public school system. This chapter will examine the phenomenon that has come to be called Monkey Trials: how they came about, who has pursued them and why, and what the outcome of such efforts has been. The most famous of these court cases was the first. Also known as the Scopes Trial it captured the world's attention and made public a growing rift in modern culture between religion and reason. Though it occurred in 1925 the forces the Scopes Trial unleashed had been brewing for as much as a century or more prior to Clarence Darrow and William Jennings Bryan setting foot in the Rhea County courthouse and would continue on into the twenty-first century.

Thoughts about the mutability (the ability to change) of living things had been pondered back to the Greeks and possibly the Egyptians before them. Little hints had been seen here and there, but there were more questions than answers. On an intellectual level special creation by the gods seemed to make more sense and be a more obvious explanation for the diversity of life on earth. Enlightenment thinkers began to toy with the idea that there might be answers to questions about the workings of the universe that were natural rather than supernatural beginning in the seventeenth century. In its modern form evolution, or transmutation as it was known, first appeared in the late eighteenth and early nineteenth centuries in the Western world. By the mid-nineteenth century a number of naturalists had already reached a point where they were ready to accept a plausible explanation for biological change if one would come along. It was at this point that Charles Darwin published *On the Origin of Species* (1859) and laid out the explanation that best fit the evidence and made sense. Still, however, many naturalists as well as theologians felt transmutation something that just did not exist no matter how good of an explanation was put forward. As a result debates and arguments broke out as various camps for and against evolution started competing. While opposing sides fought and argued in England, France, Germany, and elsewhere the battle reached a special fervor in the United States because of the unique cultural conditions there. (In the early twenty-first century battles over the teaching of evolution have appeared in England, Turkey, and Russia).

There were a number of complex reasons for this. The United States was still experiencing intellectual growing pains in the late nineteenth century. Americans wanted to develop their own unique set of values separate from the Old World and all areas were open to discussion. As part of this program religion was being filtered anew as well. Americans in the mid-nineteenth century were still profoundly religious, but many grew disillusioned with traditional forms of Christian worship while others rejected the newer forms. Added to this was the growing population of non-Christians in America. People stayed away from their churches in droves yet were increasingly interested in matters of the spirit. A host of alternative philosophies from transcendentalism to spiritualism, Mormonism, and Christian Science sprang up and went in and out of vogue.

One of the more popular avenues in the pursuit of spirituality was the glorification of nature. Philosophers and writers from James Fenimore

Cooper to Ralph Waldo Emerson, Henry David Thoreau, and Herman Melville saw in nature a source of spiritual renewal. Artists like Winslow Homer, Asher B. Durand, and Frederick Edwin Church were captivated by the uniquely American landscape and its perceived spirituality. Naturalists too were drawn to the woods for many of the same reasons the writers were. In addition to wanting to prove to their European cousins that they could do original thought-provoking work and advance the cause of science, there was something else the naturalists looked for, something more personal. The naturalists and fossil hunters who fanned out over the American landscape in the nineteenth century sought not only to discover scientific facts about this alien world, but to find answers to the higher questions of life. Many nineteenth-century religious movements made clear efforts to support their beliefs with science. Spiritualism, Christian Science, and Theosophy all did and saw no contradiction, in fact they saw science and religion as compatible.

EVOLUTION IN AMERICA

Evolutionary thinking came somewhat slowly to the United States. Once it did most intellectuals and naturalists were eventually converted to a general belief in evolution, though not always to Darwin's central idea of natural selection. The dissatisfaction with natural selection caused some Americans to adopt ideas put forward earlier in the century by Frenchman Jean-Baptiste Lamarck who theorized that organisms changed through the accumulation of hereditary traits brought on by adaptation to environmental pressure and that changes to any aspect of the organism's structure were passed on to its offspring, creating a new species. For some, Lamarck's law of acquired characters was a simpler, more straightforward alternative to Darwin's selection and variation. This caused something of a definitional problem.

Many naturalists and other intellectuals called themselves or were called Darwinists, but what they actually believed about evolution varied, not only from each other, but from Darwin as well. There were naturalists who rejected evolution and theologians who accepted it. Anti-evolutionists pounced upon differing explanations as proof that evolution was a sham. Added to this the image of monkeys turned to men was a potent icon which lodged in the public eye. Technical explanations for the course of human evolution, scientists' constant referral to primates, the loose use of the term ape—as well as a good bit of eugenic and racial baggage that was attached to it both consciously and unconsciously—caused considerable confusion and resentment among lay people. To social groups steeped in the narrative tradition of Genesis, the claim of a close relationship between human and monkey, if not direct descent, was deeply offensive whether explained clearly or not. Fundamentalists gleefully exploited the monkey/man image to great effect and found it easy to turn people against the idea.

An attempt to bridge the gap between evolution and religion was in the work of George Frederick Wright (1838–1921). He studied at Oberlin College in Ohio for the ministry. The college's president was the legendary Charles G. Finney (1792–1875), a prime mover of the Second Great Awakening. Finney stressed the idea that the Bible was the ultimate authority and that religious belief was supported by evidence from nature. Becoming a minister upon his graduation, Wright began to teach himself natural history by reading Darwin and Lyell, whose books' central themes were in direct contradiction to what he had been taught as a seminarian. He also read the botanist Asa Gray and came to believe he could accept both religion and evolution without contradiction. Gray had corresponded with Darwin before 1859 and was sent a preproduction copy of *On the Origin of Species* for comment. For Gray, a Presbyterian, and a host of other Victorian naturalists, God was the first indisputable cause of evolution, with random chance left out of the equation. This theistic evolution approach gave them a way to accept evolution and Christianity together. Gray became a mentor to Wright as both believed strict religionists who denounced evolution as anti-Christian were just as wrong as strict evolutionists who denied religion. Wright thought that God created the first few species and then let natural selection take over. Wright did not view Scripture as literal or infallible, at first.

By the 1880s, however, Wright grew more conservative after taking a position at his old school, Oberlin. In that more overtly religious environment he slowly changed from defending science from religion to defending religion from science. He grew to fear that science and evolutionary thinking in particular were seeking not just to coexist with, but to undermine religious faith. He lost his liberal view and adopted an increasingly literalist stance. He wrote a number of books on natural history which received poor reviews from scientists. They accused him of badly misreading the geology he discussed and knocked him for being a theologian out of place trying to explain science. The stinging critique by geologists pushed Wright more into the traditional anti-evolutionist camp. Wright then tried to find scientific explanations for events like Noah's flood and the destruction of Sodom and Gomorrah. Ironically, he interpreted the line about man being made from the dust of the earth as meaning that after evolving from the lower primates (who could be seen as coming from "dust"), man became man when God put the spark of spirituality in him. Wright, like a number of high-profile naturalists, could accept the bodily evolution of humans, but when it came to consciousness he believed the only explanation was divine intervention.

This contradiction was symptomatic of an age where intellectuals were hard-pressed to accept two such different philosophies as Christianity and evolution but wanted to. Wright's attempt to support religion with scientific thinking and geological and archaeological evidence was a step toward what would eventually be known as Creation Science.

CHRISTIAN FUNDAMENTALISM

Spurred along by the changing intellectual landscape of America, Christian Fundamentalism was born. Scholars have disagreed over how this intense and strict interpretation of Christian doctrine came about. Christian Fundamentalism was traditionally thought of as having emerged in the United States in the early part of the twentieth century as an anti-intellectual backlash against the rise of other approaches to Christian worship collectively known as Modernism. This is now seen as an over-simplification. One school of thought argued that Fundamentalism was a development that grew out of revivalism, which itself originated in the early twentieth century, not over political or social concerns, but over what was considered the growth of false Christian doctrine due to a growing secularization of society and fragmentation among Christians. Another argued that Fundamentalism could be traced back to the millennial (end of the world) movement of the early nineteenth century begun with the Millerittes of New York. Fundamentalists considered themselves pious Christians in a world turning away from God and proper Christian belief. Modernists, on the other hand, held the Bible to be symbolic, not literal; accepted the teaching of evolution as compatible with Christianity; and followed a fluid more liberal view of Christianity that they believed had to be "modernized" by each generation in order to stay viable (an idea the Fundamentalists vehemently rejected).

It can also be argued that the roots of Christian Fundamentalism can be traced back to the early nineteenth century and even possibly to the post-colonial period. Many of the groups of European, mostly Christian, settlers who came to North America came for reasons of religious freedom. With the freedom to flourish some groups achieved political power and influence. Once other groups began to catch up in significance, whether politically or numerically, the older established groups began to react as if threatened. The Anti-Catholic movement of the nineteenth century, for example, was in large part due to Protestant fears that the growing number of Catholics in the country threatened their hegemony over the corridors of cultural and political power. As a result the dark paranoid side of religious power came out (Jews would later replace Catholics as the enemy of choice in religious fervor, to be followed in turn by Muslims).

Part of this growing antagonism can be found in the reaction to the intangible notion of national self-hood. Nations and peoples often have religious or unique ethnicities as cultural glue. These things give the nation a focus and sense of shared community. The United States, however, began as a nation which intentionally did not have a state religion, and as a society was made up of multiethnic immigrants. As a result, other things were looked to in order to supply the kind of notions that would unify the nation and its people. With no national religion—though Christians of varying denominations predominated—public worship came to rest upon things like the Constitution, shared patriotism, the unique natural landscape, and even technological advances. These things were looked at to supply the focus for a national sublime (that thing which transcends the everyday and gives someone special pride in themselves and their home). American public society became increasingly secularized with religious worship shifting to the private sphere. Most religious groups in America found this situation rather acceptable as it allowed them full access to society yet allowed them to hold onto their religious and ethic traditions without fear (or at least a reduced fear) of persecution or exclusion. Spirituality was important to Americans, but the majority preferred to do their own theological thinking without the intrusion of the government or other churches. The government's role was to protect religion not promote it. This compartmentalization had its benefits beyond simple protection of a person's right to worship in their own way. When a citizen left their ethnic or religious community to venture out into the wider society of different religious affiliations it was easier for them to interact with others because their interface was not through religion. Pride in the Constitution, nature, technological advances, or even sporting events required no specific religion and were open to all and helped to bring down social barriers. However, some smaller Christian sects, particularly in the South and Midwest, began to feel marginalized by the increase in secularization. Religious leaders warned their congregations that society was slipping into ungodliness by not following proper Christian doctrine and by becoming too materialistic and worldly thus losing their chance at salvation and redemption. The only way to stop this from happening was, they said, to get back to the basics of Christianity; they had to go back to the fundamental tenets of their worship and belief.

In the early part of the twentieth century a group of loosely affiliated preachers and Christian intellectuals in the United States and Great Britain (where similar concerns had been percolating) joined together and wrote what became the manifesto of a new religious movement. Published between 1910 and 1915 in twelve volumes of ninety-four essays (including one from George Frederick Wright) *The Fundamentals: A Testimony of Truth* took a collection of beliefs, proofs of Biblical truth, ideas, concerns, practices, and suggestions for behavior which had been previously fragmentary notions and combined them into a codified body of knowledge. The enterprise was bankrolled by conservative Christian oilman Lyman Stewart of Union Oil of California (which in later years became UNOCAL). Printed and distributed in vast numbers *The Fundamentals* covered a wide range of topics, though mostly it rejected the growing popularity of the higher biblical criticism of the Modernists.

Taking their name from the book's title, Christian Fundamentalists held to the belief in the virgin birth of Jesus, His physical resurrection, the reality of miracles, faith and grace over good works as the road to salvation, and the inerrancy of the Bible. While *The Fundamentals* was the intellectual framework of the movement, the primary text was the King James Version of the Bible. Many augmented their studies by using the *Scofield Reference Bible* (1909). This special edition, also supported financially by Lyman Stewart, was the King James Version with numerous annotations and commentaries by lawyer-turned-Bible-exegete Cyrus Scofield (1843–1921). His quirky interpretation of scripture leaned heavily on the dispensationalism of John Nelson Darby, James Inglis, and James Brookes. Dispensationalism is the belief that there are periods where God deals with humans in different ways in the form of special tests which man must pass. Scofield also helped secure the notion of Young Earth Creationism in America by including in his commentary a reference to the calculations of Irish cleric Archbishop James Ussher (1581–1656) who placed the year of creation at 4004 B.C.E. Scofield was also particularly interested in the Book of Revelations and the concept of the Rapture (a term that goes back to fourth-century Byzantium, but is not actually in the Bible). The *Scofield Reference Bible* had, and still has, enormous influence on the growth and spread of the Christian Fundamentalist movement. It could not help but sour the relationship between science and religion because as Scofield framed it, Christianity was under threat from evolution because it undermined the dramatic and theatrical view of the human condition the Fundamentalists had become enamored of.

Not all the groups or individuals drawn to Christian Fundamentalism believed in exactly the same thing. Easy generalizations about these groups were formed in part because of the grand generalizations made by the various sides about the other. Anyone even loosely associated was blindly labeled a fundamentalist, even though it constituted a relative minority. As a grouping of diverse individuals and denominations—including Presbyterians, Baptists, and others—Fundamentalism was not a homogeneous block. Some were radical, while others were moderate. Many Fundamentalists insisted that they were making scholarly studies of the Bible and that their methodologies were more "scientific" than the theoretical work of scientists. The simplistic reading that all the

Fundamentalists were uneducated hillbillies is inaccurate. Their emphasis on gathering and studying facts in the traditional Baconian method led them in part to their opposition to evolution because they thought it was theoretically and not empirically based (something which is still an anti-evolutionist argument).

Many Fundamentalists were revivalists who wanted a return to a past Golden Age. There were millenialists who looked forward to the end of the world and their ultimate salvation by God (and several different types of millenialists on top of that); they looked forward to a utopian future for themselves. Many were evangelists who wanted to bring all this good news to the immoral and corrupt world in order that these people would be saved as well. In general, Christian Fundamentalists, regardless of denomination, were and still are haunted by the immorality of the world: promiscuity, abortion, materialism, humanism, women's rights, civil rights, gay rights, and a host of other social demons. The most insidious of these underlying demons of society as they saw it was evolution. Many in the movement felt that by killing off evolution they would excise the demons in themselves allowing them to approach salvation cleansed of the world's materialist taint.

That they are scriptural literalists is a major part of the problem for the Fundamentalists in relationship to evolution. Religious groups which did not take a literalist approach to their scriptures had a much easier time dealing with evolution because they had ways open to them to deal with science: They could simply ignore it, they could make the science fit their theology, or they could make their theology fit the science. The literalists found themselves in a corner of their own making. As most scripture was written hundreds, if not thousands, of years ago they could be contradicted by science, especially where the natural world was concerned. For example, if scripture says, "God is great!" there is not much science has to say because science is not equipped with the tools to examine such a statement. If scripture says the earth is only 10,000 years old, on the other hand, the enormous amount of scientific data showing the earth to be billions of years old becomes a problem the literalists have to deal with. And herein lays the rub. Science in general, but geology and evolution in particular undermine a literal reading of Genesis. While that is bad enough, the real trouble is that if Genesis is undermined the rest of the Bible is, especially the end. So if evolution says the world may not have begun the way religionists say it did then the world might not end the way the religionists say. If no divine creation at the beginning of times then maybe no divine salvation at the end of time.

Another problem for the fundamentalists was that creation and evolution took two very different views of the human condition. Evolution in the nineteenth century saw the human experience as an optimistic and upwardly mobile process. Humans had begun as lower forms and had climbed up to achieve success. The Christian Fundamentalist view of the human experience was a pessimistic downward spiral to oblivion that could only be stopped by the return of Jesus. For them, humans had begun at a pinnacle in the Garden of Eden and had fallen to a degraded level. As such, evolution was the antithesis of Christianity. (The later twentieth century saw evolutionary biology take a more subtle, nuanced, and non-progressive view abandoning the old way.) With their pessimistic view of the human condition and future, Christian Fundamentalists saw evolution as the thing that undermined everything they held dear. It reduced the story of Genesis to an allegory that could not be taken literally.

By the early 1920s, spurred on by the horrors of World War I some Fundamentalists decided it was time to take action and engage the evil directly. Evolution was warping and influencing the minds of America's schoolchildren and that had to be stopped. Unsatisfied with keeping their religious preferences in the private sphere, the Fundamentalists worked at projecting their influence into the public sphere. In order to turn the tide of evolutionist thinking, Fundamentalist-supported bills were put before the legislatures of twenty states to ban the teaching of evolution. The bill introduced in Kentucky was spearheaded by Nebraska politician William Jennings Bryan and strove to outlaw the teaching of human evolution in any school receiving public funds. While they were mostly unsuccessful—including the one in Kentucky—diehard supporters kept on fighting for anti-evolution legislation.

PHASE ONE

Beginning his career as a Nebraska congressman, William Jennings Bryan (1860–1925) ran unsuccessfully for president three times in 1896, 1900, and 1908. He was a nationally known political figure who had fought for the protection of the poor and especially Midwestern farmers. In the late nineteenth and early twentieth centuries a string of Republican Party electoral successes severely damaged the Democratic Party. This allowed Bryan to rise to the upper levels of what was left of the Democrats. Seeing an opportunity, Bryan gave his famous "Cross of Gold" speech in 1896 in which he called for silver to be used as the standard basis of American money. This, he said, would allow poor farmers to pay off their crushing bank loans more easily. Bryan, like many from America's Western states distrusted of the Eastern region with its sprawling cities and large foreign populations. He was a staunch temperance man and thinly veiled anti-Catholic, both characteristics that drove away immigrant ethnic voters (the Democratic Party's core constituency) but was attractive to Christian Fundamentalists.

By the 1920s, Bryan became associated with Christian Fundamentalism and stepped to the forefront of the anti-evolution movement. The same month Bryan's Kentucky anti-evolution bill was killed, he ran an editorial in the *New York Times* excoriating the teaching of Darwinism. Claiming evolution an unproven theory, Bryan said that he opposed its teaching (like many opposed to it, Bryan's understanding of the process of evolution was shaky at best). The *Times* then asked Henry Fairfield Osborn (1857–1935), president of the American Museum of Natural History in New York, to write a rebuttal. Osborn said that what Bryan was asking in his article was whether God used evolution as part of His divine plan. Osborn's simple answer was, yes, He did. Science and evolution, Osborn countered, promoted spirituality, and not undermined it. Osborn, however, was not a strict Darwinian; he believed evolution was a divinely inspired law. The universe, he said, was not the result of accident or chance, but of a divine order. Osborn tried to be civil with Bryan and discuss the matter on a professional level, but Bryan made that increasingly difficult. The discussion between the two soon turned to name calling and sarcasm. Replying to Osborn's counter-editorial, Bryan disparaged him and Princeton University biologist Edwin Grant Conklin (who wrote a companion piece to Osborn's), as "tree men," equating them to monkeys swinging by their tails and chattering nonsense. He dismissed Osborn's argument completely and claimed Osborn thought the discovery of fossils more important than the birth of Jesus Christ.

In a way it is ironic that Osborn took up the battle against Christian Fundamentalism. He was himself a devout Presbyterian whose theories of human evolution contained as much theology as it did science. He argued that a mechanism he called Aristogenesis allowed individual organisms to evolve purposefully under their own willpower. While he agreed with mainstream biologists about most of the mechanics of evolution he argued that the initial push which started a living group evolving in any direction came from the hand of God. Osborn was an evolutionary theist and so not quite as far from the Fundamentalists as might be expected. What did separate him from them was his unshakable belief that teaching schoolchildren science and evolution was not only good for them, but good for the country as well. Osborn felt that knowledge of evolution made a person better at worshiping God.

Bryan was not the only fundamentalist who had it in for evolution, or for Osborn. John Roach Straton, despite his father being a pastor, did not become a practicing Christian until he attended a revival meeting in his teens. Embracing the life with zeal he became a pastor himself and served at a series of churches in Chicago, Norfolk, and finally the Calvary Baptist Church in New York (1918–1929). He became an ardent Christian Fundamentalist and anti-evolutionist publishing a newspaper called *The Fundamentalist*. Straton's hackles were raised considerably in

1925 when Osborn opened a ground-breaking new museum exhibit on human evolution called The Hall of the Age of Man. This was the first major museum exhibit in America on human evolution. Osborn brought in hominid fossils from around the world and installed posters and charts laying out his view that the evolution of the human race was a progressive and goal-oriented process. What really got Straton's attention, as it did for thousands of museumgoers, was a triptych of large murals depicting the lives of the Neandertals and Cro-Magnons. Painted by New York artist Charles R. Knight under Osborn's directions, the paintings depicted human ancestors with a good deal of racial baggage: the Neandertals were shown as "lower orders" while the Cro-Magnons were decidedly Aryan.

The reverend fumed that Osborn and his exhibition were poisoning the minds of New York schoolchildren. Straton wanted the exhibition dismantled or at least to include a reference to the biblical story of creation. Osborn did not want religion introduced into the exhibition because he feared that average people already did not understand the workings of evolution, and religion would only confuse the issue further. Unsatisfied with Osborn's reply, Straton began preaching anti-evolution and anti-museum sermons, even hanging a large banner outside his church lambasting the professor. Straton charged that Osborn and company were insulting God and that taxpayer money should not go to such a den of iniquity as the American Museum of Natural History. The discourse was covered extensively in the New York press as well as a number of newspapers around the country. The public tended to side with Osborn. The uproar did not diminish the crowds waiting to see the museum or the hall, and non-fundamentalist theologians came to the museum's defense. The tussling between Bryan, Osborn, and Straton turned out to be a warm-up for what would become the icon of the conflict between evolution and religion. Focus soon shifted from New York to Tennessee.

The pivotal moment in the warfare between supporters of human evolution and those of special creation in America came with what journalist H.L. Mencken called the Monkey Trial. The state of Tennessee passed its anti-evolution law, the Butler Bill, in March 1925, and the fledgling American Civil Liberties Union (ACLU) of New York moved to oppose it. The ACLU advertised in Tennessee newspapers asking for volunteers to help fight the law through a test case. The city fathers of the small town of Dayton saw a chance for national publicity and approached the ACLU, offering Dayton as the site to hold the trial. They then enticed Dayton public school physical education and general science teacher John T. Scopes to volunteer as the accused. William Jennings Bryan was asked to join the prosecution by William Bell Riley of the World Christian Fundamentals Association.

As the primary defense council the ACLU called in the legendary Clarence Darrow (1857–1938). Darrow had become infamous over his 1924 defense of the child murderers Nathan Leopold and Richard Loeb. That the two spoiled rich kids from Chicago society had brutally killed little Bobby Franks was not in question. Darrow fought to keep his charges from receiving the death penalty by arguing that society had morally corrupted them and as a result, though guilty, they did not deserve to die. His gambit worked, and Leopold and Loeb were spared. In 1925 Darrow took on the cause of defending evolution. The Monkey Trial became an epic public battle between Bryan and Darrow over the freedom to teach unpopular ideas in the public school system in the United States.

Part of Darrow's strategy was to call various well-known scientists, including Henry Fairfield Osborn, to the stand to explain evolution and

THE DARWINIAN THEORY.

A racist picture card exploiting the twin fears of North American Christian Fundamentalists: evolution and race. Undated but thought to be from the time of the Scopes Monkey Trial. (Author's collection.)

why teaching it was not a cause for worry to the people of Tennessee. Doing this was central to Darrow's case, but the judge denied the opportunity for the experts to speak before the jury, saying scientific explanations were irrelevant to the case (Bryan wanted to call experts of his own particularly the Reverend Straton and the creation "scientist" George McCready Price but neither could attend). Frustrated, Darrow took a desperate gamble and argued that he should be able to call William Jennings Bryan to the stand as an authority on the Bible. Over objections from his co-prosecution, Bryan accepted Darrow's challenge and confidently took the stand. Darrow proceeded to grill Bryan on anomalous and contradictory aspects of scripture. He was also able to have Bryan admit that the biblical "days" of the Creation might not have been twenty-four hours, but days of indeterminate length. Bryan stumbled and stammered his way through the cross-examination, on occasion even raising the eyebrows of his staunchest supporters by some of his answers. His simplistic view of creation and unlettered understanding of evolution made him an easy target for Darrow's sharper analytic mind. When it was over, Bryan was exhausted and visibly shaken. Despite Bryan's poor performance on the stand, Scopes was convicted. Losing was not a problem for Darrow as he wanted to take the case to the Supreme Court to have it declared unconstitutional. The conviction was overturned on a technicality in the lower courts so it never made it to the Supreme Court. The trial did, however, have a lasting effect on American religion and its relationship to the idea of evolution and public thinking. All other such trials would be forever compared to this one.

The popular image of the Scopes trial came in large part from H.L. Mencken's description of it in a series of articles for the *Baltimore Sun*. Mencken was a well-known syndicated journalist and political commentator. He outraged many by arguing that religion was a curse upon society that clouded people's minds and led them to believe irrational things. He wrote of the trial as a battle of forward-looking intellectual sophisticates stemming the tide of advancing hordes of backward, clawing barbarians. This picture was advanced by the appearance of the play based on the trial, *Inherit the Wind* (1955). Playwrights Jerome Lawrence and Robert E. Lee constructed a tragedy of Greek proportions that, while powerful, missed all the subtleties and underlying tensions of the actual events by rolling out the stereotypes of big city versus backwoods and the enlightened versus the closed-minded. This was in part because the authors were not actually talking about the Scopes trial, but the McCarthy hearings of the 1950s. The play was a great success and inspired a motion picture a few years later. However, this image is misleading and inaccurate. To accept it on face value makes it impossible to understand what was, and is, going on. In the history of Monkey Trials not everything is always as it seems.

Monkeys and Movies

Most people know what they think they know about the Scopes Trial from the play and film, *Inherit the Wind*. Written almost three decades after the facts by Jerome Lawrence and Robert E. Lee, the play oversimplified and greatly redrew the facts to tell a story of political power gone awry and the consequences of mass hysteria. The play's opening is completely fabricated. In the story, Bertram Cates (Scopes) is arrested against his will and vigorously hated, to the point of violence, by the local townspeople. The reality is that John Thomas Scopes was a well-liked teacher who volunteered to act as the accused in an artificially contrived test case and never spent any time in jail. (Clarence Darrow reported that far from being reviled in Dayton he was showed unending courtesy and never felt threatened or disliked, despite the fact that his views differed so dramatically from most of the people of the town.) The central dramatic action of the play is the confrontation between the two lawyers, Henry Drummond and Mathew Harrison Brady (Darrow and Bryan).

While the play uses some actual transcript dialogue for the confrontation, it is added to and heightened for dramatic effect. Brady is portrayed as a religious bigot and out-of-control inquisitor. William Jennings Bryan, while a passionate Christian who rejected Social Darwinism (a perverted form of the original idea), was also a liberal democratic politician with a long career of fighting for the rights of poor farmers and the downtrodden and against the abuses of the Robber Barons of the Gilded Age.

The authors openly acknowledged that they were less interested in the evolution trial of the 1920s then they were in commenting on the infamous McCarthy Un-American hearings of the 1950s. The title of the play was never used during the trial, but is yet another commentary on the McCarthy hearings. Using an historical event to examine a current political situation is a common mechanism of literature. The various monkey trials were never really about evolution versus science, they were about political power and the ultimate control of the school system. *Inherit the Wind* is not an accurate depiction of historical events or even really a commentary on the battle between science and religion: It too is a wider commentary on political power and its discontents. But then again . . . what isn't?

Following the Scopes trial, some anti-evolutionists stepped away from a purely scriptural point of view and began building what they thought would be a scientific refutation of evolution. Creation science was a Fundamentalist belief in a literal interpretation of the Genesis story of a six-day creation week occurring no more than 10,000 years ago, which all could be proven by scientific methods. Not all creationists took the young-earth position. Some allowed the earth might be billions of years old. This gap theory allowed that after the creation week, a period of indeterminate length passed before the historical period began. Strict literalists rejected this notion because it did not fit biblical accounts and contradicted God's word. Gap theory traces back at least to Scottish theologian Thomas Chalmers (1780–1847) who used it as a way of reconciling Scripture with the knowledge geologists were producing about the nature of the earth. Along with gap theory was day-age theory that argued that the days of Genesis were not twenty-four hour days but were of indeterminate length. Again, strict creationists saw this as a cop-out to try to get around scientific evidence.

Mainstream America and the media lost interest in creationism, but followers did not. In the 1930s the first in a series of societies was formed to keep the work going. Groups like the Religion and Science Association and the Deluge Geology Society came into being and, though small, were hotbeds of creationist activity and debate. Squabbling among believers grew intense, partly as a result of younger enthusiasts obtaining conventional science educations. With their new knowledge and what they saw in the field they found it increasingly difficult to reconcile some of the more extreme claims of older followers. The new generations were not quite armchair theorizers like their unlettered forebears. Their deeper understanding of the natural world gave them a different insight. Unable to overcome the validity of biology and geology some creationists began giving ground, accepting scientific tenets as long as they could hold on to the ultimately divine origin of humans and their souls.

PHASE TWO

In order to teach evolution, textbooks were needed. The first school-oriented science textbook in the United States to deal with evolution was Asa Gray's *First Lessons in Botany* originally published in 1857 and re-issued after 1859. Gray, despite his relationship with Darwin, did not embrace Darwinism completely. His thinking remained theistic, and his textbook reflected it. He argued that God had created a number of species that then went on to become others. Initially, Gray did not accept common descent but by the 1880s had done so. The other leading naturalist in the United States turning out textbooks was Louis Agassiz. His high school text on zoology did not take the evolutionary stance. He was an ardent opponent of transmutation and based his books on the concept of special creation. Throughout the 1860s and 1870s old-school science textbooks either adopted a theistic approach to evolution or avoided it altogether.

The first high school textbooks to accept evolution as the operating principle of biology began to appear in the 1880s written by the generation of naturalists raised on it. Charles Bessey's *Botany for High Schools and Colleges* (1880) and Joseph LeConte's *Compend of Geology* (1884) both aggressively taught evolution as the underlying process of biology. Around the turn of the century more pro-evolution texts appeared that also emphasized the part it played in human history. Authors like David Starr Jordan, Vernon Kellog, and Charles Davenport all made a point of connecting biology, heredity, and racial hierarchies. These authors were also eugenicists who believed the human race should be carefully and systematically bred in order to eliminate "inferiors." The most popular high school biology book of the first part of the twentieth century was George Hunter's *Civic Biology*, first published in 1914 and going through numerous editions. Besides having evolution as the driving force in the progression and appearance of life on earth, Hunter, like the other eugenicists, saw Nordic humans as the end product of evolution. This was the book that was the standard-issue biology text in the Dayton, Tennessee, school system.

In the latter half of the twentieth century anti-evolutionists saw text-books as the best delivery method for their ideas as well. Textbook publishers became the targets for the wrath of the Fundamentalists. Texas and California are the country's largest consumers of school textbooks. As such publishers sometimes allowed themselves to be influenced by these states and rearranged their books to suit the sensibilities of creationists in those states who exerted pressure on the publishers to remove "objectionable" materials before they would buy them. As many other states followed Texas and California's lead, considerable sums were involved. Fearful of losing lucrative multimillion-dollar contracts, publishers often reduced or excised discussion of evolution in order to curry favor.

The state of Arkansas followed Tennessee by passing its own anti-evolution law in 1928. Similar to the Butler Bill the Arkansas law prohibited the teaching of evolution in public schools, colleges, and universities. Though harsh and put in to placate the powerful Christian Fundamentalist presence in the government and populace, the Arkansas law was never enforced. No opposition to the law materialized and so little was made of it one way or the other. This was helped by the fact that increasingly as the twentieth century passed few teachers in the state—even those who believed in evolution—actually taught it at the grammar or high school level for fear of retribution. Throughout the mid-twentieth century the teaching of evolution fell slowly into a period of quiet neglect around the country. With the advent of the Cold War between the capitalist West and the communist East things changed. In order to combat what Western leaders felt was a direct threat the American public school system began to overhaul its curriculum in order to better prepare the country's children to meet this threat. New teaching methods and curriculum standards were enacted across the country to produce citizens with the requisite technical and academic tools needed to thwart the spread of communism. Science curricula in particular were changed following the Soviet launch in 1957 of the first artificial satellite in earth orbit, *Sputnik*. The National Science Foundation and other public science advocacy groups produced guidelines for improving public school science education. With these developments the teaching of evolution was brought back from obscurity to take a central role in biology classes across the country. The renewed emphasis on evolution education combined with the growing social unrest and turmoil of the 1960s brought on by the war in Vietnam, the civil rights movement, and the women's liberation movement drove some Fundamentalists to despair over the same sort of social breakdown and secularism which had so upset their forebears in the 1920s. A new wave of anti-evolution sentiment and legislative efforts began. The growing tension was augmented by the re-emergence of the creation science movement.

The man often credited with having provided the framework for the revival of the creation science movement of the 1960s is Henry Morris (1918–2006) and his book *The Genesis Flood* (1961). A trained engineer with a doctorate in hydrodynamics Morris, like many of the more scientifically trained creationists, came to his Fundamentalism later in life. Following a series of teaching positions Morris decided to turn his technical training on the problem of explaining Noah's flood. He came to believe that traditional geology and evolution were the enemies of religion. His literalist interpretation of the Bible became increasingly inflexible—though on a personal level he was a gracious adversary—and refused to concede to anything which did not directly support scripture. To this end he joined forces with theologian John Whitcomb to produce *The Genesis Flood*. Abandoning uniformitarian geology for catastrophism Morris argued that an envelope of water vapor surrounded the earth in biblical times and that God piercing it is what provided the waters for the deluge. After going through great lengths to employ the language of fluid dynamics to explain the flood, he reached an impasse which he surmounted through the use of miracles. Morris stepped into the same hole many literalist creation science proponents did. Wanting to beat the heathen scientists at their own game they could only go so far down that road before being forced by simple logic to abandon science and untie the Gordian knot with a miracle, thus undermining their effort.

Morris's thesis was not particularly new. He was in large part repeating the work of George McCready Price from the early part of the twentieth century. Price (1870–1963) was likely the most influential scientific creationist of the early twentieth century and helped pave the way for later writers like Morris. Price grew up a millenarian in a Seventh-Day Adventist sect in New Brunswick, Canada. The leader of the Adventist church was the charismatic Ellen White (1827–1915) who claimed God passed information to her followers through her while she was in a trance. She stressed a literal belief in Genesis without any of the nuanced interpretations, like the day-age and gap theories.

In the 1890s Price briefly attended a small New Brunswick school, taking a few classes in natural history (these were the closest he would ever get to formal training in anything approaching science). By the turn of the century he began to fixate on the geologic aspects of evolution as the key to the whole system. He paid particular attention to stratigraphy and the explanation for why younger layers were occasionally found below older ones. Geologists had long argued that the strata of the earth went in a progression from youngest at the top to oldest at the bottom. The concept of uplift was used to explain why the sequence was not always so simple: Geologic forces could sometimes lift older layers above younger ones, making them seem out of sequence. Price found this explanation unacceptable. As to the initial formation of the layers, Price found

Charles Lyell's steady-state, uniform-action explanation also unacceptable. His Adventist upbringing led him to Georges Cuvier's catastrophism. Price modified Cuvier's many periods of catastrophic upheaval and reduced them down to just Noah's flood as Morris would do decades later. It was the flood that accounted for the geology of the earth and the fossils that were remnants of those animals that did not get in the ark. He called this system flood geology and published it as *The New Geology* (1923).

Henry Morris had the good fortune to address these same issues at the moment when the resurgent creationist movement was looking for a secular explanation to support their position. Morris rode the wave of his success by forming first the Creation Science Research Center (CSRC) and then the Institute for Creation Research (ICR), both of which became influential centers of young-earth creationism and the heart of the creation science movement. They took on the task, not so much of convincing unbelievers in the mainstream scientific community, but in influencing sympathetic members of local school boards to begin working creation science into their curricula. It was this creation science movement and its organizations that fostered the second phase of the assault on the public school system.

In 1965 the school board in Little Rock, Arkansas, adopted a biology textbook which included a chapter on evolution. Biology teacher Susan Epperson felt pushed into a corner as a result. A believer in evolution who was ready to teach it, she was conflicted by the fact that while she was expected to teach the curriculum she was given, the state said teaching evolution was against the law. The contradictory rules led her to bring a suit for clarification and to protect herself from being jailed by the state for upholding a rule given her by her employer; paradoxically also the state. The case of *Epperson v. Arkansas* revolved around the argument that a state could not outlaw evolution nor could it tailor its public school education curriculum to cater to a religious majority. In 1968 the Supreme Court agreed with that sentiment and Epperson won the case.

The simplistic prohibition against teaching evolution at the bottom of both the Scopes and Epperson cases violated the separation clause of the U.S. Constitution that disallowed the government to make laws promoting or restricting a particular religion. This meant that defeating these types of laws was relatively easy. Anti-evolution forces, however, learned valuable lessons from these two cases. Following a learning curve, they realized that in order to defeat—or kill—evolution they would need a more sophisticated strategy. In the early 1980s constitutional lawyer, Christian Fundamentalist, and director of the Institute for Creation Research Wendell Bird, along with colleague Paul Ellwanger, a "repertory therapist" and president of a group calling itself Citizens for Fairness in

Education, pioneered a new tactic. Bird and Ellwanger put together a version of creationism they believed was not religious in nature. Bird called it the "theory of abrupt appearances" and argued that organisms had appeared suddenly without the need for a common ancestor and with no mention of a deity. It was an innovative way of presenting their case because instead of pushing the theological aspect of their belief, the authors would stress the idea of intellectual "fairness." If evolution were to be taught, then it would be fair to give creationism equal treatment in the classroom, allowing the students to decide which to believe. Ellwanger publicly stated that his bill was about fairness in education not religion. However, he told his supporters privately that they should not bring up religion, even though this was what the entire exercise was about, because it would hurt their efforts to get the laws passed. They would mask their religiosity with calls for intellectual freedom and try to sneak them past the Constitution.

The first test of this new approach came in Arkansas in 1981. Awkwardly titled the "Balanced Treatment for Creation-Science and Evolution-Science Act" (but commonly known as Act 590) it mandated that whenever evolution was taught in an Arkansas school creationism had to be taught as well. This wording allowed anti-evolutionists to say with a straight face that they were not outlawing evolution only giving "fair and balanced" treatment to all intellectual ideas relating to biological origins. The proposal was brought to a sympathetic "born-again" senator who, impressed by what he read, bypassed the state Department of Education and the attorney general and put it forward as a bill. A sympathetic governor then signed it into law in March 1981without reading it or initiating any debate. By the following December the ACLU launched a suit to repeal it. In *McLean v. Arkansas* the Supreme Court again held for the plaintiffs saying that what the state was calling Creation Science was in fact not science at all, but a theological belief.

Although theoretically viable, this approach of removing God from a religious precept had its downside. In order to rework their beliefs into a more constitutionally palatable form, supporters had to systematically water down, at least publicly, their religious beliefs. Ironically, they argued that in their version of creation, God was not God. They tried to explain that just because they used the word *creator* it did not necessarily imply a deity. The gambit failed, and the law was defeated. In a final irony, the attorney general's office that had been given the unenviable task of defending an unconstitutional law was accused of wrongdoing, not by the secular press but by creationists. Televangelist Pat Robertson and moral-majority pundit Jerry Falwell accused the attorney general's office of siding with the ACLU and the forces of darkness. Even though the Arkansas case fell apart for the creationists, they did not give up.

While these cases put a damper on the anti-evolution movement the case that had the most impact was *Edwards v. Aguillard*. Like Arkansas,

the state of Louisiana passed its "Creation-Science Act" claiming it was
in support of academic freedom and intellectual fairness. Opponents
said it was a blatant attempt to introduce Christian Fundamentalism
into the school's curriculum. A court case ensued and the lower Louisiana
court held that the act was a promotion of religion. The state appealed
to the Supreme Court in June of 1987. The Supreme Court ruled in
support of the lower court's ruling that the Creation-Science Act
was unconstitutional because it violated the separation clause. Justice
Brennan who cited the "Lemon test" to support his position wrote the
majority decision. (This came from *Lemon v. Kurtzman* and has become
a yardstick for First Amendment cases.) The Lemon test says that a law
must pass three steps in order to be constitutional: It must have a secular
purpose, it must not promote or retard religious expression, and it must
not "entangle" the government with religion. Brennan and six other justices felt the Louisiana act did violate these points making it unconstitutional. They argued the act restricted teacher's ability to teach science
unencumbered by religious precepts. The two dissenting justices were
Antonin Scalia and William Rehnquist. Scalia shocked a number of
observers when in his opinion he questioned whether the Constitution
really did contain a separation clause. In addition to saying that creationism could not be taught in a science class the court also added that
they were not giving special treatment to evolution. They stated that any
scientifically valid alternative to evolution could be taught in science
class. This seemingly innocuous statement formed the basis of the next
phase of anti-evolution Monkey Trials. In the late twentieth and early
twenty-first centuries a number of cases came along which attempted to
use a new way of thinking about creationism and an improved model of
Paul Ellwanger's idea.

PHASE THREE

In the later part of the twentieth century America's political landscape
had taken a shift toward the right of the political spectrum. The forces
of social and political conservatism had been growing since the 1950s.
The leaders of the political and religious right saw the liberal ideals of
Franklin Roosevelt, the Democratic Party and the New Deal, not as
great benefits, but as undermining their "values." Under the Reagan and
first Bush administrations conservatives took heart that they were re-
gaining their ascendancy. This would allow them to realize their dream
of taking the country back to an earlier era. The period of brief liberal
reemergence under the Clinton administration only served to bolster
conservatives' determination to eliminate the political left and liberalism
in general. In order to win the 2000 presidential election George W.
Bush courted the world of Christian Fundamentalism even though he

was not himself a fundamentalist. He and his political operatives and strategists saw in the religious right as a base of support ripe for exploitation. They brought in conservative religious leaders and made high-profile showings of support for their favorite issues like abortion, law and order, gay and immigrant issues, the breaking down of the barriers between church and state, and of course evolution. All this led zealous Christian leaders as well as a number of related and sympathetic organizations to think they now had a real chance not to just weaken the hold of evolution training in schools but to eliminate it. Creationists began to explore new ways of bringing their message—particularly Intelligent Design (see chapter 24)—into the classroom. Anti-evolution, pro-Creationist legislation began reappearing in front of state legislatures across the union.

In 1999 the Kansas Board of Education decided to delete mention of evolution from the state's science curriculum and standardized tests. On 11 August of that year the board voted six to four to remove most references to evolution from the state's educational standards. Although it was not an outright ban, it eliminated the need to teach the subject by removing any evolution questions from the school system's standardized tests. Grade school and high school teachers regularly concentrated on subjects that were tested on all-important state exams. With evolution not on the tests, teachers would be less likely to spend time on it. Although creationists across the state vowed they would ensure the ruling was obeyed, many districts continued to teach evolution regardless. Their feeling was that in an attempt to give their students the best education they could, evolution would be taught. Not wanting to antagonize voters for his upcoming presidential bid, Vice President Al Gore, who had been an outspoken proponent of science education, refused to condemn the ban. In Texas, another presidential candidate wanting to curry favor with the religious right, Governor George W. Bush, said he thought creationism should be taught in public schools.

The Kansas situation came about partly as a result of a set of standards for teaching science put out by the prestigious National Academy of Science (NAS) in 1995. In a move to bring the state up to those standards, as many states did, increased emphasis was placed on the teaching of evolution in the Kansas curriculum. A recently elected Kansas Board of Education implemented most of the requirements, but some members balked at including evolution in the curriculum considering it too speculative an idea. Pundits and science advocates called the board backward and anti-intellectual. Some suggested not accepting Kansas high school science credits when students applied to college. The NAS along with the American Association for the Advancement of Science (AAAS) disallowed the state to use their copyrighted materials. By 2001, several of the more conservative board members were voted off, largely as a result of the publicity generated by the evolution ban. The new board reversed the ruling and brought evolution education in Kansas back from extinction, albeit temporarily. A series of elections back and forth alternately brought in and out board members from both sides.

As was becoming the pattern, just because a decision had been made did not end the struggle. By February 2005 Intelligent Design supporters were pushing to have it included—or at least wording which allowed it being included—in the school curriculum and at the same time downplaying evolution. They were also eager to redefine the very meaning of science as a way around the U.S. Constitution. A new more conservative board decided to take up this cause and proposed a plan to allow for criticism of evolution be included in the science classes. In preparation for a vote on the topic the board held public hearings to discuss what they were proposing to do. Much of the hearings were to be given over to a discussion of the differences between evolution and intelligent design. A number of board members were openly antagonistic to evolution calling it a "fairy tale." A pro-intelligent design lawyer named John Calvert, acting as a sort of unofficial council to the Kansas board, arranged to have a number of intelligent design proponents come to the hearing to explain what ID was all about (at taxpayer expense). Calvert was not a teacher or board member, but the head of a local organization called the Intelligent Design Network that reportedly had ties to the Discovery Institute. On his Web site Calvert asks worried parents if they would like "your school to change over from Evolution only to a more objective approach—an approach that promotes critical thinking, open minds and permits teachers to explain the scientific controversies about our origins?" Most of his answers to that question seemed to involve buying his intelligent design publications. Pro-evolution scientists and organizations like the American Association for the Advancement of Science turned down invitations to attend the meetings arguing the entire exercise was meant as a way of making them look bad or engaging in a nonexistent controversy, which would give scientific validation to something (intelligent design) that had none.

In June, after the contentious hearings, a preliminary vote approved the new science standards that would allow intelligent design to be taught. Biology teachers would be expected to read a strongly worded statement to their classes saying that evolution was not supported by the fossil record the way scientists said it was, and that there were other explanations for the origins of life that were not simply natural. Worst of all it included a new definition of science—just as the Discovery Institute argued there should be—which allowed for the inclusion of metaphysical explanations for natural phenomena. Teachers were also to be given a list of specific "weaknesses" in evolutionary theory. In a full vote in November 2005 the Kansas School Board once again came out against evolution and voted to implement the new anti-evolution standards. By 2007 the tables turned yet again and evolution was reintroduced into the Kansas curriculum.

DOVER

Probably the most influential Monkey Trial of the early twenty-first century was the Dover, Pennsylvania–based *Kitzmiller v. Dover*. It is significant because, like the Scopes Trial, it brought together powerful forces from both sides of the issue to engage under the relentless gaze of the media. It came out of social forces that had been building for some time concerning the state of American politics, religion, and science and their role in the public schools. A sign of things to come appeared a few years before the court case unfolded. A Dover High School student painted a mural on evolution for his science class. Proud of his work he donated it to the school and it was hung permanently in the biology lab. A school janitor began to quietly fume over the depiction of humans evolving. Waiting until the summer after the student had graduated the janitor— without the permission of the student, teacher, or school administration— tore down the mural and burned it. Later asked why he had done such a thing he replied that he had a niece in the school system and he did not want her seeing pictures of evolution.

Creationist organizations like the Institute for Creation Research and the Discovery Institute, who had worked in close concert with the Kansas school board, had been noticing sympathetic school boards. At the Dover school district, the board of education was choosing textbooks for the coming year. In September 2004 several members of the board voiced concerns over a biology text that they interpreted as teaching humans came from monkeys. The next month, while agreeing to adopt the book, they mandated that ninth-grade biology teachers would be expected to recite a formal, paragraph-long disclaimer to students saying that evolution was a theory only and not a fact and that alternative concepts for life's origins existed. This made them the first school board in America to do so. To check out these "alternatives" students were directed to the school library to read a copy of the book *Of Pandas and People*, a stack of which had been conveniently donated for use anonymously.

When the board decided to put forward the statement two members resigned in protest. During discussions conservative board members made remarks about how they thought evolution did not exist and that it should not be taught at all. The members who resigned claimed there had been pressure from fundamentalists to have the board do something about introducing creationism into the classroom. Just before the

Thanksgiving holiday a town meeting was called to discuss the new ruling as well as the board seats left vacant by the resigning members. Board
member William Buckingham—a retired local policeman—raised eyebrows when he proceeded to ask pro-evolution attendees to the meeting
who wanted to speak if they had ever been accused of child molesting.
Throughout the lead up to the trial Buckingham would accuse pro-
evolution parents and their defenders of being illiterate and unpatriotic.
Alarmed at what was happening a number of parents contacted the local
chapter of the ACLU and asked about filing a suit to stop the mandated
speech against evolution. They thought it was inappropriate to have
teachers recite a scripted statement to ninth-graders about "alternative"
explanations for how life began. It was, to them, a blatant attempt to foist
a specific religious idea (intelligent design creationism) upon children in
a public school and that it was being put forward as science.

Once the lawsuit was announced battle lines were drawn and outside
groups began appearing. In February 2005 the Thomas More Law Center
(TMLC) came to the school board's defense. The TMLC is a kind of
Christian version of the ACLU though focused exclusively on Christian
concerns. Their motto is "Defending Religious Freedom for Christians"
and they are active in the anti-gay and anti-abortion movements. The
first thing the TMLC did was try to get some of the more vocal and persistent pro-evolution parents removed from the suit saying they had no
real standing in the case. The parents responded with a novel approach
saying that their standing in the case was based on the fact that if their
children were taught intelligent design in biology class they would be
hampered in their efforts to get into a good college because the scientific
community gave no legitimate status to creationism. A series of moves
and counter moves ensued. There were accusations that board members
had made anti-evolution statements during meetings, saying they wanted
creationism and religion taught in the classroom. Audiotapes of these
meeting then "disappeared." The presiding judge John E. Jones requested
reporters who had been at the meetings turn over their notes, something
they were reluctant to do. The board sent out a newsletter—reportedly
written with TMLC help—to local parents explaining their position.
Attached to the newsletter was a statement by Pennsylvania representative Rick Santorum (R) who supported the teaching of intelligent design
saying it was a legitimate science that should be taught in the schools.
Creationist parents tried to sue the school board to force them to teach
creationism. The publishers of the contentious book at the heart of the
case, *Of Pandas and People*, joined with the defendants. The Foundation for Thought and Ethics argued that if the school board lost the case
their sales would suffer. It was discovered under cross examination
that the previous school board president with board member William
Buckingham had conspired to raise money through Buckingham's church

to buy the books and then have them donated anonymously to the school, and that they lied about it under oath.

The following April, Pennsylvania state representative Thomas Creighton (R) introduced a bill for vote that would allow local school boards like Dover to teach intelligent design creationism if they wanted along with evolution. The state then held a series of subcommittee meetings to discuss the issue. Pro-evolution speakers were grilled on the finer points of evolution in ways some thought weighted against them. One committee member admitted he believed the designer of intelligent design was God. The TMLC accused the ACLU of being up to no good in its efforts to keep the separation of church and state intact. They argued the statement the Dover board wanted read to its students was nothing more than a "modest proposal" and not worth all the fuss.

The non-jury Dover trial finally began on 27 September 2005 at the Harrisburg courthouse. Expert witnesses testified to the validity of both evolution and intelligent design. A main witness for the plaintiffs was Brown University biologist Kenneth Miller—author of the textbook Dover was using and an outspoken critic of intelligent design theory. Miller complicated the case for the defense, as he was an articulate witness and a devout Catholic. The defense had been insinuating that anyone who opposed the Dover school board's actions were anti-Christian. Most of the plaintiff parents also considered themselves good, practicing Christians. University of Louisiana Professor Barbara Forrest gave contentious testimony claiming the Seattle, Washington–based Discovery Institute was operating behind the scenes to support the teaching of intelligent design as a way of promoting its own anti-evolution agenda. The defense brought in the well-known proponent of intelligent design, biochemist Michael Behe of Pennsylvania's Lehigh University. The author of *Darwin's Black Box* (1996), Behe testified that intelligent design was a sound scientific theory that argued life is too complex to have appeared according to traditional Darwinian evolution. Dover school system teachers testified they felt bullied into reading the statement and that they had objected to the original proposal but the conservative members of the board ignored them. Plaintiff witness Professor Robert Pennock of Michigan State University testified as a philosopher of science that intelligent design was more a theology than a science. He cited occasions where high-profile intelligent design proponents had admitted as much in various talks to religious groups in the past. This was powerful evidence that intelligent design had no place in a science class.

Judge Jones handed down his ruling in early January 2006. In a ruling that surprised many he held for the plaintiffs, saying the board-created mandate to read the statement violated the separation clause. He determined that intelligent design was a theological and not scientific concept and that the board members' intention was to introduce religion into the public school system. Two weeks later the newly elected school board—which saw the removal of most of the conservatives who voted for intelligent design replaced by moderates who wanted it out—voted to rescind the original mandate. They also released the Thomas More Law Center from representing them, effectively ending any chance of an appeal. The Discovery Institute, whose entire reason for existing is to eliminate evolution teaching and replace it with intelligent design, quickly condemned the judge's ruling saying it had "many, many errors." Besides the condemnation of the TMLC and the Discovery Institute Judge Jones received threatening letters from individuals. As several Republican Party spokespersons had recently made threatening statements about other judges, the U.S. Marshals Service put Judge Jones (who was himself a Republican appointed to his position by President George Bush) under protection. The verdict also meant that the school board—regardless of who the actual members were—had to pay the court costs and legal fees of the winners. Those costs, not counting how much the school board paid the TMLC for their services, were around 2 million dollars: an amount the tiny school system was ill-prepared to pay. To lessen the blow the ACLU and the organization People for the Separation of Church and State agreed to bill the school only for the most essential charges, a move that cut the bill by half. The circus that had grown around the trial was capped when following the vote to eliminate intelligent design from the science curriculum televangelist Pat Robertson released a public statement saying that the vote had been an insult to God. The self-styled holy man said to the people of Dover, "If there is a disaster in your area, don't turn to God, you just rejected Him!"

CONCLUSION

In the early years of the twenty-first century lawsuits and legislation have become an integral part of science education. States from Michigan to Maryland to Ohio all have cases and legislation pending in one form or another. There are, in fact too many such cases to have discussed them all in this one chapter. In December 2004 the Cecil County, Maryland, school board met to choose and approve textbooks. Two members asked for more time to make their decision. Board member William Herold complained that the tenth-grade science text did not contain anything on intelligent design creationism. He argued that the "conflicts" over evolution should be taught to the students. Since 1996 Maryland has had a state law that forbade the teaching of "origin of life issues" in the public schools. Herold wanted supplemental materials handed out to students in the class about intelligent design. A public hearing on the issue brought in a number of supporters who also wanted creationism taught. One of them said that "they've been trying to make my Bible a fraud for three thousand years and they haven't succeeded!" The board, including Herold, unanimously voted to accept the standard pro-evolution biology book. However, he said he voted for it with the understanding that he was going to work to repeal the state law. He also pushed to make it mandatory that teachers begin their science classes by stating that there are alternative theories of life. In Cobb County, Georgia, the local school board, in a move similar to Dover's, attached stickers to the covers of biology textbooks telling students that evolution was only a theory not a fact. The courts first ruled the stickers violated the separation clause. An appeal got the stickers approved, but the fear of a protracted lawsuit eventually prompted supporters to withdraw the stickers in December 2006 with the promise not to pursue the case any longer. This effectively ended the controversy and removed the stickers.

In the United States, those who want to uphold the separation of church and state—an idea supporters say is the best protection we have to ward off oppressive theocracy or dictatorship—and those who want to pull it down so that one religious sect can take precedence will go on. This is the delicate balance of democracy and the role of evolution in it. Both evolution and religion have powerful and important parts to play in people's lives. By their natures, however, they need to be kept separate, not to damage or repress or short change them, but to ensure their continued vitality. Attempts to mix them only dilute both and render them useless.

FURTHER READING

Conkin, Paul. *When all the Gods Trembled* (Lanham, MD: Rowman & Littlefield Publishers, 1998).

Larson, Edward. *Trial and Error* (Oxford: Oxford University Press, 2003).

Morris, Henry. *The Long War Against God* (Green Forest, AR: Master Books, 2000).

Nelkin, Dorothy. *The Creation Controversy* (San Jose, CA: ToExcel Press, 2000).

Numbers, Ronald. *The Creationists* (Berkeley: University of California Press, 1992).

Regal, Brian. *Henry Fairfield Osborn: Race and the Search for the Origins of Man* (Burlington, VT: Ashgate Publishing, 2002).

Regal, Brian. *Human Evolution: A Guide to the Debates* (New York: ABC-CLIO Press, 2004).

Scott, Eugenie. *Evolution vs. Creationism: An Introduction* (Westport, CT: Greenwood Press, 2004).